Fariborz M. Farhan
Box 36030
Ga. Tech
Atlanta, Ga. 30332

Linear Algebra

LINEAR ALGEBRA

R. R. Stoll and E. T. Wong
Department of Mathematics, Oberlin College

ACADEMIC PRESS New York and London

ACADEMIC PRESS INC.
111 Fifth Avenue, New York, New York 10003

United Kingdom Edition published by
ACADEMIC PRESS INC. (LONDON) LTD.
Berkeley Square House, London W.1

LIBRARY OF CONGRESS CATALOG CARD NUMBER: 68-16517

PRINTED IN THE UNITED STATES OF AMERICA

Preface

This book is intended to be used as a text for a one-semester course in linear algebra at the undergraduate level. We believe our treatment of the subject will be both useful to students of mathematics and those interested primarily in applications of the theory. The major prerequisite for mastering the material is the readiness of the student to reason abstractly. Specifically, this calls for an understanding of the fact that axioms are assumptions and that theorems are logical consequences of one or more axioms. Familiarity with calculus and linear differential equations is required for understanding some of the examples and exercises. Necessary concepts of intuitive set theory are outlined in an appendix.

In view of the availability of so many textbooks having the same title and essentially the same chapter headings, we feel that the most useful role this preface can play is that of indicating to one considering it for adoption those features which, in our opinion, set it apart. One, which should have appeal to the mathematician, is a dedication to the principle that, whenever possible, definitions and theorems should be stated in a form which is independent of the notion of the dimension of a vector space. The generality which is thereby achieved we regard as outweighing one possible ill effect, namely that Section 3 of Chapter 1 and Section 1 of Chapter 2 may prove to be difficult for some on first reading.

A second feature of this book which is worthy of mention is the early introduction of inner product spaces and the associated metric concepts. Students soon feel at ease with this class of spaces because they share so many properties with physical space when equipped with a rectangular coordinate system. We capitalize on this fact by using these spaces to provide meaningful examples of concepts and illustrations of theorems in later parts of the book.

More than the usual emphasis is placed on algorithms for finding all solutions of a system of linear equations, the rank of a matrix, and the minimal polynomial of a linear transformation. The same is true of the development of working methods for the orthogonal diagonalization of self-adjoint transformations (real symmetric matrices) and their complex counterpart and the triangularization of matrices. The exercises in our book reflect our belief that students should not only cope with theoretical problems but be able to grapple with "dirty" computational problems as well.

Finally, we call attention to Chapter 9 which is concerned with several applications to other fields of the theory that have been developed. For centuries scientists have devoted themselves to idealizing problems concerning the physical world so that they may be formulated mathematically. When the idealized mathematical model has been analyzed, the results, interpreted in physical terms, will provide (hopefully) insight into the original problem. More recently, social scientists (in particular, economists) and others have begun to make a determined effort to arrive at more quantitatively based theories through the use of mathematics. In idealizing their problems investigators rely heavily on linear approximations for variables. Thus the resulting mathematical models exhibit instances of linearity. So it is not surprising to find that the theorems of linear algebra have a wide variety of applications. In the examples we have selected to illustrate the fact that not only the elementary but the deeper theorems as well have applications, we have restricted our attention to the analysis of the mathematical model that has been created to investigate a particular problem. Thus it is assumed that the interested reader is sufficiently familiar with the necessary background material to view the problem in its proper setting. The text of Chapter 9 is based on material supplied for this purpose by Dr. Thomas F. Dernburg, Department of Economics, Oberlin College, Dr. Richard N. Porter, Department of Chemistry, University of Arkansas, Dr. Gerard Stephenson, Jr., Department of Physics, University of Maryland, and Dr. William E. Restemeyer, Department of Applied Mathematics, University of Cincinnati.

In conclusion we mention that we have found it possible to cover Chapters 1 through 8, with the exception of the last section in each of Chapters 7 and 8, in a semester course attended primarily by sophomores and juniors.

December 1967
Oberlin, Ohio

R. R. Stoll
E. T. Wong

Contents

Contents

Symbols

1 Vector Spaces

The introductory paragraph in each of the subsequent chapters provides a summary of its contents, plus, in some instances, inducements for examining them. Here, in addition, we insert several remarks concerning the prerequisites for the study of this book. First of all, familiarity with the algebraic properties of the system of real numbers (for which we reserve the boldface letter **R** as a name) and the system of complex numbers (for which we reserve the bold face letter **C** as a name) is presupposed. Second, we assume a working knowledge of that part of intuitive set theory outlined in Sections 1 to 3 of the Appendix. A knowledge of the calculus is necessary to understand some examples and cope with some exercises. Occasionally an example or exercise is taken from a more advanced part of mathematics. A lack of familiarity with the topic under discussion in such cases should not be the cause of alarm and will not impede an understanding of later developments.

We turn now to the contents of this chapter. The most important concept discussed is that of an abstract vector space over **R** or **C**. Later, and after a brief introduction to the notion of a field, it is observed that an arbitrary field may be used in place of either the field of real numbers or that of complex numbers. Examples are given to show that a wide variety of mathematical systems may be classified as vector spaces over some field. Further, we introduce the idea of a subspace of a vector space, the elementary algebra of subspaces, and discuss ways to generate subspaces.

1. VECTORS

The reader is familiar with the concept of a vector or vector quantity in one sense or another; for example, in elementary physics such concepts as displacement and velocity are classified as vector quantities. Statements about vectors that everyone will endorse are, first, a vector quantity can be represented by an arrow (that is, a directed line segment) and, second, such arrows have the following properties:

(a) Two arrows are equal if, and only if, they have the same length and direction (Fig. 1).

(b) Two arrows α and β (with the same initial point) determine an arrow called the sum of α and β (symbolized by $\alpha + \beta$) by way of the parallelogram law (Fig. 2).

(c) A real number (scalar) c and an arrow α determine an arrow called the scalar product of α by c (symbolized by $c \cdot \alpha$) in the following way. The length of $c \cdot \alpha$ is $|c|$ times the length of α; if $c > 0$, then the direction of $c \cdot \alpha$ is that of α while if $c < 0$, then the direction of $c \cdot \alpha$ is opposite that of α. Finally, if $c = 0$, then $c \cdot \alpha$ is the "zero" arrow and its direction is not defined (Fig. 3).

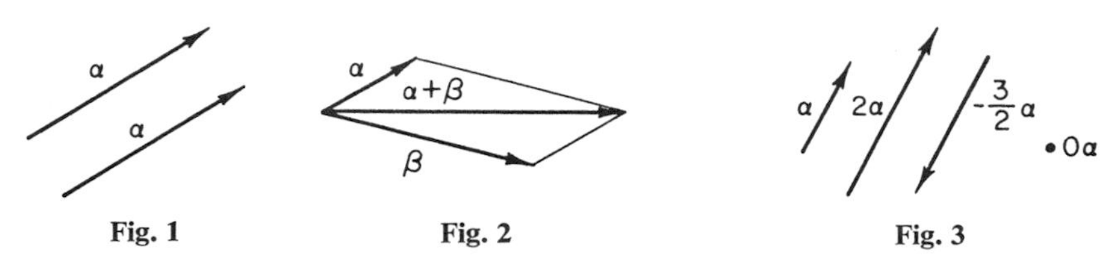

Fig. 1 Fig. 2 Fig. 3

The study of the properties of addition of arrows and the multiplication of an arrow by a scalar can be transformed from a geometric to an algebraic setting by the introduction of a coordinate system in the space at hand; for example, suppose that we are considering arrows of E_3, euclidean 3-space. Let us fix our attention on a cartesian coordinate system for this space. We can then determine, in a well-known way, a 1-1 correspondence between the points of E_3 and the set R^3 of all ordered triples of real numbers. If the point P of E_3 has the ordered triple (x, y, z) as image under the 1-1 correspondence at hand, then x, y, and z are called the coordinates of P. Consider now the arrow with initial point A and terminal point B; this arrow we shall designate by $\overrightarrow{AB}$. If a_1, a_2, a_3, and b_1, b_2, b_3 are the coordinates of A and B, respectively, then $\overrightarrow{AB}$ is equal to the arrow $\overrightarrow{OP}$, where O is the origin of the coordinate system and

P has coordinates $b_1 - a_1$, $b_2 - a_2$, $b_3 - a_3$. Thus each arrow is equal to one having the origin as its initial point. Two such arrows are equal if and only if they are identical. Hence these arrows are in 1-1 correspondence with the points of E_3: to the point P corresponds the arrow $\overrightarrow{OP}$ (the position arrow of P). In turn, a 1-1 correspondence of the set of position arrows is induced with the set R^3. If the point P has coordinates x, y, and z, we shall identify the position vector $\overrightarrow{OP}$ with the ordered triple (x, y, z). Parenthetically, we note that in any discussion involving both points and arrows the context will make clear whether a triple (x, y, z) denotes a point or an arrow. For number triples serving as names of position arrows, properties (a), (b), and (c) can be restated as follows (the proofs are left as an exercise):

(a)′ Equality. $(x, y, z) = (x', y', z')$ if and only if $x = x'$, $y = y'$, and $z = z'$.

(b)′ Addition. $(x, y, z) + (x', y', z') = (x + x', y + y', z + z')$.

(c)′ Scalar multiplication. $c \cdot (x, y, z) = (cx, cy, cz)$.

One who is intent on generalizations of the forcgoing will see immediately the possibility of *defining* (i) a point in n-space (n a positive integer) as an n-tuple $(x_1, \ldots, x_n)$ of numbers, (ii) an arrow in n-space as an ordered pair (A, B) of points (the initial and terminal points, in that order, of the arrow) and, (iii) equality of arrows $((a_1, \ldots, a_n), (b_1, \ldots, b_n))$ and $((a_1', \ldots, a_n'), (b_1', \ldots, b_n'))$ as meaning that $b_i - a_i = b_i' - a_i'$, $1 \leq i \leq n$. Extending the definition (a)′ of equality of ordered triples to that of n-tuples in the obvious way, we find that an arrow is equal to one whose initial point is $O = (0, \ldots, 0)$ and, in turn, that such arrows (let us call them position arrows), may be identified with n-tuples of real numbers. Next, operations of addition and scalar multiplication can be defined for n-tuples by imitating the results given in (b)′ and (c)′. These definitions read as follows:

$$(x_1, \ldots, x_n) + (y_1, \ldots, y_n) = (x_1 + y_1, \ldots, x_n + y_n)$$

$$c \cdot (x_1, \ldots, x_n) = (cx_1, \ldots, cx_n)$$

In turn, the system consisting of R^n and the operations of addition and scalar multiplication defined above is susceptible to generalizations. The one we wish to discuss now has its origin in the observation that the definitions of addition of n-tuples and multiplication of an n-tuple by a scalar employ only the operations of addition and multiplication of real numbers. Thus these definitions are meaningful when the coordinates of n-tuples and scalars are elements of any set of objects for which operations of addition and multiplication are defined, for instance, the system C of complex numbers or the system of rational numbers. Actually, there are many systems consisting of a set and two operations which have sufficient properties in common with the systems R and C that they can be used in place of R or C in what follows. The technical term used as a name for such systems is *field.* Later we shall give a precise definition of this concept. Until then, when we use the term "a field F," the reader should think of F as R or C, each equipped with its familiar operations.

The concept of a vector space given in the next section is an abstraction of the systems that consist of the set F^n of all n-tuples (for some one positive integer n) of elements of a field F and the operations of addition of n-tuples and multiplication of an n-tuple by a scalar (an element of F), as already defined. We comment later on the phrase "is an abstraction of."

2. DEFINITION OF A VECTOR SPACE

The reader who has had no experience with abstract mathematics may find this definition somewhat formidable. In that event the comments and examples that follow it should be carefully studied.

□ **Definition.** Let $F = \{a, b, \ldots\}$ be a field. A nonempty set $V = \{\alpha, \beta, \ldots\}$ is called a *vector space over* F if and only if

(a) there is defined in V a binary operation (called *addition* and symbolized by $+$) with the following properties:

(1) $\alpha + \beta = \beta + \alpha$ for all α, β in V;
(2) $\alpha + (\beta + \gamma) = (\alpha + \beta) + \gamma$ for all α, β, γ in V;
(3) there is a unique element, 0, in V (called the zero vector) such that $\alpha + 0 = \alpha$ for all α in V;
(4) for each α in V there is a unique element, $-\alpha$, in V such that $\alpha + (-\alpha) = 0$;

(b) there is defined an operation (called *scalar multiplication* and symbolized by $\cdot$) that assigns to each ordered pair (a, α), with a in F and α in V, a unique element of V, to be denoted by $a \cdot \alpha$, such that the following relations hold for all elements a, b of F and all elements α, β of V:

(5) $$a \cdot (\alpha + \beta) = (a \cdot \alpha) + (a \cdot \beta);$$
(6) $$(a + b) \cdot \alpha = (a \cdot \alpha) + (b \cdot \alpha);$$
(7) $$(ab) \cdot \alpha = a \cdot (b \cdot \alpha);$$
(8) $$1 \cdot \alpha = \alpha$$ where 1 is the identity element for multiplication in F.

The elements of a vector space V over a field F will be called *vectors* and denoted by lower-case Greek letters in all theoretical considerations. The elements of the field F will be called *scalars* and denoted by lower-case Latin letters. The field F, of course, is equipped with operations of addition ($+$) and multiplication ($\cdot$); the sum of scalars a and b is written as $a + b$ and their product as ab. Instances occur on the left-hand sides of relations (6) and (7). Following standard practice, we also symbolize the binary operation defined in V (addition of vectors) by $+$ and the operation of scalar multiplication by $\cdot$. We use the same symbol for the zero vector [see (3)] and the zero scalar; this will cause no confusion.

We turn now to comments of another kind. The definition of a vector space V over a field F focuses attention on a set whose elements are called vectors. No mention is made of *what* objects are vectors; hence "vector" is an undefined term. Since the same definition does not specify which element of V is the zero vector, "zero vector" is another undefined term. Moreover, the definition does not specify which function of $V \times V$ into V is called addition in V; so the "addition operation" is a further undefined term. Similar remarks apply to the terms "field" and "scalar multiplication." Turning now to items (1) to (8) of the definition, we note that they constitute a set of assumptions (axioms) about the undefined terms. To continue our discussion, let us agree to use the term *model* (of the set of undefined terms in the definition of a vector space) to refer to any system consisting of a set V, a binary operation $+$ in V, an element 0 of V, a field F, and a function $\cdot$ on $F \times V$ into V. The notation

$$(V, +, 0, F, \cdot)$$

is convenient for designating such a system. If items (1) to (8) of our definition are satisfied by a model, we call it a *model of the theory of vector spaces* or simply *a vector space*.

It is a simple matter to show that

$$(\mathsf{R}^n, +, (0, \ldots, 0), \mathsf{R}, \cdot),$$

where $+$ and $\cdot$ are the operations defined in Section 1, is a vector space.† If we had taken the time to investigate properties of this system with applications to geometric problems in mind, we would have found that those properties which may now be summarized by the statement, "R^n is

† Incidentally, whenever R^n is construed to be a vector space, it is understood that the operations of addition and scalar multiplication are those defined in Section 1 unless it is stated otherwise.

a vector space over R," are basic. This is the reason for an earlier remark to the effect that our definition is an abstraction of the system consisting of R^n, etc.

Why do we choose to study vector spaces from an abstract point of view? As a preliminary to our answer, we shall make several remarks about axiomatic theories in general. Once an axiomatic theory has been formulated by way of a specified set of undefined terms and a set of statements (the axioms of the theory) about the undefined terms, we attempt to derive from the assumed properties of these terms further properties in a way that may be described as follows. By a *proof* of the theory at hand we mean a finite sequence of statements formulated in terms of the undefined terms and, possibly, set-theoretic notions, each of which can be classified as an instance of an axiom or an instance of one of the assumed set-theoretic principles (see the Appendix) or a logical consequence of one or more of the earlier statements in the sequence. The last statement of a proof is called a *theorem* of the theory. Those statements of the theory that qualify as theorems are the properties of the undefined terms of interest. Since they have been derived from the axioms by logic alone (therefore without an assignment of meaning to the undefined terms), it follows that they determine a true statement about *every* model of the theory, In particular, if a model of the undefined terms of the theory of vector spaces is actually a vector space, then the statement about that vector space which results from the assignment of meaning to the occurrences of the undefined terms in a theorem (as specified by the model) is a true statement about that vector space. Now it happens that a wide variety of concrete mathematical systems are vector spaces. Thus the axiomatic (or abstract) approach to the study of vector spaces is an efficient one.

We turn now to our initial set of examples of vector spaces. Proofs that the systems discussed are indeed vector spaces are to be provided by the reader.

EXAMPLES

2.1. A *polynomial function* f over a field F is a function on F into F for which there exists a nonnegative integer n and an $(n + 1)$-tuple $(a_0, \ldots, a_n)$ of elements of F such that

$$f(x) = a_0 + a_1x + \cdots + a_n x^n \tag{9}$$

for all $x \in F$. The set of all polynomial functions over F will be symbolized by

$$P_\infty(F).$$

Define the sum $f + g$ of two polynomials f and g by $(f + g)(x) = f(x) + g(x)$ and the product $a \cdot f$ of a scalar a and a polynomial f by $(a \cdot f)(x) = af(x)$. Then $(P_\infty(F), +, 0, F, \cdot)$, where 0 is the *zero polynomial* (that is, the constant function 0) is a vector space or, more simply, $P_\infty(F)$ is a vector space over F.

2.2. Let f be a polynomial function over F and not the zero polynomial. Suppose that n is the least nonnegative integer such that $f(x)$ can be written in the form of (9). Then n is called the *degree* of f. We symbolize the set of all polynomials over F of degree less than or equal to n, together with the zero polynomial, by

$$P_n(F).$$

The definition of addition of polynomials given in Example 2.1 is an operation in $P_n(F)$; that is, the sum of two polynomials of degree less than or equal to n is a polynomial of degree less than or equal to n. Also the scalar product of an element of F and an element of $P_n(F)$ is an element of $P_n(F)$. In brief, $(P_n(F), +, 0, F, \cdot)$ is a model of the undefined terms in the definition of a vector space. Actually, it is a vector space.

2.3. By a system of m linear homogeneous equations in the n unknowns $x_1, x_2, \ldots, x_n$ over a field F is meant a set of m equations of the form

$$\begin{aligned} a_{11}x_1 + a_{12}x_2 + \cdots + a_{1n}x_n &= 0, \\ a_{21}x_1 + a_{22}x_2 + \cdots + a_{2n}x_n &= 0, \\ &\vdots \\ a_{m1}x_1 + a_{m2}x_2 + \cdots + a_{mn}x_n &= 0, \end{aligned}$$

where each coefficient a_{ij} is in F. A solution of the system is any n-tuple $(c_1, \ldots, c_n)$, with $c_1, \ldots, c_n \in F$, such that on setting $x_1 = c_1, \ldots, x_n = c_n$, each equation is satisfied. The set S of all solutions is nonempty [for $(0, \ldots, 0) \in S$] and is a subset of the set F^n. Addition of n-tuples, as defined for elements of F^n, is an operation in S and a scalar multiple of an n-tuple in S is an element of S (proofs are required). Then S is found to be a vector space over F. We call this space the *solution space* of the system of equations.

2.4. Let us define addition of n-tuples of complex numbers as in the case of n-tuples of real numbers. Clearly, this is an operation in C^n. Further, define the scalar product of a real number a and $(c_1, \ldots, c_n) \in \mathsf{C}^n$ by

$$a \cdot (c_1, \ldots, c_n) = (ac_1, \ldots, ac_n).$$

This is a function on $\mathsf{R} \times \mathsf{C}^n$ into C^n. As the reader may suspect (and should prove), C^n is a vector space over R.

2.5. Let V be the set of all functions that are continuous on some closed interval $[a, b]$. Let $f, g \in V$ and $a \in \mathsf{R}$. Defining $f + g$ and $a \cdot f$ as in Example 2.1, we find that $f + g \in V$ and $a \cdot f \in V$, since the sum of two continuous functions and a constant multiple of a continuous function are continuous functions. Using just elementary properties of functions, we find that V is a vector space over R. Hereafter we denote the set (space) of all functions which are continuous in the interval $[a, b]$ by

$$C[a, b].$$

2.6. Suppose $c \in [a, b]$ and consider

$$S = \{f \in C[a, b] \mid f(c) = 0\}.$$

Defining $f + g$ and $a \cdot f$ as above, we observe that if $f, g \in S$ and $a \in R$ then $f + g$ and $a \cdot f$ are in S. Indeed, S is a vector space over R.

2.7. The reader who has studied differential equations should convince himself that the set of all solutions of an nth-order linear homogeneous differential equation determines a vector space.

We list next several computational rules that hold in any vector space V over a field F. Here we begin to write simply $a\alpha$ for $a \cdot \alpha$.

(10) If $\alpha \in V$, $a \in F$, and $a \neq 0$, then $a\alpha = 0$ if and only if $\alpha = 0$.
(11) If $\alpha \in V$, $\alpha \neq 0$, and $a \in F$, then $a\alpha = 0$ if and only if $a = 0$.
(12) If $\alpha \in V$ and $a \in F$, then $(-a)\alpha = a(-\alpha) = -(a\alpha)$.

In the next rule $\alpha - \beta$ is introduced as an abbreviation for $\alpha + (-\beta)$.

(13) If $\alpha, \beta \in V$ and $a, b \in F$, then $a(\alpha - \beta) = a\alpha - a\beta$ and $(a - b)\alpha = a\alpha - b\alpha$.

To prove (10) we begin with the identity $\alpha + 0 = \alpha$ given in (3). If $a \in F$, then $a(\alpha + 0) = a\alpha$. Using (5), we deduce that $a\alpha + a0 = a\alpha$. Now

$a\alpha \in V$, and so there is in V the element $-(a\alpha)$ such that $a\alpha + (-a\alpha) = 0$ according to (4). Adding $-(a\alpha)$ to each side of the equation under consideration we obtain

$$(a\alpha + a0) - (a\alpha) = a\alpha - (a\alpha) = 0.$$

It follows that $a0 + 0 = 0$, and hence, using (3), that $a0 = 0$. Thus, if $\alpha = 0$, then $a\alpha = 0$. To prove the converse (with $a \neq 0$), assume that $a\alpha = 0$. Then a has a reciprocal $1/a$, and multiplying the equation $a\alpha = 0$ by $1/a$ gives us

$$\left(\frac{1}{a}\right)(a\alpha) = \left(\frac{1}{a}\right)0.$$

Using (7), then the identity $a(1/a) = 1$, and then (8), we have

$$\left(\frac{1}{a}\right)(a\alpha) = \left(\frac{1}{a}\right)a\alpha = 1\alpha = \alpha.$$

Further, by the result proved above, $(1/a)0 = 0$. Thus the equation $(1/a)(a\alpha) = (1/a)0$ reduces to $\alpha = 0$ and the proof is complete.

To prove (11) we begin with the identity $a + 0 = a$, where 0 is the zero scalar. If $\alpha \in V$, then $(a + 0)\alpha = a\alpha$. Using (6), we deduce that $a\alpha + 0\alpha = a\alpha$. Adding $-(a\alpha)$ to each member of this equation, we infer that $0\alpha = 0$. To prove, conversely, that if $a\alpha = 0$ where $\alpha \neq 0$, we need observe merely that if a were not 0 then, by (10), α would be 0, contrary to hypothesis.

Proofs of (12) and (13) are left as exercises. Before stating other rules, we insert some remarks. The associative law of addition of vectors [property (2)] asserts that the two possible ways to add three vectors α, β, and γ, namely

$$(\alpha + \beta) + \gamma \quad \text{and} \quad \alpha + (\beta + \gamma),$$

yield the same vector. From this assumption it can be proved by mathematical induction that all possible ways of adding n vectors $\alpha_1, \ldots, \alpha_n$, in that order, to form a single sum yield a uniquely determined vector which we denote by

$$\alpha_1 + \cdots + \alpha_n \quad \text{or} \quad \sum_{i=1}^{n} \alpha_i.$$

The commutative law of addition of vectors [property (1)] states that the sum of two vectors is independent of their order. This, too, can be extended (again the proof is by mathematical induction) to the case of n vectors; the generalization asserts that $\alpha_1 + \cdots + \alpha_n$ is independent of the order of the summands. We state as our next two rules similar extensions of (5) and (6).

(14) If $a \in F$ and $\alpha_1, \ldots, a_n \in V$, then

$$a \sum_{i=1}^{n} \alpha_i = \sum_{i=1}^{n} a\alpha_i .$$

(15) If $a_1, \ldots, a_n \in F$ and $\alpha \in V$, then

$$\left(\sum_{i=1}^{n} a_i\right)\alpha = \sum_{i=1}^{n} a_i \alpha.$$

These extensions are also established by mathematical induction as is next rule.

(16) $$\sum_{i=1}^{n} a_i \alpha_i + \sum_{i=1}^{n} b_i \alpha_i = \sum_{i=1}^{n} (a_i + b_i)\alpha_i .$$

EXERCISES

2.1. Referring to the definition of a vector space, show that properties (3) and (4) can be weakened to
(3)′ there exists an element, 0, such that $\alpha + 0 = \alpha$ for all α in V,
(4)′ for each α in V there exists an element β such that $\alpha + \beta = 0$,
by proving that if (3)′ and (4)′ hold then (3) and (4) necessarily follow.

2.2. Prove the computational rules (12) and (13).

2.3. Prove that if 0 is the zero vector, then $a0 = 0$ for any scalar a, without using either the fact that if $a \neq 0$ then it has a reciprocal or the assumption that $1\alpha = \alpha$ for any vector α.

2.4. For the set $\mathbf{R}^2$ define operations of addition and multiplication by a scalar as follows:

$$(x, y) + (x', y') = (3y + 3y', \ -x - x'),$$

$$a(x, y) = (3ay, \ -ax).$$

Is this system a vector space over $\mathbf{R}$? Justify your answer.

2.5. Show that $P_\infty(F)$ and $P_n(F)$ are vector spaces over F.

2.6. Supply proofs of the assertions made in Example 2.3.

2.7. Supply proofs of the assertions made in Examples 2.5 and 2.6.

2.8. Let V be a vector space over R. In the set $W = V \times V$ define addition and scalar multiplication by a complex number as follows:

$$(\alpha_1, \beta_1) + (\alpha_2, \beta_2) = (\alpha_1 + \alpha_2, \beta_1 + \beta_2),$$

$$(a + ib)(\alpha, \beta) = (a\alpha - b\beta, a\beta + b\alpha).$$

Show that this system is a vector space over C.

2.9. Let R^+ be the set of all positive real numbers. Define addition in R^+ by

$$x + y = xy \text{ (ordinary multiplication)}$$

and scalar multiplication by a real number a as follows:

$$a \cdot x = x^a.$$

Prove that this system is a vector space over R.

3. SUBSPACES AND THEIR ALGEBRA

Earlier, as a matter of convenience, we subsumed the notion of a theorem of an axiomatic theory to that of a proof of the theory. But in practice it is the class of theorems of a theory that occupies the central position—we formulate statements that we have some reason to believe might be theorems and then attempt to construct a proof. How does a research worker acquire "some reason to believe" that a statement of a theory might be a theorem? In the case of vector spaces he relies heavily on models of the theory for guidance; specifically, he looks for true statements of a model that translate into true statements about other models with which he is familiar. Such statements become candidates for theoremhood! We also rely on models for guidance in defining concepts that might be explored with profit. Geometry is a natural source of helpful models because the theory of vector spaces is an outgrowth of algebraic notions and methods that were devised to study certain aspects of geometry. Let us consider an illustration.

In Section 1 we showed that in relation to some one cartesian coordinate system for E_3 there is determined a 1-1 correspondence f, let us say, between the set of ordered triples of real numbers and that of position arrows. The mapping f has the following additional properties. If $f(x, y, z) = \overrightarrow{OP}$ and $f(x', y', z') = \overrightarrow{OP}'$, then

$$f[(x, y, z) + (x', y', z')] = \overrightarrow{OP} + \overrightarrow{OP}'$$

and

$$f[c(x, y, z)] = c \cdot \overrightarrow{OP};$$

that is, f is a 1-1 mapping of the vector space R^3 onto the vector space of position vectors which preserves sums and scalar multiples. We shall call such a function an *interpretation* of R^3. Another interpretation of R^3 may be obtained as follows. Let us define addition of points of E_3 and multiplication of a point by a scalar via the corresponding operations for the associated position vectors. The result is a vector space of points. The function g which maps each element (x, y, z) of R^3 onto the points having x, y, and z as coordinates is clearly an interpretation of R^3.

In a study of E_3 those sets of points called lines and planes play an important role. From the standpoint of vector spaces, lines and planes that contain the origin are distinguished because, with the operations of addition and scalar multiplication mentioned above, they qualify as vector spaces. It is the analogue of these spaces that we shall consider for vector spaces in general.

□ **Definition.** Suppose that W and V are vector spaces over the same field F such that

(i) $W \subseteq V$,
(ii) the restriction of the operation of addition in V to elements of W is the operation of addition in W,
(iii) the restriction of the operation of scalar multiplication for V to elements of W is the operation of scalar multiplication for W.

Then W is called a *subspace* of V.

Let us call a subset S of a vector space V over F *closed under addition* if and only if $\alpha + \beta \in S$ for all $\alpha, \beta \in S$. Similarly, we shall call S *closed under scalar multiplication* if and only if for $a \in F$ and $\alpha \in S$ we have $a\alpha \in S$. Then the subspaces of V are precisely those nonempty subsets of V that are closed under both operations and satisfy the definition of a vector space.

EXAMPLES

3.1. The subsets $\{0\}$ and V of a vector space V are subspaces of V. They are called its trivial subspaces. In Chapter 2 we shall consider the question of whether every vector space has a basis.

3.2. The vector space S of Example 2.3 is a subspace of F^n. The vector space $P_n(F)$ is a subspace of $P_\infty(F)$.

3.3. In this example is an illustration of the remark that a subset of a vector space V over a field F may be a vector space over F yet not a subspace of V. By virtue of the first statement in Example 2.3 (with $m = 1$ and $n = 3$) the set

$$S = \{(x, y, z) \in \mathsf{R}^3 \mid x + y + z = 0\}$$

is a subspace of R^3. Let $\alpha \in \mathsf{R}^3 - S$ [for example, (1, 1, 1) is such an element] and define W by

$$W = \{\alpha + \beta \mid \beta \in S\}.$$

In W we define addition, symbolized $\oplus$, by

$$(\alpha + \beta_1) \oplus (\alpha + \beta_2) = \alpha + (\beta_1 + \beta_2),$$

and scalar multiplication, symbolized $\odot$, by

$$a \odot (\alpha + \beta) = \alpha + a\beta.$$

It is left as an exercise for the reader to show that $(W, \oplus, \alpha, R, \odot)$ is a vector space and that it is not a subspace of R^3.

This example becomes meaningful when we introduce the interpretation g of R^3 which assigns to the vector (x, y, z) the point in E_3 having x, y, and z as coordinates. The reader is asked to convince himself that the images of the nontrivial subspaces of R^3 under g are the totality of lines and planes through the origin and, in particular, that the image of S in E_3 is the plane P defined by the equation $x + y + z = 0$. Then it will be clear that the image of W is the plane in the position that P would occupy after a translation by the arrow (1, 1, 1), if for α we choose the vector (1, 1, 1) of R^3. Finally, it should be noted that the technique for fabricating a vector space out of W is based on forcing α to play the role of the zero vector.

Suppose now that W is a subspace of the vector space V. The reader is asked to prove that the zero vector of W is the zero vector 0 of V and, consequently, that if the vector α of V is in W then $-\alpha$ is in W. Thus a subspace of V is closed under addition and scalar multiplication, contains 0, and contains $-\alpha$ if it contains α. We contend that, conversely, a subset S of V with these properties is a subspace of V. Indeed, the only clauses of the definition of a vector space that do not appear in this set of properties are those which hold throughout V (for example, the commutative law of addition) and, hence, automatically hold for elements of a subset of V. According to the next theorem, this set of conditions can be simplified further.

☐ **Theorem 3.1.** A nonempty subset S of a vector space V over F is a subspace of V if and only if $a\alpha + b\beta \in S$ whenever $a, b \in F$ and $\alpha, \beta \in S$.

PROOF. Assume that S is a subset of V that satisfies the given conditions. Then, if $a, \beta \in S$, so is $1\alpha + 1\beta = \alpha + \beta$. If $a \in F$ and $\alpha \in S$, then so is $a\alpha + 0\alpha = a\alpha$, and S has the necessary closure properties. Further, $0 \in S$, for since there exists a vector α in S by assumption, $(-1)\alpha + 1\alpha = 0$ in S. Finally, if $\alpha \in S$, then so is $(-1)\alpha = -\alpha$. Thus by the remarks above S is a subspace.

Conversely, if S is a subspace of V, then certainly the conditions stated in the theorem hold for S in view of its closure properties. ◇

☐ **Theorem 3.2.** The intersection of a nonempty collection of subspaces of a vector space V is a subspace of V.

PROOF. Let $\mathscr{C}$ be a nonempty collection of subspaces of V and let $S = \cap\mathscr{C}$. It is a straightforward exercise to show that S satisfies the conditions of Theorem 3.1. ◇

With the position arrows of E_3 as a model of R^3 in mind, the reader can easily convince himself that in contrast to the intersection of a collection of subspaces the union of a collection of subspaces of a vector space need not be a subspace. However, there does exist an analogue of Theorem 3.2. for unions. For it a definition is required.

☐ **Definition.** The sum $S_1 + \cdots + S_p$ of the subspaces $S_1, \ldots, S_p$ of a vector space V is $\{\alpha_1 + \cdots + \alpha_p \mid \alpha_i \in S_i\}$.

☐ **Theorem 3.3.** The sum $S = S_1 + \cdots + S_p$ of the subspaces $S_1, \ldots, S_p$ of a vector space V is a subspace of V and the smallest subspace of V that includes each S_i.

PROOF. The proof that S is a subspace of V is left as an exercise. It is clear that each $S_i \subseteq S$, since the zero vector is in each subspace. Finally, S is the smallest such subspace in the sense that it is included in each subspace that includes every S_i. For proof suppose that S' is a subspace that includes every S_i. Let $\alpha_i \in S_i$, $1 \le i \le p$. Then each α_i is an element of S'; hence $\alpha_1 + \cdots + \alpha_p \in S'$ since S' is closed under the operation of addition. Thus $S \subseteq S'$. ◇

Suppose that S is merely some subset of a vector space V. There is a subspace of V, namely V itself, that includes S. Hence the intersection S' of the set of all subspaces of V that includes S is defined. It is a subspace of V that includes S and is clearly the smallest. If $S = \varnothing$, we notice that $S' = \{0\}$, but, in general, our definition of S' does not provide us with a description of its elements. We proceed to remedy this matter.

□ **Definition.** Let S be a nonempty subset of a vector space V. A vector α in V is called a *linear combination* of vectors of S if and only if there exists a positive integer n, scalars $a_1, \ldots, a_n$, and vectors $\alpha_1, \ldots, \alpha_n$ in S such that

$$\alpha = \sum_{i=1}^{n} a_i \alpha_i .$$

We denote the set of all linear combinations of vectors of S by $[S]$. If $S = \varnothing$, we define $[S]$ to be $\{0\}$. If S is a finite set, say $S = \{\alpha_1, \ldots, \alpha_p\}$, we may write $[\alpha_1, \ldots, \alpha_p]$ in place of $[S]$ and refer to a linear combination of elements of S as a linear combination of $\alpha_1, \ldots, \alpha_p$.

EXAMPLES

3.4. If $S = \{\alpha\}$, then $[S] = \{a\alpha \mid a \in F\}$. More generally, if $S = \{\alpha_1, \ldots, \alpha_p\}$, then

$$[S] = \{a_1\alpha_1 + \cdots + a_p\alpha_p \mid a_1, \ldots, a_p \in F\}.$$

3.5. If, in R^3, $S = \{\epsilon_1, \epsilon_2, \epsilon_3\}$, where

$$\epsilon_1 = (1, 0, 0),$$

$$\epsilon_2 = (0, 1, 0),$$

$$\epsilon_3 = (0, 0, 1),$$

then $[S] = \mathsf{R}^3$.

3.6. If, in R^3, $S = \{(1, 1, 0), (0, 1, 1)\}$, then

$$\begin{aligned}[S] &= \{a(1, 1, 0) + b(0, 1, 1) \mid a, b \in \mathsf{R}\} \\ &= \{(a, a + b, b) \mid a, b \in \mathsf{R}\} \\ &= \{(x, y, z) \mid x - y + z = 0\}.\end{aligned}$$

3.7. In the space $P_\infty(\mathsf{R})$ let $S = \{1, x^2, \ldots, x^{2n}, \ldots\}$; then $[S]$ is the space of all polynomials p over R such that $p(-x) = p(x)$.

3.8. In the space $C[a, b]$ let

$$S = \{\sin nx \mid n = 1, 2, \ldots\} \cup \{\cos mx \mid m = 0, 1, 2, \ldots\}.$$

What is $[S]$? What useful role does $[S]$ play in its parent space?

□ **Theorem 3.4.** Let S be a subset of a vector space V. Then $[S]$ is the smallest subspace of V that includes S. We call $[S]$ the subspace *spanned* (or *generated*) by S.

PROOF. This is left as an exercise. ◇

□ **Definition.** A subset S of a vector space V is called a *set of generators* of V if and only if $[S] = V$. A vector space V is called *finite-dimensional* if and only if there exists a finite set of generators of V; if not, then V is called *infinite-dimensional.* A set B of generators of V is called a *basis* of V if and only if no proper subset of B is a set of generators of V.

EXAMPLES

3.9. In R^n let $S = \{\epsilon_1, \ldots, \epsilon_n\}$, where ϵ_i is the n-tuple with 1 as its ith coordinate and 0 as each of its other coordinates. Since $(a_1, \ldots, a_n) = \sum_1^n a_i \epsilon_i$, S is a set of generators of R^n. Since S is finite, R^n is finite-dimensional. Finally, S is a basis of R^n, for otherwise it would be possible to express some ϵ_i as a linear combination of the others. This is impossible as the reader can show.

3.10. In the space $P_\infty(\mathsf{R})$ let $S = \{1, x, \ldots, x^n, \ldots\}$. Then S is a set of generators and indeed is a basis of $P_\infty(\mathsf{R})$. (To show that S is a basis, the reader could begin with the assumption that for some nonnegative integer m the polynomial function x^m is equal to a polynomial function in which x^m does not appear and derive a contradiction.) Further, $P_\infty(\mathsf{R})$ is infinite-dimensional.

3.11. A finite-dimensional vector space has a basis; for if V is such a space, then, by definition, there exists a finite subset S of V such that $V = [S]$. If S is not a basis, then some proper subset S' of S spans V. If S' is not a basis, the foregoing step can be repeated. Proceeding in this fashion, we obtain a chain $S \supset S' \supset S'' \supset \cdots$ of subsets that must break off with some $S^{(k)}$, for S is finite and each succeeding subset has fewer elements than its predecessor. Then $S^{(k)}$ is a basis of V.

We continue our discussion of subspaces of a vector by introducing a further concept.

☐ **Definition.** If S is a subspace of a vector space V, then a subspace $\bar{S}$ of V is called a *complementary subspace* of S or a *complement*† of S if and only if

$$S \cap \bar{S} = \{0\} \quad \text{and} \quad S + \bar{S} = V.$$

☐ **Theorem. 3.5** Let S be a subspace of a vector space V. A subspace S' of V is a complementary subspace of S if and only if each vector α in V has a unique representation in the form $\alpha = \beta + \beta'$, where $\beta \in S$ and $\beta' \in S'$.

PROOF. Assume that S' is a complement of the subspace S of V. Then $V = S + S'$, which means that each α in V is the sum of a vector β in S and a vector β' in S'. Suppose that we also have $\alpha = \gamma + \gamma'$, where $\gamma \in S$ and $\gamma' \in S'$. Then $\beta + \beta' = \gamma + \gamma'$ or $\beta - \gamma = \gamma' - \beta'$, where, since S and S' are subspaces, $\beta - \gamma \in S$ and $\gamma' - \beta' \in S'$. Hence, since $S \cap S' = \{0\}$, we have $\beta - \gamma = 0$ and $\gamma' - \beta' = 0$. It follows that $\beta = \gamma$ and $\beta' = \gamma'$, which proves the uniqueness of the representation of α.

For the converse we assume that S and S' are subspaces of V such that each α in V has a unique representation in the form $\alpha = \beta + \beta'$ with $\beta \in S$ and $\beta' \in S'$. Then $V = S + S'$. Next, let $\alpha \in S \cap S'$. Then α is a member of both subspaces and consequently has the representations $\alpha = \alpha + 0$ (where $\alpha \in S$ and $0 \in S'$) and $\alpha = 0 + \alpha$ (where $0 \in S$ and $\alpha \in S'$). By the assumed uniqueness property it follows that $\alpha = 0$. Thus $S \cap S' = \{0\}$ and the proof is complete. ◇

If S and $\bar{S}$ are subspaces of a vector space V and $\bar{S}$ is a complement of S, then clearly S is a complement of $\bar{S}$; that is, the relation "is a complement of" is symmetric. The distinguishing features of such a pair of subspaces is generalized in the following definition.

† The notion of a complement of a subspace is distinct from the set-theoretic concept having the same name.

☐ **Definition.** A vector space V is said to be the *direct sum* of subspaces $S_1, \ldots, S_k$, symbolized

$$V = S_1 \oplus \cdots \oplus S_k,$$

if and only if $V = S_1 + \cdots + S_k$ and each vector α in V has a unique representation in the form $\alpha = \beta_1 + \cdots + \beta_k$ with $\beta_i \in S_i$, $1 \le i \le k$.

EXAMPLES

3.12. Let us find a complement of the subspace $S = \{(x, y, z) \mid 3x - 2y + z = 0\}$ of R^3. With the elements of R^3 interpreted as points of E_3, the image of S is the plane P which has $3x - 2y + z = 0$ as an equation. A plausible choice for a complement of S is the inverse image of a line through the origin of E_3 and which does not lie in P. The line $L: x = y = z$ meets these requirements. It is the image of $\bar{S} = [(1, 1, 1)]$ in R^3. We shall prove that $\bar{S}$ is a complement of S. To show that $S \cap \bar{S} = \{(0, 0, 0)\}$ we notice that a vector (x, y, z) in this intersection simultaneously has the form $(x, y, -3x + 2y)$ and (z, z, z). Thus x, y, and z must satisfy the following requirements:

$$\begin{aligned} x \qquad - z &= 0, \\ y - z &= 0, \\ -3x + 2y - z &= 0. \end{aligned}$$

This system of linear homogeneous equations has $(0, 0, 0)$ as its only solution. Hence $S \cap \bar{S}$ consists of the zero vector alone. Further, the system of equations

$$\begin{aligned} x \qquad + z &= a, \\ y + z &= b, \\ -3x + 2y + z &= c, \end{aligned}$$

has a solution for each choice of a, b, and c. Hence $S + \bar{S} = \mathsf{R}^3$. It is clear that any line that, like L, contains the origin and does not lie in P may be used to construct a complement of S. Thus S has infinitely many complements.

3.13. The subset $S = \{p \in P_\infty(\mathsf{R}) \mid p(0) = 0\}$ of $P_\infty(\mathsf{R})$ is a subspace. The subspace of $P_\infty(\mathsf{R})$ which consists of the zero polynomial and all polynomials of degree 0 is a complement of S.

The question may come to mind as to whether there is some systematic method for constructing a complement of a given subspace. Actually, the first question which should be considered is that of the existence of a complement. In Chapter 2 we dispose of both questions for subspaces of a finite-dimensional vector space by describing a constructive method for obtaining some complement of a given subspace. For subspaces of infinite dimensional spaces we can prove only the existence of a complement; this is our next theorem. Its proof employs the version of a deep set theoretical principle known as "Zorn's lemma." The beginner may find this lemma (which is stated in Section 4 of the Appendix) to be more abstruse than the theorem whose proof requires the lemma. In that event we suggest that the reader simply accept the statement of the theorem since thereby nothing will be lost.

□ **Theorem 3.6.** For each subspace of a vector space V there exists a complement.

PROOF. Let S be a subspace of V. If $S = V$, then $\{0\}$ is a complement of V. Assume that $S \neq V$ and let $\alpha \in V - S$. Then $S \cap [\alpha] = \{0\}$. Suppose $a\alpha = \delta$, where a is a scalar and $\delta \in S$. If $a \neq 0$, then $a^{-1}(a\alpha) = a^{-1}\delta$, $1\alpha = a^{-1}\delta$, and $\alpha = a^{-1}\delta$. Hence $\alpha \in S$; but, by assumption, $\alpha \notin S$. Hence $a = 0$ and 0 is the only vector common to S and $[\alpha]$.

Now consider the nonempty (by the observation above) collection $\mathcal{S}$ of all subspaces of V whose intersection with S is $\{0\}$. Clearly $\mathcal{S}$ is partially ordered by the relation of inclusion. Let $\mathcal{C}$ be a subchain of $\mathcal{S}$. Then $M = \cup\mathcal{C}$ is easily seen to be a subspace of V and, since $M \cap S = \{0\}$, it is in $\mathcal{S}$. Further, M is an upper bound for $\mathcal{C}$. Since the hypotheses of Zorn's lemma are satisfied, we may infer the existence of a maximal element $\bar{S}$ in $\mathcal{S}$. We shall prove that $\bar{S}$ is a complement of S. Since $\bar{S} \in \mathcal{S}$, $S \cap \bar{S} = \{0\}$. To show that $S + \bar{S} = V$, suppose, to the contrary, that there exists an element β in $V - (S + \bar{S})$. Then, as in the first paragraph of this proof, we may conclude that $[\beta] \cap (S + \bar{S}) = \{0\}$. In turn, we may infer that $([\beta] + \bar{S}) \cap S = \{0\}$ as follows. Suppose that γ is a member of the intersection. Then $\gamma \in S$ and there exists a scalar a and an element $\bar{\delta}$ in $\bar{S}$ such that $a\beta + \bar{\delta} = \gamma$. If $a \neq 0$, we may conclude that $\beta \in \bar{S} = S$, contrary to the choice of β. Thus $a = 0$ and γ is a vector in both S and $\bar{S}$; therefore $\gamma = 0$.

At this point we have shown that if $V - (S + \bar{S}) \neq \varnothing$ then there is a subspace $([\beta] + \bar{S})$ of S that properly includes $\bar{S}$ (note that $\beta \neq 0$). This is impossible since $\bar{S}$ is a maximal element of S. Thus $V-(S+\bar{S})=\emptyset$ and so $V=S+\bar{S}$. ◇

EXERCISES

3.1. Which of the following sets of vectors $\alpha=(x_1, \ldots, x_n)$ in R^n $(n>3)$ are subspaces of R^n?
(a) All α such that $x_1>0$.
(b) All α such that $x_1+3x_2=x_3$.
(c) All α such that $x_1x_2=0$.
(d) All α such that $x_2=x_1{}^2$.

3.2. Let V be the real vector space of all continuous functions f from R into R. Which of the following subsets of V are subspaces?
(a) All f such that $f(x^2)=(f(x))^2$.
(b) All f such that $f(0)=f(1)$.
(c) All f such that $f(3)=1+f(-5)$.
(d) All f such that $f(a)=0$, where a is fixed.

3.3. Let S_1 and S_2 be subspaces of a vector space V. Show that $S_1\cup S_2$ is a subspace of V if and only if one of the given subspaces is included in the other.

3.4. For each of the following choices of the set S of vectors in R^3 determine $[S]$.
(a) $S=\{(1, 1, 1)\}$.
(b) $S=\{(1, 0, 1), (2, 1, 0), (1, 1, -1)\}$.
(c) $S=\{(0, 1, -1), (2, 0, 1), (0, -1, 0)\}$.

3.5. For nonzero vectors α and β of a vector space V, show that $[\alpha]=[\beta]$ if and only if $\beta=a\alpha$ for some scalar a.

3.6. Suppose that $S_1=\{\alpha_1, \ldots, \alpha_m\}$ and $S_2=\{\beta_1, \ldots, \beta_n\}$ are two finite subsets of a vector space. Show that $[S_1]=[S_2]$ if and only if $\alpha_i\in[S_2]$, $1\le i\le m$, and $\beta_j\in[S_1]$, $1\le j\le n$.

3.7. The functions f and g defined by

$$f(x)=\begin{cases}0 & \text{if } 0\le x\le\frac{1}{2}\\ x-\frac{1}{2} & \text{if } \frac{1}{2}\le x\le 1\end{cases}\qquad g(x)=\begin{cases}-x+\frac{1}{2} & \text{if } 0\le x\le\frac{1}{2}\\ 0 & \text{if } \frac{1}{2}\le x\le 1\end{cases}$$

are elements of $C[0, 1]$. Find $[f]$, $[g]$, and $[f, g]$.

3.8. Does the set of functions $\{\sin^2 x, \cos^2 x, \sin x, \cos x\}$ span the same subspace of $C[a, b]$ as $\{1, \sin 2x, \cos 2x\}$?

3.9. Let $S=\{(x, y, z)\in\mathsf{R}^3 \mid 3x-y+z=0\}$. Show that S is a subspace of R^3 and find a basis of S.

3.10. Show that $\{(1, 0, 0)\ (1, 1, 0), (1, 2, 1)\}$ is a basis of R^3.

3.11. Let $\mathscr{B}$ be a basis for vector space V. Prove that
(a) for any finite subset $\{\alpha_1, \ldots, \alpha_n\}$ of distinct elements of $\mathscr{B}$, if $a_1\alpha_1 + \cdots + a_n\alpha_n = 0$, then $a_1 = \cdots = a_n = 0$;
(b) every nonzero vector in V can be written uniquely as a linear combination of distinct elements of $\mathscr{B}$, assuming that only nonzero summands are included.

3.12. Let $W = \{p \in P_n(\mathsf{R}) \mid p(0) = p(1) = 0\}$.
(a) Prove that W is a subspace of $P_n(\mathsf{R})$ and find a basis of W.
(b) Extend the basis found for W in (a) to a basis of $P_n(\mathsf{R})$.

3.13. Show that a subspace S of a vector space V has a unique complement if and only if S is a trivial subspace.

3.14. Find a complementary subspace of S if
(a) S consists of all polynomials p such that $p(-x) = p(x)$, in the space $P_\infty(\mathsf{R})$.
(b) S consists of all functions f such that $f(0) = 0$ in the space of all functions defined in $[-a, a]$.

3.15. Let S be a nontrivial subspace of a vector space V and $\bar{S}$ be a complement of S. Show that if $\mathscr{B}_1$ and $\mathscr{B}_2$ are bases of S and $\bar{S}$, respectively, then their union is a basis of V.

3.16. (a) Let S_1, S_2, and S_3 be subspaces of a vector space V. Show that $V = S_1 \oplus S_2 \oplus S_3$ if and only if $V = S_1 + S_2 + S_3$ and $S_1 \cap (S_2 + S_3) = \{0\}$, $S_2 \cap (S_1 + S_3) = \{0\}$, $S_3 \cap (S_1 + S_2) = \{0\}$. Generalize this to the case of n summands.
(b) Find an example of three subspaces S_1, S_2, and S_3 of R^3 such that $V = S_1 + S_2 + S_3$ and $S_1 \cap S_2 = \{0\}$, $S_1 \cap S_3 = \{0\}$, $S_2 \cap S_3 = \{0\}$, but V is not the direct sum of S_1, S_2, and S_3.

3.17. Prove that a vector space V is the direct sum of subspaces $S_1, \ldots, S_p$ if and only if $\mathscr{B}_1 \cup \cdots \cup \mathscr{B}_p$ is a basis of V, where $\mathscr{B}_i$ is a basis of S_i, $1 \leq i \leq p$, and the $\mathscr{B}_i$'s are pairwise disjoint.

4. VECTOR SPACES OVER ARBITRARY FIELDS

A set F in which two binary operations, + (addition) and $\cdot$ (multiplication), are defined is called a field if and only if the following conditions are satisfied for all a, b, and c in F.

A1. $a + b = b + a$.

A2. $a + (b + c) = (a + b) + c$.

A3. There is a unique element 0 (called zero) in F such that $a + 0 = a$.

A4. For each a in F there is a unique element $-a$ such that $a + (-a) = 0$.

M1. $a \cdot b = b \cdot a$.
M2. $a \cdot (b \cdot c) = (a \cdot b) \cdot c$.
M3. There is a unique element 1 (called one) which is $\neq 0$ and such that $a \cdot 1 = a$.
M4. For each $a \neq 0$ there is a unique element a^{-1} (the multiplicative inverse of a) such that $a \cdot a^{-1} = 1$.
D. $a \cdot (b + c) = a \cdot b + a \cdot c$.

We list below several further properties of a field,† which are immediate consequences of the defining properties.

(1) If $a + c = b + c$, then $a = b$. If $ac = bc$ and $c \neq 0$, then $a = b$.
(2) An equation of the form $a + x = b$ has a unique solution, $b + (-a)$.

We write this element as $b - a$. An equation of the form $ax = b$ with $a \neq 0$ has a unique solution, namely, ba^{-1}. We write this element as b/a.

$$a0 = 0, \tag{3}$$
$$a(-b) = -(ab), \tag{4}$$
$$-(-a) = a \quad \text{and} \quad (a^{-1})^{-1} = a, \tag{5}$$
$$(-1)a = -a, \tag{6}$$
$$(-a)(-b) = ab. \tag{7}$$

On the basis of merely an intuitive understanding of the system R of real numbers and of the system C of complex numbers it is clear that they are fields when equipped with their familiar operations of addition and multiplication. Properties (1) to (7) indicate that various famliair computation rules for these two fields stem solely from their field properties.

In mathematics many other examples of fields can be found. Our reason for mentioning this is that all results concerning vector spaces obtained so far and most of those we derive in later chapters are valid when the field of scalars F is *any* field, not just R or C. (When we state any result that is based in any way on a peculiarity of the field of scalars, we shall note this fact). Thus, in view of the existence of a great variety of fields, our results have a much wider range of validity than merely for $F = \mathsf{R}$ or $F = \mathsf{C}$.

Further examples of fields can be constructed from various subsets of R by using the familiar operations of addition and multiplication. Some illustrations appear in the following examples.

† Henceforth we write ab for $a \cdot b$.

EXAMPLES

4.1. The set Q of rational numbers with the operations of ordinary addition and multiplication is a field.

4.2. The same is true of the set of all real numbers of the form $a + b\sqrt{3}$ $(a, b \in \mathsf{Q})$.

4.3. The set Z of integers with ordinary addition and multiplication is *not* a field.

4.4. Let p be a prime and let $\mathsf{Z}_p = \{0, 1, \ldots, p-1\}$. If $a, b \in \mathsf{Z}_p$, let $a + b$ be the least positive remainder obtained by dividing the ordinary sum of a and b by p. Let $a \cdot b$ be the least positive remainder obtained by dividing the ordinary product of a and b by p; for example, if $p = 7$, then $3 + 6 = 2$ and $3 \cdot 6 = 4$. Then Z_p with these operations is a field; the reader should attempt to prove it.

EXERCISES

4.1. Derive properties (1) to (7) of fields.

4.2. Let F be a field. Prove that the result of adding 1 to itself successively is either always different from 0 or else the first time it is equal to 0 occurs when the number of summands is a prime p. In the latter case show that for each $a \in F$, $a + \cdots + a$ (p summands) $= 0$. Note that the *characteristic* of F is defined as zero in the first case and p in the second.

4.3. (a) Show that $\mathsf{Z}(\sqrt{3}) = \{a + b\sqrt{3} \mid a, b \in \mathsf{Q}\}$ is a field.
(b) Show that $\mathsf{Z}(\sqrt{3}) = \{a + b\sqrt{3} \mid a, b \in \mathsf{Q}\}$ is not a field.

4.4. In $\mathsf{R} \times \mathsf{R}$ let us define operations by

$$(a, b) + (c, d) = (a + c, b + d)$$

$$(a, b) \cdot (c, d) = (ac, bd).$$

Is $\mathsf{R} \times \mathsf{R}$ with these operations a field?

4.5. Show by an example that if in Example 4.4 we use a composite number (for example, 6) in place of a prime, the resulting system is not a field.

4.6. Let V be a vector space over a field of p elements. Suppose V has a basis of n elements. Show that V has exactly p^n elements.

2 Further Properties of Vector Spaces

This chapter begins with the derivation of properties of a basis of a vector space. The initial group of properties is independent of the dimension of the space; additional properties are found for bases of finite-dimensional spaces. The concepts of homomorphism and isomorphism for vector spaces are then introduced. In Sections 3 and 4 several computational procedures applicable to finite-dimensional spaces are presented. In this connection the definition of a matrix is given along with a method for reducing a matrix to a form that has many applications. One that is described yields a mechanical process for finding all solutions of a system of linear equations. In the last section, which is optional, a class of distinguished subsets of a vector space is defined. These subsets are the abstract analogue of arbitrary lines and planes in E_3.

1. BASES AND DIMENSION

We recall the definition of a basis of a vector space V as a set of generators of V such that no proper subset has this property. In order to explore this concept, we introduce further definitions.

□ **Definition.** A subset S of a vector space V over F is said to be *linearly dependent* if and only if there exists a finite subset $\{\alpha_1, \ldots, \alpha_p\}$ of distinct vectors of S and scalars $a_1, \ldots, a_p$ in F, not all of which are 0, such that

$$\sum_{i=1}^{p} a_i \alpha_i = 0.$$

A subset S of V is called *linearly independent* if and only if it is not linearly dependent.

Observe that a subset S of V is linearly independent if and only if every finite subset of S is linearly independent; that is, if and only if for every set $\{\alpha_1, \ldots, \alpha_p\}$ of distinct vectors of S an equation of the form $\sum_1^p a_i \alpha_i = 0$ implies that $a_1 = \cdots = a_p = 0$. In a trivial way the empty set (of vectors!) is seen to be linearly independent. Other easy consequences of the definition include the following facts:

(1) If S is a subset of V and $0 \in S$, then S is linearly dependent.
(2) If S and S' are subsets of V, $S \supseteq S'$, and S' is linearly dependent, then so is S.
(3) If $\alpha \in V$ and $\alpha \neq 0$, then $\{\alpha\}$ is linearly independent.
(4) A subset of a linearly independent subset of V is linearly independent.

For reference we state the next consequence of our definition as a theorem.

□ **Theorem 1.1.** A finite subset $S = \{\alpha_1, \ldots, \alpha_p\}$ of a vector space is linearly dependent if and only if either $\alpha_1 = 0$ or for some j, $2 \leq j \leq p$,

$$\alpha_j = a_1\alpha_1 + \cdots + a_{j-1}\alpha_{j-1}.$$

PROOF. If the set S is linearly dependent and $\alpha_1 \neq 0$, let j be the least positive integer such that $\{\alpha_1, \ldots, \alpha_{j-1}\}$ is linearly independent and $\{\alpha_1, \ldots, \alpha_{j-1}, \alpha_j\}$ is linearly dependent. Then clearly $2 \leq j \leq p$ and there exist scalars $b_1, \ldots, b_j$, not all 0 such that $\sum_1^j b_i \alpha_i = 0$. The scalar b_j is not 0, for otherwise every b_i would be 0. Hence $\alpha_j = -(b_1/b_j)\alpha_1 - \cdots -(b_{j-1}/b_j)\alpha_{j-1}$.

Conversely, if either $\alpha_1 = 0$ or $\alpha_j = a_1\alpha_1 + \cdots + a_{j-1}\alpha_{j-1}$ for some j, $2 < j < p$, it is immediate that S is linear dependent. ◇

Before turning to some examples we remark that in discussing a finite subset $S = \{\alpha_1, \ldots, \alpha_p\}$ of a vector space we shall sometimes say that $\alpha_1, \ldots, \alpha_p$ are linearly dependent (or independent) instead of saying that S is linearly dependent (or independent).

EXAMPLES

1.1. In R^3 the vectors $\epsilon_1 = (1, 0, 0)$, $\epsilon_2 = (0, 1, 0)$, and $\epsilon_3 = (0, 0, 1)$ are linearly independent. On the other hand, the vectors

$$\alpha_1 = (1, -1, 0), \alpha_2 = (3, 2, 1), \alpha_3 = (1, 4, 1)$$

are linearly dependent because $2\alpha_1 - \alpha_2 + \alpha_3 = (0, 0, 0)$ [or $\alpha_3 = -2\alpha_1 + \alpha_2$].

Is $\{(2, -1, 1,) (1, 0, 0), (0, 2, 1)\}$ linearly independent or linearly dependent? A systematic procedure for obtaining an answer is to consider scalars x, y, and z such that

$$x(2, -1, 1) + y(1, 0, 0) + z(0, 2, 1) = (0, 0, 0).$$

Thus, x, y, and z satisfy the conditions

$$\begin{aligned} 2x + y &= 0, \\ -x + 2z &= 0, \\ x + z &= 0. \end{aligned}$$

The space of solutions of this system of equations is $\{(0, 0, 0)\}$, and so the set in question is linearly independent.

1.2. In the space $P_\infty(\mathsf{R})$ the set $\{1, x, \ldots, x^n, \ldots\}$ is linearly independent (see Example 1.3.9).†

1.3. Let S be a linearly independent subset of a vector space V. Suppose $V - [S] \neq \varnothing$. Then, if $\beta \in V - [S]$, the set $S \cup \{\beta\}$ is linearly independent. To prove this it is sufficient to show that any finite subset of $S \cup \{\beta\}$ that has β as a member is linearly independent. Let $\{\beta, \alpha_1, \ldots, \alpha_p\}$ be such a set and suppose that

$$b\beta + \sum_{i=1}^{p} a_i \alpha_i = 0.$$

Then $b = 0$, for otherwise $\beta \in [S]$, contrary to the hypothesis. Hence $\sum_1^p a_i \alpha_i = 0$ and, in turn, each a_i is equal to 0, since the α's are linearly independent. Hence the given set is linearly independent.

The contrapositive of the result just derived is also of interest. It reads: if S is a linearly independent subset of V, β is an element of V, and $S \cup \{\beta\}$ is a linearly dependent set, then $\beta \in [S]$.

□ **Theorem 1.2.** Let S be a subset of a vector space V. If $S \neq \varnothing$ and $S \neq \{0\}$, then S is linearly independent if and only if S is a basis of $[S]$.

† To refer to a theorem, example, exercise, or section in the chapter in which it appears we use only the number by which it is identified in the text. When a reference is made to one of these items in another chapter, we prefix its identifying number with a numeral that identifies the chapter; for instance, in Chapter 1 we shall refer to the third example in Section 2 as Example 2.3 and in another chapter we shall refer to the same example as Example 1.2.3.

PROOF. To establish that if such an S is linearly independent then S is a basis of $[S]$ we prove the contrapositive statement: if S is not a basis of $[S]$, then S is linearly dependent. Assume that S is not a basis of $[S]$. Then some proper subset S' of S generates $[S]$. Let α be an element of $S - S'$. Then $\alpha \in [S']$ and so $\alpha = \sum_1^n a_i \alpha_i$, where each α_i is in S'. Thus $\{\alpha, \alpha_i, \ldots, \alpha_n\}$ is a linearly dependent subset of S and S is linearly dependent.

To establish the converse we again turn to the contrapositive: if S is linearly dependent, then S is not a basis of $[S]$. Assume that S is linearly dependent. Then there exist $\alpha_1, \ldots, \alpha_n \in S$ and scalars $a_1, \ldots, a_n$, not all 0, such that $\sum_1^n a_i \alpha_i = 0$. If $n = 1$, it follows that $0 \in S$. Since $S \neq \{0\}$ by assumption, $S - \{0\}$ is a proper subset of S, which clearly generates $[S]$. Thus S is not a basis of $[S]$ in this case. Next, suppose that $n > 1$. Let a_k be one of the nonzero scalars among $a_1, \ldots, a_n$. Then α_k is a linear combination of the other α's and $S - \{\alpha_k\}$ generates $[S]$. So in all cases S is not a basis of $[S]$. ◇

□ **Corollary.** A subset S of a vector space $V \neq \{0\}$ is a basis of V if and only if $[S] = V$ and S is linearly independent.

Observe that this corollary is false if $V = \{0\}$. The space $\{0\}$ has $\{0\}$ as a set of generators and as a basis. Since there remains nothing more to say about this space, we often ignore it henceforth.

□ **Theorem 1.3.** A set S of vectors in a vector space V is linearly independent if and only if each α in $[S]$ has a unique representation as a linear combination of elements of S.

PROOF. We shall consider the case where S is a finite set. It is left as an exercise for the reader to modify our proof so that it is applicable to any S.

Let $S = \{\alpha_1, \ldots, \alpha_p\}$ and assume that S is linearly independent. Suppose that α in $[S]$ has the representations

$$\alpha = \sum_{i=1}^{p} a_i \alpha_i \quad \text{and} \quad \alpha = \sum_{i=1}^{p} b_i \alpha_i.$$

Then $\sum_1^p a_i \alpha_i = \sum_1^p b_i \alpha_i$ and $\sum_1^p (a_i - b_i)\alpha_i = 0$. It follows that $a_i - b_i = 0$ or $a_i = b_i$, $1 \leq i \leq p$.

Conversely, assume that each α in $[S]$ has a unique representation in the form $\sum_1^p a_i \alpha_i$. Suppose $a_1, \ldots, a_p$ are scalars such that $\sum_1^p a_i \alpha_i = 0$.

Since $0\alpha_1 + \cdots + 0\alpha_p = 0$, we conclude that $a_i = 0$, $1 \leq i \leq p$. Hence S is linearly independent. ◇

☐ **Corollary.** If $\{\alpha_1, \ldots, \alpha_n\}$ is a basis of V, then each vector α in V has a unique representation of the form

$$\alpha = \sum_{i=1}^{n} a_i \alpha_i.$$

☐ **Definition.** A linearly independent set S of vectors of a vector space V is called a *maximal linearly independent set* of V if and only if S is not properly included in any linearly independent subset of V. In other words, a maximal linearly independent set is a maximal element in the collection of all linearly independent subsets of V, partially ordered by inclusion.

☐ **Theorem 1.4.** A linearly independent subset $\mathscr{B}$ of a vector space V is a maximal linearly independent set if and only if $\mathscr{B}$ is a basis of V.

PROOF. Assume that $\mathscr{B}$ is a maximal linearly independent subset of V. Then for each $\alpha \in V - \mathscr{B}$, $\{\alpha\} \cup \mathscr{B}$ is linearly dependent; hence $\alpha \in [\mathscr{B}]$ (see Example 1.3). Thus $\mathscr{B}$ is a basis of V by the corollary to Theorem 1.2.

Conversely, assume that $\mathscr{B}$ is a basis of V. Then $\mathscr{B}$ is linearly independent by the corollary to Theorem 1.2. Further, if $\alpha \in V - \mathscr{B}$, then $\alpha \in [\mathscr{B}]$ and no set that properly includes $\mathscr{B}$ is linearly independent. Thus $\mathscr{B}$ is a maximal linearly independent set. ◇

☐ **Theorem 1.5.** There exists a basis for each vector space.

PROOF.† We have already observed that the space consisting of the zero vector alone has a basis. Suppose $V \neq \{0\}$. Then there is a nonzero vector α in V, hence a linearly independent subset $\{\alpha\}$ of V. So the collection $\mathscr{S}$ of all linearly independent subsets of V is nonempty. Now $\mathscr{S}$ is partially ordered by the relation of inclusion, and we contend that this (nonempty) partially ordered set satisfies the hypothesis of Zorn's lemma.

† Once again we are forced to use Zorn's lemma, and once again we suggest that the proof be omitted by the uninitiated.

For proof consider a subchain $\mathscr{C}$ of $\mathscr{S}$. It is immediate that $\cup\mathscr{C} \in \mathscr{S}$ and obvious that $\cup\mathscr{C}$ is an upper bound of $\mathscr{C}$. This being the case, we can infer the existence of a maximal element in $\mathscr{S}$, that is, the existence of a maximal linearly independent set in V. Such a set is a basis according to Theorem 1.3. ◇

□ **Theorem 1.6.** If S is a linearly independent subset of a vector space V, then there exists a basis of V that includes S.

PROOF. Let C be a basis of a complement of $[S]$. Then $C \cup S$ is a basis of V. ◇

Let us emphasize that Theorems 1.5 and 1.6 merely assert the *existence* of bases; that is, neither offers a method for *constructing* a set of the kind whose existence is asserted. Later such methods are provided for the case of finite dimensional spaces.

EXAMPLES

1.4. The vectors $\alpha = (1, 1, 0)$ and $\beta = (-1, 0, 1)$ in R^3 are linearly independent. The subspace S generated by the pair consists of all triples of the form $(a - b, a, b)$ so that $S = \{(x, y, z) \mid x - y + z = 0\}$. A complement of S is $[(1, 1, 1)]$. Thus $\{(1, 1, 0), (-1, 0, 1), (1, 1, 1)\}$ is a basis of R^3, according to the proof of Theorem 1.6.

1.5. In the vector space $P_\infty(\mathsf{R})$ the pair $\{f, g\}$, where $f(x) = x + 1$ and $g(x) = x^2 - x + 1$, is linearly independent. Further, $[f, g] = \{(a + b) + (a - b)x + bx^2 \mid a, b \in \mathsf{R}\}$. A complement $\bar{S}$ of this subspace is $[\{x, x^3, x^4, \ldots, x^n, \ldots\}]$. Since the indicated set of generators for the complement is a basis of $\bar{S}$, $\{f, g, x, x^3, \ldots, x^n, \ldots\}$ is a basis of $P_\infty(\mathsf{R})$, according to the proof of Theorem 1.6.

1.6. Clearly the set $S = \{(x, y, z)) \mid ax + by + cz = 0$ and $a \neq 0\}$ is a proper subspace of R^3. Using results from analytic geometry, we infer that $\bar{S} = [(a, b, c)]$ is a complement of S. Now $[(-b, a, 0), (-c, 0, a)] = S$ and so $\{(-b, a, 0), (-c, 0, a), (a, b, c)\}$ is a basis of R^3.

□ **Theorem 1.7.** If a vector space V has a basis of n elements, then every basis of V has exactly n elements.

PROOF. For convenience we introduce the notation

$$\#(A)$$

for the number of elements of a finite set A. Assume that $\mathscr{B}$ and $\mathscr{B}'$ are bases of V and that $\#(\mathscr{B}) = n$. It is to be shown that $\#(\mathscr{B}') = n$. Let $m = \#(\mathscr{B} \cap \mathscr{B}')$; then, clearly $0 \leq m \leq n$. Consider first the case $m = n$. In this event $\mathscr{B}' = \mathscr{B}$ because our assumptions imply that $\mathscr{B} \subseteq \mathscr{B}'$, $\mathscr{B}'$ is a linearly independent set, and $\mathscr{B}$ is a maximal linearly independent set (see Theorem 1.3). With $\mathscr{B}' = \mathscr{B}$, we may conclude that $\#(\mathscr{B}') = n$; so the proof is complete for the case $m = n$.

Assume next that $m < n$. We shall show the existence of a basis $\mathscr{B}_1$ of V such that $\#(\mathscr{B}_1) = n$ and $\#(\mathscr{B}_1 \cap \mathscr{B}') = m + 1$. Suppose that $\mathscr{B} = \mathscr{C} \cup \{\alpha_{m+1}, \ldots, \alpha_n\}$, where $\mathscr{C} = \mathscr{B} \cap \mathscr{B}'$. Since $\mathscr{C} \cup \{\alpha_{m+2}, \ldots, \alpha_n\}$ is a proper subset of $\mathscr{B}$ it is not a basis of V. Hence there exists a vector β in $[\mathscr{B}']$ such that $\beta \notin [\mathscr{C} \cup \{\alpha_{m+2}, \ldots, \alpha_n\}]$. We prove next that the set

$$\mathscr{B}_1 = \{\beta\} \cup \mathscr{C} \cup \{\alpha_{m+2}, \ldots, \alpha_n\}$$

of n elements is a basis of V. When β is expressed as a linear combination of elements of $\mathscr{B}$, the scalar coefficient of a_{m+1} is not zero, since otherwise β would be a member of $[\mathscr{C} \cup \{\alpha_{m+2}, \ldots, \alpha_n\}]$. Thus $\alpha_{m+1} \in [\mathscr{B}_1]$ and it follows that $\mathscr{B} \subseteq [\mathscr{B}_1]$. Hence $\mathscr{B}_1$ spans V. Since, in addition, it is clear that $\mathscr{B}_1$ is linearly independent, we conclude that $\mathscr{B}_1$ is a basis of V. Thus $\mathscr{B}_1$ is a basis such that $\#(\mathscr{B}_1) = n$ and $\#(\mathscr{B}_1 \cap \mathscr{B}') = m + 1$.

By repeating the foregoing "construction" $n - m$ times we obtain a basis $\mathscr{B}_{n-m}$ of V such that $\#(\mathscr{B}_{n-m}) = n$ and $\#(\mathscr{B}_{n-m} \cap \mathscr{B}') = n$. From the first part of the proof it follows that $\#(\mathscr{B}') = n$. ◇

The *invariance of dimensionality*, as the foregoing result for finite-dimensional vector spaces is called, extends to the case of infinite-dimensional spaces in the sense that there exists a 1-1 correspondence between any two bases of an infinite-dimensional vector space. Since a proof employs considerably more set theory than we presuppose, we shall not pursue the matter.

□ **Definition.** If V is a vector space different from $\{0\}$ and has a basis of n elements, we call it an *n-dimensional vector space* or say simply that V has *dimension n*. The vector space $\{0\}$ we call the *0-dimensional space*. In either case we denote the dimension of V by *dim V*.

EXAMPLES

1.6. The subset S of F^n, which consists of the vectors

$$\epsilon_1 = (1, 0, 0, \ldots, 0),$$
$$\epsilon_2 = (0, 1, 0, \ldots, 0),$$
$$\vdots$$
$$\epsilon_n = (0, 0, \ldots, 0, 1),$$

is a basis of F^n. Hence dim $F^n = n$. This particular basis we shall call the *standard basis* of F^n.

1.7. The vector space $P_n(\mathsf{R})$ has dimension $(n + 1)$. The vector space of all solutions of a differential equation of the form

$$\frac{d^n y}{dx^n} + p_1(x)\frac{d^{n-1}y}{dx^{n-1}} + \cdots + p_{n-1}(x)\frac{dy}{dx} + p_n(x)y = 0,$$

where $p_1, \ldots, p_n \in C[a, b]$ has dimension n.

1.8. Since a subset B of a vector space V is a basis if and only if B is a maximal linearly independent subset of V (see Theorem 1.4), any set of more than n vectors of an n-dimensional space is linearly dependent.

We turn now to the derivation of several basic properties of finite-dimensional vector spaces. First of all we call attention to the fact that a vector space is finite-dimensional in the sense of the definition given in Chapter 1 if and only if it has dimension n for some nonnegative integer n. Suppose now that V is a finite-dimensional vector space other than $\{0\}$. If it is spanned by a set S of k (nonzero) vectors, then dim $V \leq k$. Indeed, if S is a linearly independent set, then S is a basis of V (corollary to Theorem 1.2) and dim $V = k$. Otherwise S is linearly dependent. In this event we make use of Theorem 1.1 to construct a linearly independent (proper) subset $S' = \{\alpha_1, \ldots, \alpha_n\}$ of S which is maximal in the sense that if $\beta \in S - S'$ then $\{\alpha_1, \ldots, \alpha_n, \beta\}$ is linearly dependent. It follows that $S \subseteq [S']$, hence that $V = [S']$. According to the corollary to Theorem 1.2, S' is a basis of V and dim $V < k$.

Now suppose that V is an n-dimensional vector space and S is a subset of n vectors that spans V. By the remarks above S must be a basis and therefore a linearly independent set. On the other hand, if dim $V = n$

and S is a linearly independent subset of n vectors, then S is a basis (Theorem 1.6) and therefore spans V. We state these results as our next theorem.

□ **Theorem 1.8.** If V is a vector space of dimension n, then (a) any linearly independent set of n vectors is a basis and (b) any set of n vectors that spans V is a basis.

The proofs of some of our earlier existence theorems can be recast for finite-dimensional vector spaces to provide constructive methods for obtaining the object whose existence is asserted. As an illustration, consider Theorem 1.6. Suppose that S is a linearly independent set of p elements in the n-dimensional space V; then $p \leq n$ (Example 1.8), and if $p = n$ there is nothing to prove (Theorem 1.8). So assume that $p < n$; then $V - [S]$ is nonempty. Let α be one of its elements; then $S \cup \{\alpha\}$ is linearly independent (Example 1.3). If $p + 1 = n$, we are finished. If $p + 1 < n$, we repeat the procedure, and after $n - p$ steps we have a basis of V that includes S.

An alternative constructive method for extending a nonempty linearly independent set $\{\alpha_1, \ldots, \alpha_p\}$ of elements of an n-dimensional space V to a basis of V employs a known basis $\{\beta_1, \ldots, \beta_n\}$ of V. Consider the ordered set

$$(\alpha_1, \ldots, \alpha_p, \beta_1, \ldots, \beta_n).$$

It is linearly dependent and therefore (Theorem 1.1) one of its members is a linear combination of preceding members. It follows that there is a first such element which, since the α_i's are linearly independent, must be one of $\beta_1, \ldots, \beta_n$. Suppose it is β_k. Now consider the ordered set

$$(\alpha_1, \ldots, \alpha_p, \beta_1, \ldots, \beta_{k-1}, \beta_{k+1}, \ldots, \beta_n).$$

It generates V and, if linearly independent, is a basis of V of the required kind. Otherwise we repeat the foregoing routine. After a finite number of such steps a basis that includes $\{\alpha_1, \ldots, \alpha_p\}$ is obtained.

A constructive method for extending a given linearly independent set to a basis provides a method of determining a complement of a subspace (see Theorem 1.3.6). Suppose that V is a finite-dimensional vector space, that S is a subspace of V, and that $\mathscr{B}$ is a basis of S. Let $\mathscr{B}'$ be an extension of $\mathscr{B}$ that is a basis of V. Then clearly $\mathscr{B}' - \mathscr{B}$ is a basis of a complement of S.

The foregoing discussion tacitly assumes, insofar as its usefulness is concerned, that a subspace of a finite dimensional vector space is finite-dimensional. We prove this next.

□ **Theorem 1.9.** A subspace S of an n-dimensional vector space V is finite-dimensional and $\dim S \leq n$. Further, $\dim S < \dim V$ if and only if $S \subset V$.

PROOF. We shall establish the first assertion and leave the second as an exercise. If $S = \{0\}$, then S is 0-dimensional and the proof is complete. Otherwise there is a nonzero vector α_1 in S. Then $S_1 = [\alpha_1]$ is included in S; if $S = S_1$, then $\dim S = 1$ and we are finished. If $S_1 \subset S$, let $\alpha_2 \in S - S_1$ and define $S_2 = [\alpha_1, \alpha_2]$, etc. After no more than n steps of this kind a basis of S is obtained, according to the result in Example 1.8. ◇

We turn next to the derivation of an important arithmetical result that relates the dimensions of subspaces.

□ **Theorem 1.10.** Let S_1 and S_2 be finite-dimensional subspaces of a vector space V. Then

$$\dim (S_1 + S_2) = \dim S_1 + \dim S_2 - \dim (S_1 \cap S_2).$$

PROOF. Let $m = \dim S_1$ and $n = \dim S_2$. Since the union of a basis of S_1 and a basis of S_2 generates $S_1 + S_2$, it is clear that $S_1 + S_2$ is finite-dimensional. Further, since $S_1 \cap S_2$ is a subspace of S_1 (or, equally well, of S_2), it is also finite-dimensional. Thus the alleged equality stated in the theorem is meaningful. Its correctness is immediate if $S_1 \cap S_2 = \{0\}$ since then $\dim (S_1 \cap S_2) = 0$ and $\dim (S_1 + S_2) = m + n$.

Next suppose that $S_1 \cap S_2 \neq \{0\}$ and let T be a complement of $S_1 \cap S_2$ in S_1. We shall prove that $S_1 + S_2 = T + S_2$. Let $\alpha \in S_1 + S_2$; then $\alpha = \alpha_1 + \alpha_2$, where $\alpha_1 \in S_1$ and $\alpha_2 \in S_2$. In turn, $\alpha_1 = \gamma + \delta$, where $\gamma \in S_1 \cap S_2$ and $\delta \in T$. Hence $\alpha = (\gamma + \delta) + \alpha_2 = \delta + (\gamma + \alpha_2) \in T + S_2$ and therefore $S_1 + S_2 \subseteq T + S$. Since the reverse inclusion obviously holds, equality follows and we may infer that

$$\dim (S_1 + S_2) = \dim (T + S_2).$$

Next, we note two instances of the special case of the theorem considered in the first paragraph. Since $T \cap S_2 = \{0\}$,

$$\dim (T + S_2) = \dim T + \dim S_2.$$

Also, since $(S_1 \cap S_2) \cap T = \{0\}$ and $(S_1 \cap S_2) + T = S_1$,

$$\dim S_1 = \dim (S_1 \cap S_2) + \dim T.$$

Combining appropriately the three equations displayed above gives the desired conclusion. ◇

Although we have labeled certain methods in the latter part of this section constructive, they are basically descriptive in nature and merely assure us that certain entities can be constructed in a finite number of steps. What is missing are computational procedures, or algorithms, for such chores as finding a basis of a subspace spanned by a given set of vectors and extending a given linearly independent set of vectors to a basis. In Section 3 we take steps to remedy this deficiency. Possibly the reader can anticipate some of the techniques given there in working out the numerical exercises included in the following list.

EXERCISES

1.1. Find three vectors in R^3 that are linearly dependent and such that any two of them are linearly independent.

1.2. If α, β, and γ are linearly independent elements of a vector space, prove that $\{a + \beta, \beta + \gamma, \gamma + \alpha\}$ is a linearly independent set.

1.3. Let $S = \{\alpha_1, \ldots, \alpha_m\}$, where $\alpha_i = (a_{i1}, \cdots, a_{in})$, $1 \leq i \leq m$. Show that this subset of R^n is linearly dependent if and only if the system of n linear homogeneous equations in m unknowns,

$$a_{11}x_1 + a_{21}x_2 + \cdots + a_{m1}x_m = 0,$$

$$a_{12}x_1 + a_{22}x_2 + \cdots + a_{m2}x_m = 0,$$

$$\vdots$$

$$a_{1n}x_1 + a_{2n}x_2 + \cdots + a_{mn}x_m = 0,$$

has a nontrivial solution, that is, a solution other than $x_1 = 0, \ldots, x_m = 0$. From this result deduce that if the number of unknowns exceeds the number of equations in a given system of linear homogeneous equations, the system will have a nontrivial solution.

1.4. Is $\{\sin nx \mid n = 1, 2, \ldots, p\}$ a linearly independent subset of the vector space of all continuous functions on R into R?

1.5. Let V be the space of all functions that are continuous on $[0, 1]$. Let f_1, f_2, f_3 be three elements of V such that there exist $a_1, a_2, a_3 \in [0, 1]$ for which $f_i(a_i) = 0$ and $f_i(a_j) \neq 0$ if $i \neq j$ $(i, j = 1, 2, 3)$. Show that (a) each of $\{f_1, f_2\}$,

$\{f_1, f_3\}$, and $\{f_2, f_3\}$ is a linearly independent set and (b) if $\{f_1, f_2, f_3\}$ is linearly dependent then

$$f_1(a_2)f_2(a_3)f_3(a_1) + f_1(a_3)f_2(a_1)f_3(a_2) = 0.$$

1.6. Test each of the following subsets of R^4 for linear dependence. For each set S that is linearly dependent find a subset that is a basis of the subspace $[S]$.
(a) $S = \{(1, 0, 1, 1), (0, 1, 2, 3), (-2, 3, 4, 7)\}$.
(b) $S = \{(1, 1, 0, 1), (2, 1, 1, 2)\}$.
(c) $S = \{(2, 1, 0, 1), (-1, 2, 0, -1), (1, 1, 2, 3)\}$.

1.7. Extend each of the following linearly independent sets in R^4 to a basis of R^4.
(a) $S = \{(1, 1, 1, 1), (0, 2, 1, 1)\}$.
(b) $S = \{(1, 0, 1, 1), (-1, -1, 0, 0), (0, 1, 1, 0)\}$.

1.8. Find a basis for each of the following subspaces of R^3.
(a) $S = \{(x, y, z) \mid 3x - 4y + z = 0\}$.
(b) $S = \{(x, y, z) \mid x + y - z = 0 \text{ and } 2x - y + z = 0\}$.

1.9. In the space P_n(R) show that each of $\{1, t, \ldots, t^n\}$ and $\{1, t - a, \ldots, (t - a)^n\}$, where a is a fixed number, is a basis. Express the members of the first basis in terms of the second.

1.10. Find a basis for the subspace of R^3 spanned by the set of vectors (x, y, z) such that $x^2 + y^2 + z^2 = 2$.

1.11. Suppose that V is a vector space of dimension n over the field C of complex numbers. Show that V may be considered as a vector space over R. What is the dimension of V over R?

1.12. For the subspaces S_1 and S_2 of R^3 defined by

$$S_1 = \{(x, y, z) \mid 3x + y - 5z = 0\}$$

$$S_2 = \{(x, y, z) \mid x - y + 3z = 0\}$$

find $\dim (S_1 + S_2)$ by using Theorem 1.9.

1.13. Complete the proof of Theorem 1.10.

2. ISOMORPHISM

By this time the reader has encountered pairs of vector spaces that are indistinguishable from each other if the nature of the vectors is ignored. The pair of spaces P_2(R) and R^3 provides an illustration. Indeed, if it is agreed that elements of P_2(R) will be written in the form

$$a_0 + a_1 x + a_2 x^2,$$

then, insofar as vector space calculations with these objects is concerned, we may as well calculate with ordered triples of real numbers, that is, elements of R^3 in the usual way. In more detail, the obvious pairing, $a_0 + a_1x + a_2x^2 \to (a_0, a_1, a_2)$ of elements of $P_2(\mathsf{R})$ and R^3 is preserved under addition [i.e., if $a_0 + a_1x + a_2x^2 \to (a_0, a_1, a_2)$ and $b_0 + b_1 + b_2x^2 \to (b_0, b_1, b_2)$, then the sum of the two polynomials is paired with the sum of the corresponding triples] and scalar multiplication [i.e., if $a_0 + a_1x + a_2x^2 \to (a_0, a_1, a_2)$, then $c(a_0 + a_1 + a_2x^2) \to c(a_0, a_1, a_2)$]. Thus the two spaces are indistinguishable insofar as vector space calculations are concerned. The mathematical term for this kind of indiscernibility is *isomorphism*. It is defined below for arbitrary vector spaces in terms of a more general concept.

□ **Definition.** Let V and W be vector spaces over the same field F. A mapping T of V into W is called a *homomorphism* of V into W if and only if for all α, β in V and all a in F

$$(\alpha + \beta)T = \alpha T + \beta T$$

and

$$(a\alpha)T = a(\alpha T).$$

If the range of T is W, then T is called a *homomorphism* of V *onto* W. A 1-1 homomorphism of V onto W is called an *isomorphism* of V *onto* W.

Symbolizing the value of a function T at α by αT may be confusing to the reader who is in the habit of writing the value of function f at x as $f(x)$. Our choice is not merely a whim; circumstances will arise in which it has some advantages. It should be noticed that the sum and scalar multiple indicated on the left side of these equations are computed in V, while those on the right side are computed in W. Since the notion of homomorphism is defined only for vector spaces over the same field, we may, on occasion, safely omit mentioning the system of scalars.

An isomorphism T of V onto W as a 1-1 and onto mapping has an inverse T^{-1} which is a 1-1 mapping of W onto V. It is left as an exercise to prove that T^{-1} is an isomorphism of W onto V. Because of this symmetry, which is present when there exists an isomorphism T of V onto W, we shall refer to V and W as *isomorphic vector spaces* and often call T an *isomorphism between* V and W.

A homomorphism T of a vector space V into a vector space W maps the zero vector of V onto the zero vector of W [since $0T = (0 + 0)T = 0T + 0T$] and $-\alpha$(in V) onto $-(\alpha T)$ [for $0 = 0T = (\alpha + (-\alpha))T = \alpha T + (-\alpha)T$, hence $(-\alpha)T = -(\alpha T)$]. If S is a subspace of V, then

$$(S)T = \{\alpha T \mid \alpha \in S\}$$

is a subspace of W. Also

$$K_T = \{\alpha \in V \mid \alpha T = 0\}$$

is a subspace of V called the *kernel* of T.

□ **Theorem 2.1.** If V and W are vector spaces and T is a homomorphism of V into W, then T is 1-1 if and only if the kernel K_T of T is $\{0\}$.

PROOF. Assume that $K_T = \{0\}$; then, if $\alpha T = \beta T$, we have in turn $(\alpha - \beta)T = 0$, $\alpha - \beta \in K_T$, and $\alpha = \beta$. Conversely, assume that T is 1-1; then, if $\alpha T = 0$ it follows that $\alpha = 0$, since $0T = 0$. ◇

EXAMPLES

2.1. Let $V = P_\infty(\mathsf{R})$ and define $T: V \to V$ by $p(x)T = p'(x)$, the derivative of $p(x)$. The (true) statement that T is a homomorphism summarizes several familiar properties of the operation of differentiation of polynomial functions. Notice that T is onto but not 1-1.

2.2. In the same space define $T: V \to V$ by $p(x)T = \int_0^x p(t)\, dt$. Then T is seen to be a 1-1 homomorphism of V into V. Notice that this T is not onto.

2.3. Let T: $\mathsf{R}^3 \to \mathsf{R}^2$, where $(x, y, z)T = (x + y, y + z)$. Then T is a homomorphism. Since $(1, 0, 0)T = (1, 0)$ and $(0, 0, 1)T = (0, 1)$, the range of T is R^2, that is, T is onto R^2. It is not 1-1, since the kernel of T is $[(1, -1, 1)]$.

2.4. Let $T: \mathsf{R}^3 \to \mathsf{R}^3$, where $(x, y, z)T = (x + 2z, y + z, x + z)$. A straightforward computation shows that T is a homomorphism. Since the system of equations

$$\begin{aligned} x \quad\; + 2z &= a,\\ y + \; z &= b,\\ x \quad\; + \; z &= c \end{aligned}$$

has a unique solution $(x = 2c - a,\ y = b + c - a,\ z = a - c)$ for all choices of a, b, and c, T is an isomorphism of R^3 onto itself. It follows that $K_T = \{0\}$. Looking ahead to part (b) of Theorem 2.2, it follows that $\{\epsilon_1 T, \epsilon_2 T, \epsilon_3 T\} = \{(1, 0, 1), (0, 1, 0), (2, 1, 1)\}$ is another basis of R^3.

We continue with several additional elementary results.

□ **Theorem 2.2.** If T is a 1-1 homomorphism of the vector space V into the vector space W and S is a nonempty subset of V, then

(a) S is linearly independent if and only if $(S)T$ is a linearly independent subset of W;

(b) S is a basis of $[S]$ if and only if $(S)T$ is a basis of $[(S)T]$;

(c) $\alpha \in [S]$ if and only if $\alpha T \in [(S)T]$.

PROOF. Part (a) follows from the fact that for $\alpha_1, \ldots, \alpha_p \in S$ and scalars $a_1, \ldots, a_p$, $a_1(\alpha_1 T) + \cdots + a_p(\alpha_p T) = 0$ if and only if $a_1\alpha_1 + \cdots + a_p\alpha_p = 0$.

For (b) assume that S is a basis of $[S]$. Then $(S)T$ is linearly independent and spans $[(S)T]$. Hence $(S)T$ is a basis of $[(S)T]$. Conversely, assume that $(S)T$ is a basis of $[(S)T]$. Then, in turn, $(S)T$ is a linearly independent subset of W, S is a linearly independent subset of V, and S is a basis of $[S]$.

For (c), assume that $\alpha \in [S]$. Then α is a linear combination of elements of S, so that αT is a linear combination of elements of $(S)T$. Thus $\alpha T \in [(S)]$. Conversely, assume that $\alpha T \in [(S)T]$. Then $\alpha T = a_1(\alpha_1 T) + \cdots + a_p(\alpha_p T)$ for some choice of scalars $a_1, \ldots, a_p$ and vectors $\alpha_1, \ldots, \alpha_p \in S$; that is, $\alpha T = (\sum_1^p a_i\alpha_i)T$ and therefore $\alpha = \sum_1^p a_i\alpha_i \in [S]$, since $K_T = \{0\}$. ◇

□ **Theorem 2.3.** Two finite-dimensional vector spaces V and W over F are isomorphic if and only if $\dim V = \dim W$.

PROOF. Assume that $\dim V = \dim W = n$. Let $\mathscr{B} = \{\alpha_1, \ldots, \alpha_n\}$ be a basis of V and $\mathscr{B}' = \{\beta_1, \ldots, \beta_n\}$ be a basis of W. Define a mapping T of V into W as follows: if $\alpha \in V$ and $\alpha = \sum_1^n a_i\alpha_i$, then set $\alpha T = \sum_1^n a_i\beta_i$. It is left as an exercise to show that T is an isomorphism.

For the converse assume that T is an isomorphism of V onto W, both finite-dimensional spaces. Let $\mathscr{B} = \{\alpha_1, \ldots, \alpha_n\}$ be a basis of V. We shall prove that $\{\alpha_1 T, \ldots, \alpha_n T\} = \mathscr{B}'$ is a basis of W. First, $\mathscr{B}'$ is linearly independent by Theorem 2.2(a). Further, $[\mathscr{B}'] = [(\mathscr{B})T] = [\mathscr{B}]T = (V)T = W$, since T is onto. Hence $\mathscr{B}'$ is a basis of W with the same number of elements as a basis of V. ◇

Observe that the isomorphism constructed in the proof of Theorem 2.3 may be described as follows: pair the elements of a basis $\{\alpha_1, \ldots, \alpha_n\}$ of V with those of a basis $\{\beta_1, \ldots, \beta_n\}$ of W and if $\alpha_i \to \beta_i$, $1 \leq i \leq n$, then let $(\sum_1^n a_i\alpha_i)T = \sum_1^n a_i\beta_i$. We show next that each isomorphism T of two n-dimensional vector spaces V amd W over F can be constructed in this way. Indeed, by the second part of the proof of Theorem 2.3, if $\{\alpha_1, \ldots, \alpha_n\}$ is a basis of V, then $\{\alpha_1 T, \ldots, \alpha_n T\}$ is a basis of W and $(\sum_1^n a_i\alpha_i)T = \sum_1^n a_i(\alpha_i T)$; that is, T is an isomorphism of the type constructed above.

□ **Corollary.** An n-dimensional vector space over the field F is isomorphic to F^n.

Proof. We have seen that $\dim F^n = n$. ◇

An isomorphism between a given n-dimensional vector space V over F and F^n can be constructed by using the recipe given in the first part of the proof of Theorem 2.3. Let us construct one by using the standard basis of F^n. The image of a vector α in V is an n-tuple whose coordinates are the (uniquely determined) scalars that occur in the representation of α as a linear combination of the elements of a basis of V. Once a basis of V is specified, the image of α can be determined by assigning an order to the elements of that basis and choosing as the first coordinate of the image n-tuple the scalar that accompanies the first member of this basis, and so on. Such a basis we call an *ordered basis*. Ordered bases are written as n-tuples. If $(\alpha_1, \ldots, \alpha_n)$ is an ordered basis of V and if for α in V we have $\alpha = \sum_1^n x_i\alpha_i$, then the image of α under *the* isomorphism T such that $\alpha_i T = \epsilon_i$ is $(x_1, \ldots, x_n)$. We call x_i the ith *coordinate* of α relative to the ordered basis $(\alpha_1, \ldots, \alpha_n)$. If an ordered basis of V is designated by a letter, say $\mathscr{B}$, we find it convenient to write, on occasion,

$$[\alpha]_{\mathscr{B}}$$

for the n-tuple of coordinates of the vector α relative to this basis.

We conclude this section with several remarks about the corollary to Theorem 2.3. We might infer from it, together with the fact that isomorphic vector spaces are indistinguishable from the point of view of linear computations, that it is foolish to continue our study of abstract finite-dimensional vector spaces—that there is no loss of generality if we restrict it to F^n. This inference is faulty for the following reasons. The most important properties of vectors and vector spaces are those that are independent of bases, that is, unchanged by isomorphism (see the remark preceding the corollary to Theorem 2.3), and an isomorphism between an n-dimensional vector spacc V over F and F^n involves choosing a basis of V. Thus a study of just F^n would amount to being tied down to one particular basis of V or showing that definitions and theorems are independent of that basis. Accordingly, for the most part, we shall ignore the corollary in question and continue to discuss vector spaces as entities, independent of any basis. There is still a further reason for doing this; many specific vector spaces would lose much of their intuitive content if they were regarded as spaces of n-tuples, and we rely on intrinsic features of specific vector spaces to gain insight into the structure of the theory as a whole.

EXERCISES

2.1. Let V and W be vector spaces over F and let T be an isomorphism of V onto W. Prove that T^{-1}, the inverse of T, is an isomorphism of W onto V.

2.2. Prove that the image of a subspace of a vector space under a homomorphism is a subspace.

2.3. Prove that the kernel of a homomorphism T of a vector space V into a vector space W is a subspace of V.

2.4. Prove that if T is a homomorphism of a vector space V into a vector space W and S is a subspace of W, then $\{\alpha \in V \mid \alpha T \in S\}$ is a subspace of V.

2.5. Prove that if a vector space is isomorphic to one of its proper subspaces it is infinite-dimensional.

2.6. Prove that a homomorphism of a vector space V into a vector space W is $1-1$ if and only if it maps linearly independent sets onto linearly independent sets.

2.7. Prove that a homomorphism of a vector space V into a vector space W is an isomorphism if and only if it maps a basis $\mathscr{B}$ of V onto a basis $\mathscr{B}'$ of W in such a way that distinct elements of $\mathscr{B}$ have distinct images.

2.8. Prove that if V and W are finite-dimensional vector spaces over F with $\dim V = \dim W$ and T is a homomorphism of V into W the following statements are equivalent to one another: "T is an isomorphism," "T is 1-1," "the range of T is W."

2.9. Find vector spaces V and W, a homomorphism T_1 of V into W, and a homomorphism T_2 of W into V such that $(\alpha T_1)T_2 = \alpha$ for all $\alpha \in V$ but $(\beta T_2)T_1 \neq \beta$ for all $\beta \in W$.

2.10. Let V be a vector space over F, and W, a set equipped with operations of addition (that is, for α, $\beta \in W$, $\alpha + \beta$ is defined and in W) and scalar multiplication with F (that is, for $a \in F$ and $\alpha \in W$, $a\alpha$ is defined and in W). Suppose T is a 1-1 mapping of V onto W such that $(\alpha + \beta)T = \alpha T + \beta T$ and $(a\alpha)T = a(\alpha T)$ for all $a \in F$ and $a, \beta \in V$. Prove that W is a vector space over F.

2.11. Complete the proof of Theorem 2.3 by showing that the mapping T is an isomorphism of V onto W.

2.12. Let T be the mapping of R^3 into itself such that

$$(x, y, z)T = (2x - y - 4z,\ 5y + 10z,\ x + 3y + 6z).$$

(a) Show that T is a homomorphism.
(b) Find the kernel of T and a basis for the range of T.

2.13. Let T be the mapping of R^3 into itself such that

$$(x, y, z)T = (x + y,\ y + 2z,\ -x + y + z).$$

(a) Show that T is an isomorphism of R^3 onto itself.
(b) Find T^{-1}.

2.14. Let T be a homomorphism of R^3 into itself. Show that there exist scalars a_{ij}, $1 \leq i \leq 3$, $1 \leq j \leq 3$ such that $(x, y, z)T = (a_{11}x + a_{21}y + a_{31}z,$ $a_{12}x + a_{22}y + a_{32}z,$ $a_{13}x + a_{23}y + a_{33}z)$. Extend your result to R^n.

2.15. A mapping T of $P_\infty(\mathsf{R})$ into itself is defined by

$$(a_0 + a_1x + \cdots + a_nx^n)T = a_0 + a_1(x - c) + \cdots + a_n(x - c)^n,$$

where c is a fixed real number. Is T a homomorphism? An isomorphism?

2.16. Considering R as a vector space over itself, find an explicit description of all homomorphisms of R into R. Which of these are isomorphisms of R onto R? Do the same for C as a vector space over R.

3. CALCULATION METHODS

In this section we discuss some computational procedures for systematizing certain types of calculation that arise in a natural way in the study of finite-dimensional vector spaces. Questions that crop up repeatedly and require computations for their answers center about a given finite set S of vectors in an n-dimensional space and fit into one of the following categories.

(a) Is S a linearly independent or linearly dependent set?

(b) If S is a linearly dependent set, what is a basis (or the dimension) of $[S]$?

(c) What is a basis of $[S]$ consisting of elements of S?

(d) Is a given vector (of the space at hand) a member of $[S]$?

We may restrict our attention, in discussing methods of attacking such questions, to subsets of F^n in view of the isomorphism T between an n-dimensional space V over F and F^n which is at hand as soon as an ordered basis of V is selected. For in view of Theorem 2.2, a set S in V is linearly independent if and only if its image $(S)T$ in F^n is linearly independent; the inverse image of V of a basis of $[(S)T]$ is a basis of $[S]$; dim $[S] = \dim [S(T)]$; and so on. Expressed in another way, once an ordered basis of V is selected, we may designate each vector by its n-tuple of coordinates relative to that basis and calculate in F^n.

Suppose, then, that $S = \{\alpha_1, \ldots, \alpha_m\}$, where $\alpha_i \in F^n$, $1 \leq i \leq m$. Temporarily, we shall regard S as the (ordered) m-tuple

$$S = (\alpha_1, \ldots, \alpha_j, \ldots, \alpha_m)$$

and introduce (i) the m-tuple

$$S_{ij} = (\alpha_1, \ldots, \alpha_j, \ldots, \alpha_i, \ldots, \alpha_m),$$

obtained from S by the interchange of ith and jth coordinates of S, (ii) the m-tuple

$$S_i(a) = (\alpha_1, \ldots, a\alpha_i, \ldots, \alpha_m),$$

obtained from S by replacing α_i with $a\alpha_i$, where a is a nonzero scalar, and (iii) the m-tuple

$$S_{ij}(c) = (\alpha_1, \ldots, \alpha_i + c\alpha_j, \ldots, \alpha_j, \ldots, \alpha_m),$$

obtained from S by replacing α_i with $\alpha_i + c\alpha_j$, $j \neq i$. It is an easy exercise to prove that

$$[S_{ij}] = [S_i(a)] = [S_{ij}(c)] = [S] \tag{1}$$

and that each of S_{ij}, $S_i(a)$, $S_{ij}(c)$, and S is a linearly independent set if and only if one of them is linearly independent. Next, assuming that

$$\alpha_i = (a_{i1}, \ldots, a_{in}), \qquad 1 \leq i \leq m,$$

let us arrange the mn scalars a_{ij} as the rectangular array

$$A = \begin{bmatrix} a_{11}a_{12} \cdots a_{1n} \\ a_{21}a_{22} \cdots a_{2n} \\ \cdot \\ \cdot \\ \cdot \\ a_{m1}a_{m2} \cdots a_{mn} \end{bmatrix}$$

This object is called a *matrix* with entries in F or simply a *matrix over F*. Let us digress for a moment to comment on this notion. A matrix over F is a function whose domain is a set of the form

$$\{(i, j) \mid 1 \leq i \leq m, 1 \leq j \leq n\},$$

and whose values are in F. In the contexts in which matrices arise (we shall encounter several) it is quite helpful to represent them in the form already displayed.† The n-tuple

$$(a_{i1}, \ldots, a_{in})$$

is called the ith *row* of A and the m-tuple

$$\begin{bmatrix} a_{ij} \\ \cdot \\ \cdot \\ \cdot \\ a_{mj} \end{bmatrix}$$

is called the jth *column* of A. The element a_{ij} of A is referred to as the (i, j)th entry of A. We call A an $m \times n$ (read m by n) matrix to indicate the fact that it has m rows and n columns. If a row of A is the zero n-tuple $(0, \ldots, 0)$, we shall call it a *zero row*. A *zero column* is defined similarly.

We return now to the introduction of A as the matrix whose rows are the n-tuples $\alpha_1, \ldots, \alpha_m$. In this role we call A the *matrix of S*. The subspace $[S]$ of F^n spanned by S is called the *row space* of A. Clearly, the matrices corresponding to S_{ij}, $S_i(a)$, and $S_{ij}(c)$, respectively, can be obtained from A by the following operations on A.

(I) The interchange of the ith and the jth rows of A.
(II) The replacement of the ith row, α_i, of A by $a\alpha_i (a \neq 0)$.
(III) The replacement of the ith row, α_i, of A by $\alpha_i + c\alpha_j (j \neq i)$.

We call these manipulations (elementary) *row operations* of types I, II, and III, respectively, on A.

† This type of notation for a function emphasizes the value of the function of each argument. It is similar to that used for denoting sequences.

Our earlier observation summarized in (1) means that the matrix obtained from A by the application of a finite number of row operations on A has the same row space as S, that is, $[S]$. With a judicious choice of operations, A can be transformed into a matrix that may give a much simpler description of $[S]$. Let us turn to an illustration.

EXAMPLE

3.1. Let $S = (\alpha_1, \alpha_2, \alpha_3, \alpha_4)$, where

$$\alpha_1 = (0, 1, -1, 2, 5), \qquad \alpha_2 = (0, 2, -2, -1, 0),$$
$$\alpha_3 = (0, 1, -1, 1, 3), \qquad \alpha_4 = (0, 2, -2, 0, 2).$$

The matrix of S is

$$A = \begin{bmatrix} 0 & 1 & -1 & 2 & 5 \\ 0 & 2 & -2 & -1 & 0 \\ 0 & 1 & -1 & 1 & 3 \\ 0 & 2 & -2 & 0 & 2 \end{bmatrix}.$$

Using row operations of the kind indicated, we transform A successively as follows:

$$A \xrightarrow{\text{III}} \begin{bmatrix} 0 & 1 & -1 & 2 & 5 \\ 0 & 0 & 0 & -5 & -10 \\ 0 & 1 & -1 & 1 & 3 \\ 0 & 2 & -2 & 0 & 2 \end{bmatrix}, \quad \xrightarrow{\text{III}} \begin{bmatrix} 0 & 1 & -1 & 2 & 5 \\ 0 & 0 & 0 & -5 & -10 \\ 0 & 0 & 0 & -1 & -2 \\ 0 & 2 & -2 & 0 & 2 \end{bmatrix},$$

$$\xrightarrow{\text{III}} \begin{bmatrix} 0 & 1 & -1 & 2 & 5 \\ 0 & 0 & 0 & -5 & -10 \\ 0 & 0 & 0 & -1 & -2 \\ 0 & 0 & 0 & -4 & -8 \end{bmatrix}, \quad \xrightarrow{\text{II}} \begin{bmatrix} 0 & 1 & -1 & 2 & 5 \\ 0 & 0 & 0 & 1 & 2 \\ 0 & 0 & 0 & -1 & -2 \\ 0 & 0 & 0 & -4 & -8 \end{bmatrix},$$

$$\xrightarrow{\text{III}} \begin{bmatrix} 0 & 1 & -1 & 2 & 5 \\ 0 & 0 & 0 & 1 & 2 \\ 0 & 0 & 0 & 0 & 0 \\ 0 & 0 & 0 & -4 & -8 \end{bmatrix}, \quad \xrightarrow{\text{III}} \begin{bmatrix} 0 & 1 & -1 & 2 & 5 \\ 0 & 0 & 0 & 1 & 2 \\ 0 & 0 & 0 & 0 & 0 \\ 0 & 0 & 0 & 0 & 0 \end{bmatrix},$$

$$\xrightarrow{\text{III}} \begin{bmatrix} 0 & 1 & -1 & 0 & 1 \\ 0 & 0 & 0 & 1 & 2 \\ 0 & 0 & 0 & 0 & 0 \\ 0 & 0 & 0 & 0 & 0 \end{bmatrix}.$$

From the last matrix we infer that (i) S is linearly dependent, (ii) dim $[S] = 2$, and that $S' = \{(0, 1, -1, 0, 1), (0, 0, 0, 1, 2)\}$ is a basis of $[S]$. Further, since dim $[S] = 2$ and $\{\alpha_1, \alpha_2\}$ is a linearly independent subset of S, this pair of vectors is a basis of $[S]$ consisting of elements of S.

Using the set S' as a basis of S, we have

$$[S] = [S'] = \{a(0, 1, -1, 0, 1) + b(0, 1, 0, 1, 2) \mid a, b \in \mathsf{R}\}$$
$$= \{(0, a, -a, b, a + 2b) \mid a, b \in \mathsf{R}\}$$

Thus, a vector $(x_1, \ldots, x_5)$ of R^5 is in $[S]$ if and only if

$$x_1 = 0$$
$$x_2 = a$$
$$x_3 = -a$$
$$x_4 = b$$
$$x_5 = a + 2b$$

Eliminating the parameters a and b, we conclude that a vector of R^5 is in $[S]$ if and only if it is a solution of the following system of linear homogeneous equations.

$$\begin{aligned} x_1 \qquad\qquad\quad &= 0 \\ x_2 + x_3 \qquad\qquad &= 0 \\ x_2 \qquad + 2x_4 - x_5 &= 0 \end{aligned}$$

Equivalently, $[S]$ may be characterized as the solution space of this system.

The method used in this example to derive a basis for the space spanned by the given set S can be formulated as an algorithm. Once this algorithm (let us call it the *basic* algorithm temporarily) is available, we then have a computation method for answering questions in each of the four categories mentioned at the beginning of this section. Indeed, with the basic algorithm we can dispose of a question of type (a) as well as of type (b). Once dim $[S]$ is known, we can cope with a question of type (c) as follows: specify an order for the elements of S; let us say $(\alpha_1, \ldots, a_m)$. Clearly, there is no loss of generality if we assume that each $\alpha_i \neq 0$. If dim $[S] = 1$, then $\{\alpha_1\}$ is a basis of $[S]$. If dim $[S] > 1$, test the pair $\{\alpha_1, \alpha_2\}$ for linear dependence, using the basic algorithm. If this pair is linearly dependent, consider in turn $\{\alpha_1, \alpha_3\}$, $\{\alpha_1, \alpha_4\}$, ... until a linearly independent pair $\{\alpha_1, \alpha_k\}$ is obtained. If dim $[S] = 2$, then $\{\alpha_1, \alpha_k\}$ is a basis. If dim $[S] > 2$, consider in a similar way $\{\alpha_1, \alpha_k, \alpha_{k+1}\}$, $\{\alpha_1, \alpha_k, \alpha_{k+2}\}$, ..., until a linearly independent set of three vectors is obtained. Continuing in this mechanical way, we eventually obtain a basis of $[S]$ consisting of elements of S.

Finally, consider a question of type (d). Is a given vector β of the space at hand a member of $[S]$? This can be answered by adjoining β to S', a subset of S which is a basis of $[S]$, and applying the basic algorithm to determine whether $S' \cup \{\beta\}$ is linearly independent or linearly dependent. If it is linearly dependent, then $\beta \in [S]$, and if it is linearly independent then $\beta \notin [S]$. The last of the examples illustrates a more sophisticated approach to a question of type (d). Once $[S]$ is characterized as the solution space of a system of linear homogeneous equations, there is an obvious algorithm for deciding whether a given vector is a member of $[S]$. We recommend this method to the reader, but we shall not provide an algorithm for obtaining such a system of equations.

We turn now to a description of an algorithm for determining a basis of the space spanned by a given finite set S of vectors in F^n. It consists of a recipe for reducing a matrix to a particular form by using row operations. When applied to the matrix of S, it places a basis of $[S]$ in evidence.

☐ **Definition.** An $m \times n$ matrix E is called an *echelon matrix* if and only if

(a) the leftmost nonzero coordinate in each nonzero row of E is 1;

(b) each column that contains the leftmost 1 of some nonzero row has 0 as each of its other coordinates;

(c) each zero row appears below every nonzero row;

(d) if rows $1, \ldots, r$ are the nonzero rows of E and if the first 1 in row i occurs in column c_i $(i = 1, \ldots, r)$, then $c_1 < \cdots < c_r$.

The last matrix obtained in the example is an echelon matrix. By imitating the method used to obtain it from the matrix A we can show that *any $m \times n$ matrix can be transformed to an echelon matrix by row operations.*

STEP 1. Locate the leftmost nonzero column (if there is none, all entries are zero and the matrix is in echelon form). If the first coordinate of that column is 0, apply a type I operation to obtain a nonzero entry there. If this entry is $\neq 1$, change it to 1 with an operation of type II.

STEP 2. Change each other nonzero coordinate of the column under consideration to 0 by an operation of type III.

Thus the given matrix A is now transformed to the form

$$A_1 = \begin{bmatrix} 0\cdots 0 & 1 & *\cdots * \\ 0\cdots 0 & 0 & *\cdots * \\ \vdots & & \\ 0\cdots 0 & 0 & *\cdots * \end{bmatrix}.$$

STEP 3. Repeat Step 1 on the matrix whose rows are the rows below that row whose initial entry has just been made 1.

STEP 4. Repeat Step 3 until each zero row appears below every nonzero row.

STEP 5. Replace each nonzero entry above the initial 1 of a nonzero row by 0 using type III operations.

In practice we may choose to depart from this recipe for the reduction of a matrix to an echelon matrix in order to take advantage of peculiar features that a given matrix may have. The distinguished feature of an echelon matrix is given in the next theorem. Its proof is left as an exercise.

□ **Theorem 3.1.** Let E be a nonzero echelon matrix. The nonzero rows of E form a basis of the row space of E.

EXAMPLES

3.2. Suppose that

$$A = \begin{bmatrix} 1 & 0 & 2 & 0 & 1 & 7 \\ -2 & 0 & -1 & -4 & -13 & -7 \\ 2 & 0 & 5 & 1 & 3 & 14 \\ 1 & 0 & 1 & 1 & 4 & 5 \end{bmatrix}$$

is the matrix of a set S of four vectors of R^6. It can be reduced to the following echelon matrix:

$$E = \begin{bmatrix} 1 & 0 & 0 & 0 & 3 & 5 \\ 0 & 0 & 1 & 0 & -1 & 1 \\ 0 & 0 & 0 & 1 & 2 & -1 \\ 0 & 0 & 0 & 0 & 0 & 0 \end{bmatrix}.$$

Thus S is linearly dependent, dim $[S] = 3$, and the three nonzero rows of E form a basis of $[S]$. A vector $(x_1, \ldots, x_6) \in [S]$ if and only if

$$\begin{aligned} x_2 \qquad\qquad\qquad\qquad &= 0 \\ 3x_1 \quad - x_3 + 2x_4 - x_5 \qquad &= 0 \\ 5x_1 \quad + x_3 - x_4 \qquad - x_6 &= 0 \end{aligned}$$

for arbitrary choices of x_1, x_3, and x_4.

3.3. Let us outline another application of the reduction of a matrix to an echelon matrix. A system of linear homogeneous equations

$$(H) \qquad \begin{aligned} a_{11}x_1 + a_{12}x_2 + \cdots + a_{1n}x_n &= 0 \\ &\vdots \\ a_{m1}x_1 + a_{m2}x_2 + \cdots + a_{mn}x_n &= 0 \end{aligned}$$

determines a matrix A in a natural way, namely, the matrix whose (i, j)th entry is the coefficient of x_j in the ith equation. We call A the matrix of the system (H). It is clear that performing row operations on A of types I, II, and III corresponds, respectively, to interchanging pairs of equations, multiplying an equation by a nonzero scalar, and adding a multiple of one equation to another. It is easily seen that each such operation on a system of linear equations yields an *equivalent system*, that is, one having the same set of solutions as the original, Thus, if A, the matrix of (H), can be reduced to the echelon matrix E, the solutions of the associated system of linear equations are precisely those of (H). From a system of linear equations whose matrix is an echelon matrix a description of its solutions (hence that of any equivalent system) that is as explicit as one can hope for can be written down immediately.

As an illustration, consider the linear homogeneous system of four equations in $x_1, \ldots, x_6$ whose matrix is that at the beginning of Example 3.2. An equivalent system, whose matrix is the echelon matrix displayed in that example is

$$\begin{aligned} x_1 \qquad + 3x_5 + 5x_6 &= 0 \\ x_3 \quad - x_5 + x_6 &= 0 \\ x_4 + 2x_5 - x_6 &= 0. \end{aligned}$$

Thus the solutions consist of all $(x_1, \ldots, x_6)$ such that

$$\begin{aligned} x_1 &= -3x_5 - 5x_6, \\ x_3 &= x_5 - x_6, \\ x_4 &= -2x_5 + x_6; \end{aligned}$$

that is, x_2, x_5, and x_6 may be assigned arbitrary values, and for each assignment of values to x_2, x_5, and x_6 the corresponding values of the other x's are computed by the above recipes.

Alternatively, the form of a solution vector is

$$\begin{aligned} &(-3x_5 - 5x_6, x_2, x_5 - x_6, -2x_5 + x_6, x_5, x_6) \\ &\qquad = x_2(0, 1, 0, 0, 0, 0) + x_5(-3, 0, 1, -2, 1, 0) \\ &\qquad\quad + x_6(-5, 0, -1, 1, 0, 1). \end{aligned}$$

Since x_2, x_5, and x_6 may be assigned arbitrary values,

$$\{(0, 1, 0, 0, 0, 0), (-3, 0, 1, -2, 1, 0), (-5, 0, -1, 1, 0, 1)\}$$

is a basis of the solution space.

One final remark is in order. As the reader may have conjectured already, a matrix can be reduced by row operations to exactly one echelon matrix. Because we have not needed this theorem, a proof has not been given. A weaker result, which is obvious and which suffices for our present purposes, reads as follows: the row spaces of two echelon matrices derived from a matrix by row operations have equal dimension.

EXERCISES

3.1. Let $S = \{\alpha_1, \ldots, \alpha_m\}$ and $S' = \{\beta_1, \ldots, \beta_m\}$ be subsets of F^n such that $\beta_j = \alpha_j$ for all j's except $j = i$ and $\beta_i = \alpha_i + c\alpha_k (k \neq i)$. Show that $[S] = [S']$.

3.2. Reduce the following matrices to echelon matrices:

$$\text{(a)} \begin{bmatrix} -1 & 2 & 3 \\ 0 & 1 & 4 \\ 1 & 2 & -2 \end{bmatrix}, \quad \text{(b)} \begin{bmatrix} -1 & 1 & 0 & 1 \\ 5 & 2 & -1 & 0 \\ 0 & 0 & 1 & 1 \\ 0 & 2 & 3 & 5 \end{bmatrix}, \quad \text{(c)} \begin{bmatrix} 0 & 0 & -1 & 2 & 3 \\ -1 & 1 & -3 & 0 & 2 \\ 0 & 2 & 1 & 4 & 3 \end{bmatrix}.$$

3.3. Let $S = (\alpha_1, \alpha_2, \alpha_3)$, where $\alpha_1 = (1, 0, 2, 1)$, $\alpha_2 = (-1, 1, -1, 2)$, and $\alpha_3 = (1, 3, 3, 4)$.
(a) Reduce the matrix of S to an echelon matrix.
(b) Determine dim $[S]$ and a subset of S that is a basis of $[S]$.

3.4. Show that the calculations made in Example 3.1 yield the following result. The set

$$\begin{array}{ll} x - x^2 + 2x^3 + 5x^4, & 2x - 2x^2 - x^3, \\ x - x^2 + x^3 + 3x^4, & 2x - 2x^2 + 2x^4, \end{array}$$

of four polynomials of $P_4(\mathrm{R})$ is linearly dependent and $\{x - x^2 + x^4, x^3 + 2x^4\}$ is a basis of the subspace spanned by the given set.

3.5. In R^3 let $S_1 = \{(-1, 1, 3), (4, 1, 0)\}$ and $S_2 = \{(2, 1, 1), (-1, 2, 3)\}$.
(a) Determine the dimensions of $[S_1 \cup S_2]$ and $[S_1] \cap [S_2]$.
(b) Find a basis of $[S_1] \cap [S_2]$.

3.6. If A is the matrix of an ordered basis of F^n, describe an echelon matrix to which A can be reduced.

3.7. Solve the following systems of equations:

$$\text{(a)} \quad \begin{aligned} x_1 + 2x_2 + x_3 &= 0, \\ -x_1 + 3x_3 + 5x_4 &= 0, \\ x_1 - 2x_2 + x_3 + x_4 &= 0, \end{aligned}$$

$$\text{(b)} \quad \begin{aligned} x_2 - 2x_3 + x_4 + x_6 + 2x_8 &= 0, \\ x_2 - 2x_3 - 2x_4 - x_6 + 3x_8 &= 0, \\ 2x_2 - 4x_3 + 3x_6 - 7x_7 + 7x_8 &= 0, \\ x_4 - x_6 + 5x_7 - 2x_8 &= 0. \end{aligned}$$

3.8. (a) Revise the method described in Example 3.3 to solve a system of linear homogeneous equations in order to be able to cope with systems of non-homogeneous equations like

$$a_{ii} x_1 + a_{i2} + \cdots + a_{in} x_n = b_i, \qquad 1 \le i \le m,$$

by considering the "augmented" matrix

$$\begin{bmatrix} a_{11} & \cdots & a_{1n} & b_1 \\ \vdots & & & \\ a_{m1} & \cdots & a_{mn} & b_m \end{bmatrix}.$$

In particular, describe how to recognize a system that has no solutions.

(b) Solve the following systems of equations:

$$\text{(i)}\quad \begin{aligned} x+2y+u+v&=6,\\ 2y+u+v&=0,\\ x\qquad +u-v&=0,\\ -2x+y+u\qquad &=0, \end{aligned} \qquad \text{(ii)}\quad \begin{aligned} x-y+2z&=5,\\ 2x-2y-z&=0,\\ x-y+z&=3,\\ 2x-2y\qquad &=1, \end{aligned}$$

$$\text{(iii)}\quad \begin{aligned} x+y+z&=2,\\ -x\qquad +4z&=5,\\ 2x+2y-z&=4. \end{aligned}$$

(c) For what values of a, b, c, and d has the following system of equations a solution? Does the set of all such quadruples (a, b, c, d) form a vector space?

$$\begin{aligned} 3x+2y+5u+4v&=a,\\ 5x+3y+2u+v&=b,\\ 11x+7y+12u+9v&=c,\\ 4x+3y+13u+11v&=d. \end{aligned}$$

3.9. Prove Theorem 3.1.

3.10. Let A be the matrix of a system of m linear homogeneous equations in n unknowns. Suppose that an echelon matrix to which A can be reduced has exactly r nonzero rows. Show that the dimension of the solution space of the system of equations is $n-r$.

4. CHANGE OF BASIS

We continue with our discussion of calculation methods. Now we consider the relation between the coordinates of a vector in relation to different bases of the space in question. Let V be an n-dimensional space over F with

$$\mathscr{B} = (\alpha_1, \ldots, \alpha_n) \quad\text{and}\quad \mathscr{B}' = (\alpha_1', \ldots, \alpha_n')$$

as ordered bases. Let $\beta \in V$ and suppose that

$$[\beta]_{\mathscr{B}} = (x_1, \ldots, x_n) \quad\text{and}\quad [\beta]_{\mathscr{B}'} = (x_1', \ldots, x_n').$$

We shall find the relation between the x_i's and the x_i''s by using the relation that exists between the two bases. Each element of the $\mathscr{B}$-basis, as an element of V, is expressible as a unique linear combination of elements of the $\mathscr{B}'$-basis. Suppose that

$$\begin{aligned} \alpha_1 &= a_{11}\alpha_1' + \cdots + a_{1n}\alpha_n' \\ &\vdots \\ \alpha_n &= a_{n1}\alpha_1' + \cdots + a_{nn}\alpha_n. \end{aligned}$$

Once again a matrix is at hand. We call

$$(a_{ij}) = \begin{bmatrix} a_{11} \cdots a_{1n} \\ \vdots \\ a_{n1} \cdots a_{nn} \end{bmatrix}$$

the *matrix of the* (ordered) *basis* $\mathscr{B}$ *relative to the* (ordered) *basis* $\mathscr{B}'$ (or simply the matrix of $\mathscr{B}$ to $\mathscr{B}'$).

Substitution of these expressions for $\alpha_1, \ldots, \alpha_n$ in the representation of β as $\sum_1^n x_i \alpha_i$, yields

$$\begin{aligned} \beta &= x_1\left(\sum_1^n a_{1j}\alpha_j'\right) + \cdots + x_i\left(\sum_1^n a_{ij}\alpha_j'\right) + \cdots + x_n\left(\sum_1^n a_{nj}\alpha_j'\right) \\ &= \left(\sum_1^n x_i a_{i1}\right)\alpha_1' + \cdots + \left(\sum_1^n x_i a_{ij}\right)\alpha_j' + \cdots + \left(\sum_1^n x_i a_{in}\right)\alpha_n'. \end{aligned}$$

Comparing this representation of β in terms of the $\mathscr{B}'$-basis with the given one, $\beta = \sum_1^n x_i' \alpha_i'$, we deduce that

$$\begin{aligned} x_1' &= \sum_1^n x_i a_{i1} = x_1 a_{11} + \cdots + x_n a_{n1}, \\ &\vdots \\ x_j' &= \sum_1^n x_i a_{ij} = x_1 a_{1j} + \cdots + x_n a_{nj}, \\ &\vdots \\ x_n' &= \sum_1^n x_i a_{in} = x_1 a_{1n} + \cdots + x_n a_{nn}. \end{aligned}$$

In order to summarize the foregoing results (as well as another derived later in this section) we introduce the notion of the product of two matrices.

☐ **Definition.** Let $A = (a_{ij})$ be an $m \times n$ matrix over the field F and $B = (b_{ij})$ be an $n \times r$ matrix over F. The *product* AB of A and B is the $m \times r$ matrix (c_{ij}) such that

$$c_{ij} = \sum_{k=1}^{n} a_{ik} b_{kj}, \qquad 1 \le i \le m, \quad 1 \le j \le r.$$

The product of A and B is computed by what is often called "row by column multiplication"; that is, the (i, j)th entry of AB is the sum of the products of each entry of the ith row of A and the corresponding element of the jth column of B. Notice that for the definition to be meaningful it is mandatory that the number of columns of A be equal to the number of rows of B. Following are two illustrations of the products of two matrices:

$$\underset{4 \times 3}{\begin{bmatrix} 1 & -1 & 0 \\ 0 & 2 & 1 \\ -1 & 0 & 1 \\ 0 & 1 & 0 \end{bmatrix}} \underset{3 \times 4}{\begin{bmatrix} 1 & 0 & 0 & 0 \\ 1 & 1 & -1 & 0 \\ 1 & 1 & -1 & 0 \end{bmatrix}} = \underset{4 \times 4}{\begin{bmatrix} 0 & -1 & 1 & 0 \\ 2 & 3 & -1 & -1 \\ -1 & 1 & 1 & -1 \\ 1 & 1 & -1 & 0 \end{bmatrix}}$$

$$\underset{1 \times 4}{(1, -1, 2, 3)} \underset{4 \times 3}{\begin{bmatrix} 2 & 3 & -1 \\ -1 & 0 & 1 \\ 1 & -1 & 1 \\ 2 & -1 & 0 \end{bmatrix}} = \underset{1 \times 3}{(11, -2, 0)}$$

Interpreting the n-tuples of coordinates of β in relation to the $\mathscr{B}$-basis as $1 \times n$ matrices, the foregoing system of equations may be summarized by the matrix equation

$$(x_1, \cdots, x_n)(a_{ij}) = (x_1', \ldots, x_n'),$$

$$[\beta]_{\mathscr{B}}(a_{ij}) = [\beta]_{\mathscr{B}'}.$$

EXAMPLES

4.1. In R^2 consider, along with the standard basis $\mathscr{B} = (\epsilon_1, \epsilon_2)$, the basis $\mathscr{B}' = (\delta_1, \delta_2)$, where

$$\delta_1 = (\cos\theta, \sin\theta),$$

$$\delta_2 = (-\sin\theta, \cos\theta).$$

The earlier definition of the product of two matrices makes possible a concise description of this set of n^2 relations.

$$C = AB,$$

where A is the matrix of $\mathscr{B}$ to $\mathscr{B}'$, B is the matrix of $\mathscr{B}'$ to $\mathscr{B}''$, and C is the matrix of $\mathscr{B}$ to $\mathscr{B}''$. The following schematic diagram may clarify our result:

$$\mathscr{B} \xrightarrow{A} \mathscr{B}' \xrightarrow{B} \mathscr{B}'' \qquad (\mathscr{B} \xrightarrow{C = AB} \mathscr{B}'')$$

EXAMPLE

4.3. Each of the following pairs is a basis of R^2:

$$\mathscr{B} = ((1, 0), (0, 1)), \quad \mathscr{B}' = ((1, 1), (-1, 1)), \quad \mathscr{B}'' = ((1, 2), (-2, 1)).$$

The matrix of $\mathscr{B}$ to $\mathscr{B}'$ is

$$\begin{bmatrix} \frac{1}{2} & -\frac{1}{2} \\ \frac{1}{2} & \frac{1}{2} \end{bmatrix}.$$

The matrix of $\mathscr{B}'$ to $\mathscr{B}''$ is

$$\begin{bmatrix} \frac{3}{5} & -\frac{1}{5} \\ \frac{1}{5} & \frac{3}{5} \end{bmatrix}$$

and the matrix of $\mathscr{B}$ to $\mathscr{B}''$ is

$$\begin{bmatrix} \frac{1}{2} & -\frac{1}{2} \\ \frac{1}{2} & \frac{1}{2} \end{bmatrix} \begin{bmatrix} \frac{3}{5} & -\frac{1}{5} \\ \frac{1}{5} & \frac{3}{5} \end{bmatrix} = \begin{bmatrix} \frac{1}{5} & -\frac{2}{5} \\ \frac{2}{5} & \frac{1}{5} \end{bmatrix}.$$

EXERCISES

4.1. Let $V = P_2(\mathsf{R})$. Show that $\mathscr{B}' = (-1, 1 + x, x^2 - 1)$ is a basis. Find the matrix of $\mathscr{B} = (1, x, x^2)$ to $\mathscr{B}'$ and that of $\mathscr{B}'$ to $\mathscr{B}$. Find the coordinates of $3x^2 - x + 6$ in relation to $\mathscr{B}'$.

4.2. In R^3, $\mathscr{B} = (\epsilon_1, \epsilon_2, \epsilon_3)$,
$\mathscr{B}' = ((1, 1, 0), (0, -2, 1), (1, 2, 3))$,
$\mathscr{B}'' = ((0, -2, 1), (1, 1, 1), (-2, 0, 4))$
are bases.
(a) Find the matrix of $\mathscr{B}$ to $\mathscr{B}'$, of $\mathscr{B}'$ to $\mathscr{B}''$, and of $\mathscr{B}$ to $\mathscr{B}''$ directly. Verify that the third matrix is the product of the first and the second.
(b) Find the coordinates of (x, y, z) in R^3 relative to each basis.

4.3. Let $\mathscr{B} = (\alpha_1, \ldots, \alpha_n)$ and $\mathscr{B}' = (\alpha_1', \ldots, \alpha_n')$ be bases of an n-dimensional space V over F. For $\beta \in V$ let $[\beta]_{\mathscr{B}} = (x_1, \ldots, x_n)$ and $[\beta]_{\mathscr{B}'} = (x_1', \ldots, x_n')$. Prove that the mapping $h: F^n \to F^n$, where

$$(x_1, \ldots, x_n)h = (x_1', \ldots, x_n')$$

is an isomorphism of F^n with itself.

4.4. Generalize Example 4.2 to the space $P_n(\mathsf{R})$. In particular, show that the coefficients of $p(x) = \sum_0^n a_i x^i$, relative to the basis $(1, x - c, \ldots, (x - c)^n)$ are expressible in terms of derivatives of p at $x = c$.

4.5. Let T be a homomorphism of R^n into itself. Let $\mathscr{B} = (\epsilon_1, \ldots, \epsilon_n)$ be the standard basis of R^n and A be the $n \times n$ matrix of the ordered set $(\epsilon_1 T, \ldots, \epsilon_n T)$. Show that if $\alpha = (x_1, \ldots, x_n) \in \mathsf{R}^n$ then $\alpha T = (x_1, \ldots, x_n)A$.

4.6. Let $\mathscr{B}$ and $\mathscr{B}'$ be ordered bases of an n-dimensional vector space V. Let A be the matrix of $\mathscr{B}$ to $\mathscr{B}'$ and B be the matrix of $\mathscr{B}'$ to $\mathscr{B}$. What are the matrices AB and BA?

5. GEOMETRIC ASPECTS OF VECTOR SPACES†

In Section 1.3 we mentioned the interpretation of R^3 in E_3 by which each vector (x, y, z) of R^3 is mapped onto the point of E_3 having coordinates x, y, and z relative to some fixed rectangular coordinate system. The image of a nontrivial subspace of R^3 under this mapping is a line or a plane through the origin O of E_3. Further, each such line or plane is the image of some one such subspace. The question now arises in a natural way: what are the subsets of R^3 which have arbitrary lines or planes as images? Consider a line L in E_3. Recall that such a geometric configuration may be described by a set of parametric equations of the form

$$x = a + lt,$$

$$y = b + mt,$$

$$z = c + nt,$$

where t is a real number, $(l, m, n) \neq (0, 0, 0)$ and, if L is not through the origin, $(a, b, c) \neq (0, 0, 0)$; that is, L is the graph of

$$\{(a, b, c) + t(l, m, n) \mid t \in \mathsf{R}\}.$$

Now the graph of $\{t(l, m, n) \mid t \in \mathsf{R}\}$ is the line, L', let us say, parallel to L and through O. Since L' is the image of the one-dimensional subspace

† This section is optional.

$[(l, m, n)]$ of R^3, we conclude that L may be described as the image of the set of vectors

$$\{(a, b, c) + \delta \mid \delta \in S\}$$

where (a, b, c) is a fixed element of R^3. In a similar manner we may deduce that a plane of E_3 is the image of a subset of R^3 with the same form for some two-dimensional space S of R^3.

Before giving the definition suggested by these observations, let us agree to write

$$\alpha + S \quad \text{for} \quad \{\alpha + \delta \mid \delta \in S\},$$

where α and S are, respectively, a fixed element and a fixed nonempty subset of a vector space.

□ **Definition.** A subset M of a vector space V is called a *linear manifold* of V if and only if there exists an element α of V and a subspace S of V such that

$$M = \alpha + S.$$

□ **Theorem 5.1.** Two linear manifolds $M_1 = \alpha_1 + S_1$ and $M_2 = \alpha_2 + S_2$ of a vector space V are equal if and only if $S_1 = S_2$ and $\alpha_1 - \alpha_2 \in S_1$.

PROOF. Assume that $M_1 = M_2$. Since $0 \in S_1$, $\alpha_1 + 0 = \alpha_1 \in M_1$. Thus there exists $\delta_2 \in S_2$ such that $\alpha_1 = \alpha_2 + \delta_2$. Hence $\alpha_1 - \alpha_2 \in S_2$. Since S_2 is a subspace, it follows that $-(\alpha_1 - \alpha_2) = \alpha_2 - \alpha_1 \in S_2$.

The proof that $S_1 = S_2$ now follows easily. Assume that $\delta_1 \in S_1$. Then $\alpha_1 + \delta_1 = \alpha_2 + \delta_2$ for some δ_2 in S_2. Thus $\delta_1 = (\alpha_2 - \alpha_1) + \delta_2$ and, since $\alpha_2 - \alpha_1 \in S_2$, we conclude that $\delta_1 \in S_2$. In other words, $S_1 \subseteq S_2$.

Reversing the roles of M_1 and M_2 leads to the reverse inclusion relation. Thus $S_1 = S_2$ and in turn $\alpha_1 - \alpha_2 \in S_1(=S_2)$.

The converse statement is left as an exercise. ◇

□ **Definition.** The subspace S that is uniquely determined by a linear manifold M is called the *direction space* of M. The *dimension* of a linear manifold is taken to be the dimension of its direction space. One-dimensional linear manifolds of a vector space are called *lines* and two-dimensional linear manifolds are called *planes*. If the vector space is n-dimensional ($n \geq 2$), its ($n - 1$)-dimensional linear manifolds are called *hyperplanes*. Further, zero-dimensional linear manifolds are called *points*.

We turn now to the notion of parallelism for linear manifolds. In E_2 the euclidean space of dimension two, we can classify two lines L and L' as parallel if and only if either $L = L'$ or $L \cap L' = \varnothing$. This approach, however, breaks down in E_3, as the reader can readily ascertain. An alternative is provided by our notion of a direction space.

□ **Definition.** Two linear manifolds in a vector space are called *parallel* if and only if they have the same direction space.

□ **Theorem 5.2.** If M_1 and M_2 are parallel linear manifolds of a vector space V, then either $M_1 = M_2$ or $M_1 \cap M_2 = \varnothing$.

PROOF. Assume that M_1 and M_2 are parallel linear manifolds. Then there exists a subspace S of V such that $M_1 = \alpha_1 + S$ and $M_2 = \alpha_2 + S$ for $\alpha_1, \alpha_2 \in V$. Assume that $M_1 \cap M_2 \neq \varnothing$. Let $\alpha \in M_1 \cap M_2$. Then $\alpha = \alpha_1 + \delta_1 = \alpha_2 + \delta_2$ for elements $\delta_1, \delta_2 \in S$. Hence $\alpha_1 - \alpha_2 \in S$. By Theorem 5.1 we infer that $M_1 = M_2$. ◇

The following restricted converse of Theorem 5.2 is provable.

□ **Theorem 5.3.** If M_1 and M_2 are hyperplanes of a vector space V, such that either $M_1 = M_2$ or $M_1 \cap M_2 = \varnothing$, then M_1 and M_2 are parallel.

PROOF. This is left as an exercise.

Our next theorem for linear manifolds of a vector space corresponds to the "parallel postulate" of Euclid's geometry.

□ **Theorem 5.4.** Let α be a vector and M be a linear manifold of a vector space V. Then there exists exactly one linear manifold that contains α and is parallel to M.

PROOF. Let α and M be given. Suppose $M = \alpha' + S$. Then $M' = \alpha + S$ is a linear manifold that contains α and is parallel to M. If M'' is any linear manifold that is parallel to M, then $M'' = \alpha'' + S$. If, also, $\alpha \in M''$, $\alpha - \alpha'' \in S$, hence (Theorem 5.1) $M'' = M'$. ◇

EXAMPLE

5.1. Consider a system

$$(N) \qquad \begin{aligned} a_{11}x_1 + \cdots + a_{1n}x_n &= b_1 \\ &\vdots \\ a_{m1}x_1 + \cdots + a_{mn}x_n &= b_m \end{aligned}$$

of nonhomogeneous linear equations in $x_1, \ldots, x_n$ and the associated system of homogeneous linear equations

$$(H) \qquad \begin{aligned} a_{11}x_1 + \cdots + a_{1n}x_n &= 0 \\ &\vdots \\ a_{m1}x_1 + \cdots + a_{mn}x_n &= 0, \end{aligned}$$

where the a_{ij}'s and the b_i's are elements of a field F. We have observed that the set S of solutions of (H) is a subspace of F^n. Assume that $\alpha_0 = (x_1{}^0, \ldots, x_n{}^0)$ is a solution of (N). Then $\alpha_0 + \delta$ is a solution of (N) for each $\delta \in S$. Conversely, if α is any solution of (N), then $\alpha - \alpha_0 \in S$. Hence if (N) has solutions, the set of all solutions of (N) is the linear manifold $\alpha_0 + S$, where α_0 is any one solution of (N) and S is the solution space of (H).

The relationship of a subspace of a vector space to the linear manifolds defined by that subspace suggest a consideration of the following notion.

☐ **Definition.** A mapping T of a vector space V into itself is called a *translation* of V if and only if there exists a vector α_0 in V such that

$$\alpha T = \alpha_0 + \alpha$$

for all α in V.

It is clear that a translation determines a 1-1 correspondence of V with itself under which the images of subspaces are linear manifolds. If T is a translation of V and $\alpha T = \alpha_0 + \alpha$ for all α, then α_0 is uniquely determined by T. Indeed, $\alpha_0 = \alpha_0 + 0 = 0T$; that is, α_0 is the image of 0 under T. Conversely, each vector β of V determines a translation; namely, the translation T_β such that $\alpha T_\beta = \beta + \alpha$ for all α in V. It follows that the mapping A, such that $\beta A = T_\beta$, of V into $t(V, V)$, the set of all translations of V, is a 1-1 correspondence.

We propose to convert $t(V, V)$ into a vector space over F and then show that A is an isomorphism between V and $t(V, V)$. If T_1 and T_2 are translations of V, then we define $T_1 + T_2$ to be the map on V into V such that

$$\alpha(T_1 + T_2) = (\alpha T_1)T_2, \qquad \text{all } \alpha \in V.$$

If $T \in t(V, V)$ and $a \in F$, then we define aT to be the map on V into V such that

$$\alpha(aT) = a\alpha_0 + \alpha, \qquad \text{all } \alpha \in V,$$

where $\alpha T = \alpha_0 + \alpha$. Clearly, $T_1 + T_2$ and αT are elements of $t(V, V)$. To show that $(t(V, V) +, 0, F, \cdot)$ is a vector space over F and, simultaneously, isomorphic to V, it is sufficient (see Exercise 2.10) to prove that the mapping A has the following property:

$$(a\alpha + b\beta)A = a(\alpha A) + b(\beta A).$$

For this we study $T_{a\alpha + v\beta}$, the image of $a\alpha + b\beta$ under A. For $\gamma \in V$ we have

$$\begin{aligned}
\gamma T_{a\alpha + b\beta} &= (a\alpha + b\beta) + \gamma \\
&= (a\alpha + \gamma) + b\beta \\
&= (a\alpha + \gamma)T_{b\beta} \\
&= (\gamma T_{a\alpha})T_{b\beta} \\
&= (\gamma(aT_\alpha))(bT_\beta) \\
&= \gamma(aT_\alpha + bT_\beta).
\end{aligned}$$

Thus $(a\alpha + b\beta)A = T_{a\alpha + b\beta} = a(\alpha A) + b(\beta A)$ and the proof is complete.

Let us now consider simultaneously two interpretations of a vector space V. In one we regard a vector α of V as a translation (notation: $\bar{\alpha}$) of V, that is, as an element of $t(V, V)$. In the other we regard α as an element of a set P: in this role we call α a point (notation: $\underline{\alpha}$). Our denotations may suggest what we are up to! They are simply to substantiate that we may think of elements of an arbitrary vector space (and not merely elements of R^2 or R^3, as discussed in Chapter 1) as arrows that act on points to produce points. Indeed, if $\bar{\alpha} \in (V, V)$ and $\underline{\beta} \in P$, we picture the action of $\bar{\alpha}$ on $\underline{\beta}$ as

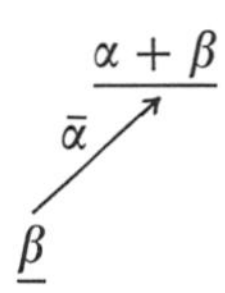

In particular, we picture $\bar{\alpha}$ as an arrow originating at the origin:

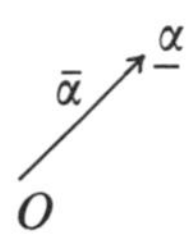

Let us list the basic properties of V regarded as a set of translations "acting" on the points of P.

(1) For each point $\underline{\beta}$ and each vector $\bar{\alpha}$, $\underline{\beta}\bar{\alpha}(= \underline{\alpha + \beta})$ is a uniquely determined point ($=$ element of P).

(2) For any point $\underline{\beta}$ and any vectors $\bar{\alpha}$ and $\bar{\alpha}'$,

$$(\underline{\beta}\bar{\alpha})\bar{\alpha}' = \underline{\beta}(\overline{\alpha + \alpha'}).$$

(3) For any point $\underline{\beta}$ and the zero vector $\bar{0}$,

$$\underline{\beta}\bar{0} = \underline{\beta}.$$

(4) For each pair of points $\underline{\beta}$ and $\underline{\gamma}$ there exists exactly one vector $\bar{\alpha}$ such that

$$\underline{\beta}\bar{\alpha} = \gamma \qquad (\bar{\alpha} \text{ is simply } \bar{\gamma} - \bar{\beta}).$$

The foregoing has been deliberately formulated to suggest the following generalization. Consider the axiomatic theory with primitive notions

(i) a set $P = \{x, y, \ldots\}$ whose elements are called points;

(ii) a vector space $V = \{\alpha, \beta, \ldots\}$ over a field, whose elements are called vectors, and

(iii) a relation from $P \times V$ to P such that

(1)′ the relation is a function [we shall write the value of this function at (x, α) as $x\alpha$],

(2)′ for $x \in P$ and $\alpha, \beta \in V$, $(x\alpha)\beta = x(\alpha + \beta)$,

(3)′ for $x \in P$ and the zero vector 0, $x0 = x$,

(4)′ for points x and y, there is exactly one vector α such that $x\alpha = y$.

Any model of this theory (that is, an interpretation of the primitive notions in which the conditions stated are true) is called an *affine space*. The subject matter known as *affine geometry* is a study of affine spaces.

EXERCISES

5.1. Complete the proof of Theorem 5.1.

5.2. Suppose that we know that a subset M of a vector space V is a linear manifold of V. Thus there is a vector α and a uniquely determined subspace S of V such that $M = \alpha + S$. Show how to recover S from M.

5.3. Find the direction subspace of each of the following linear manifolds of R^3:
(a) $M = \{(x, y, z) \in \mathrm{R}^3 | 2x - 3y + z = 2\}$.
(b) $M = \{(x, y, z) \in \mathrm{R}^3 | x = x + t,\ y = -1 + 3t,\ z = 5 + 6t\}$.

5.4. Let V be a vector space over a field F of q elements. Show that each line of V has exactly q elements.

5.5. Show that there exists a 1-1 correspondence between the points of any two lines of a vector space.

5.6. Show that for each pair of distinct elements of a vector space V there exists exactly one line that contains them.

5.7. Supply a proof of Theorem 5.3.

5.8. Construct a system of linear equations whose solution set is the linear manifold $(1, 0, 1, 0) + [(1, -1, 1, 0), (0, 1, -1, 1)]$ of R^4.

5.9. Suppose that M_1 and M_2 are linear manifolds of a vector space V and $M_1 \cap M_2 \neq \varnothing$. Show that the intersection is a linear manifold of V. What is its dimension?

3 Inner-Product Spaces

In the study of the euclidean space E_3 we find definitions of such metric concepts as the length of an arrow and the angle between two arrows. These notions can be treated efficiently in terms of the function called the *dot product* for arrows. If the points P and Q have the coordinates (x_1, x_2, x_3) and (y_1, y_2, y_3), respectively, relative to some rectangular coordinate system, the dot product of $\overrightarrow{OP}$ and $\overrightarrow{OQ}$ is defined by

$$\overrightarrow{OP} \cdot \overrightarrow{QR} = \sum_{i=1}^{3} x_i yi.$$

Then the length $|\overrightarrow{OP}|$ of $\overrightarrow{OP}$ is given by

$$|\overrightarrow{OP}| = \sqrt{\overrightarrow{OP} \cdot \overrightarrow{OP}}.$$

If θ is the measure of the angle between $\overrightarrow{OP}$ and $\overrightarrow{OQ}$, the law of cosines yields the relation

$$\overrightarrow{OP} \cdot \overrightarrow{OQ} = |\overrightarrow{OP}|\,|\overrightarrow{OQ}| \cos \theta,$$

which in turn specifies a unique value for θ in the interval $[0, \pi]$.

These facts suggest a natural way to attempt the introduction of metric concepts into the vector space R^n (and thereby into any n-dimensional vector space V over R by way of a basis of V—see Example 1.4 which follows). It is this: define the dot product of $(x_1, \ldots, x_n)$ and $(y_1, \ldots, y_n)$ in R^n to be $\sum_1^n x_i y_i$ and then (hopefully) derive as properties of this function on $\mathsf{R}^n \times \mathsf{R}^n$ into R those that make it useful (when interpreted in E_3) in the case in which $n = 3$. An examination of this case may suggest the following as being adequate.

1. The dot product is a symmetric function: $\overrightarrow{OP} \cdot \overrightarrow{OQ} = \overrightarrow{OQ} \cdot \overrightarrow{OP}$.

2. The dot product is linear in its first argument (hence in each argument): $(a\overrightarrow{OP} + b\overrightarrow{OQ}) \cdot \overrightarrow{OR} = a(\overrightarrow{OP} \cdot \overrightarrow{OR}) + b(\overrightarrow{OQ} \cdot \overrightarrow{OR})$.

3. The dot product is positive definite: $\overrightarrow{OP} \cdot \overrightarrow{OP} > 0$ if $\overrightarrow{OP}$ is not the zero arrow.

In this chapter we define metric notions and develop some consequences for a class of vector spaces that includes spaces of n-tuples of real numbers. To achieve generality as well as simplicity, however, we have not proceeded in the concrete way suggested above. Instead, we introduce numerical-valued functions on $V \times V$, where V is a vector space, which are restricted only by properties (1) to (3). Thus we prepare the ground for determining all functions that can serve our present purposes. Since we do not formulate definitions in terms of a basis of a vector space, results that we obtain are applicable to every basis and, in particular, to a basis that we may judge to be "best" in some sense.

At the outset we choose the field of scalars to be R. Later we indicate the modifications that are sufficient to extend the basic results to vector spaces over the field C of complex numbers.

1. EUCLIDEAN SPACES

□ **Definition.** An *inner product* in a vector space V over R is a function $(\,|\,)$ on $V \times V$ into R such that

(1) $\quad (\alpha \mid \beta) = (\beta \mid \alpha)$, for all $\alpha, \beta \in V$,

(2) $(a\alpha + b\beta \mid \gamma) = a(\alpha \mid \gamma) + b(\beta \mid \gamma)$, for all $\alpha, \beta, \gamma \in V$ and all $a, b \in$ R,

(3) $\quad (\alpha \mid \alpha) > 0$ if $\alpha \neq 0$.

A *euclidean space* is a vector space over R together with a specified inner product in the space.

From the symmetry (1) and the linearity in the first argument (2) of an inner product we infer its linearity in the second argument:

$$(\gamma \mid a\alpha + b\beta) = a(\gamma \mid \alpha) + b(\gamma \mid \beta)$$

for all $\alpha, \beta, \gamma \in V$ and all $a, b \in$ R. This bilinearity of an inner product implies in turn that

$$\left(\sum_{i=1}^{n} a_i \alpha_i \,\middle|\, \sum_{j=1}^{m} b_j \beta_j\right) = \sum_{i=1}^{n} \sum_{j=1}^{m} a_i(\alpha_i \mid \beta_j) b_j. \tag{4}$$

EXAMPLES

1.1. Suppose that $\alpha = (x_1, \ldots, x_n)$ and $\beta = (y_1, \ldots, y_n)$ are elements of R^n and we define

$$(\alpha \mid \beta) = \sum_{i=1}^{n} x_i y_i .$$

This is an inner product in R^n that we call the *standard inner product.*

1.2. For $\alpha = (x_1, y_1)$ and $\beta = (x_2, y_2)$ in R^2 define

$$(\alpha \mid \beta) = x_1 x_2 + 2x_1 y_2 + 2y_1 x_2 + 5y_1 y_2 .$$

The function so defined for all α and β is an inner product in R^2. That it satisfies (3) follows from the fact that $(\alpha \mid \alpha) = (x_1 + 2y_1)^2 + y_1{}^2$.

1.3. For $f, g \in C[a, b]$ define

$$(f \mid g) = \int_a^b f(x)\, g(x)\, dx.$$

This is an inner product in $C[a, b]$ that we call the *standard inner product* for this space.

1.4. Let V be any n-dimensional vector space over R and let $\mathscr{B} = (\alpha_1, \ldots, \alpha_n)$ be an ordered basis of V. The function $(\ \mid\ )$ on $V \times V$ into R defined by

$$(\alpha \mid \beta) = \sum_{i=1}^{n} x_i y_i ,$$

where $\alpha = \sum_1^n x_i \alpha_i$ and $\beta = \sum_1^n y_i \alpha_i$ is an inner product in V. Thus any finite-dimensional space over R can be converted to a euclidean space. Notice that in this euclidean space

$$(\alpha_i \mid \alpha_j) = \begin{cases} 1 & \text{if } \ i = j, \\ 0 & \text{if } \ i \neq j, \end{cases} \qquad 1 \leq i, j \leq n.$$

We interrupt our discussion to define the *Kronecker delta*, δ, as the function on the set of ordered pairs of positive integers into $\{0, 1\}$ such that δ_{ij} [the value of δ at (i, j)] is 1 when $i = j$ and 0 when $i \neq j$. Thus we may write the above result as

$$(\alpha_i \mid \alpha_j) = \delta_{ij} .$$

The same is true for the standard basis $(\epsilon_1, \ldots, \epsilon_n)$ of R^n when this space is equipped with the standard inner product; that is,

$$(\epsilon_i \mid \epsilon_j) = \delta_{ij} .$$

Later we shall prove that we can construct an ordered basis $\mathscr{B}$ of a given finite-dimensional euclidean space V such that for all vectors α and β in V the inner product at (α, β) is equal to $\sum_1^n x_i y_i$, where $(x_1, \ldots, x_n) = [\alpha]_{\mathscr{B}}$ and $(y_1, \ldots, y_n) = [\beta]_{\mathscr{B}}$.

Let V be an n-dimensional euclidean space with inner-product $(\,|\,)$ and let $\mathscr{B} = (\alpha_1, \ldots, \alpha_n)$ be an ordered basis of V. Suppose that $\alpha = \sum_1^n x_i \alpha_i$ and $\beta = \sum_1^n y_j \alpha_j$ are elements of V. Then, using (4),

$$(\alpha \mid \beta) = \sum_{i=1}^{n} \sum_{j=1}^{n} x_i(\alpha_i \mid \alpha_j) y_j. \tag{5}$$

It follows that the inner product is determined by the n^2 scalars

$$c_{ij} = (\alpha_i \mid \alpha_j), \qquad 1 \leq i, j \leq n. \tag{6}$$

These scalars define the matrix

$$\begin{bmatrix} c_{11} & \cdots & c_{1n} \\ \vdots & & \\ c_{n1} & \cdots & c_{nn} \end{bmatrix},$$

which is called the *matrix of the inner product* $(\,|\,)$ relative to the (ordered) $\mathscr{B}$-basis. By virtue of properties (1) and (3) of an inner product $c_{ij} = c_{ji}$ and $c_{ii} > 0$ for $1 \leq i, j \leq n$.

On writing the coordinates of α relative to the $\mathscr{B}$-basis as the $1 \times n$ matrix $(x_1, \ldots, x_n)$ and the coordinates of β as the $n \times 1$ matrix

$$\begin{bmatrix} y_1 \\ \vdots \\ y_n \end{bmatrix},$$

we can express the calculation rule (5) as the matrix product

$$(\alpha \mid \beta) = (x_1, \ldots, x_n)(c_{ij}) \begin{bmatrix} y_1 \\ \vdots \\ y_n \end{bmatrix}^{\dagger}$$

† The indicated product is not ambiguous, for multiplication of matrices is an associative operation as we shall prove later.

by identifying the 1×1 matrix on the right side of this equation with its single entry. Introducing the notation

$$(y_1, \ldots, y_n)^t$$

for the last factor above, we can indicate the inner product of α and β on a single line as

$$(\alpha \mid \beta) = (x_1, \ldots, x_n)(c_{ij})(y_1, \ldots, y_n)^t.$$

EXAMPLES

1.5. The matrix of the inner product defined in Example 1.4 relative to the basis at hand is

$$(\delta_{ij}) = \begin{bmatrix} 1 & 0 & \cdots & 0 \\ 0 & 1 & \cdots & 0 \\ \vdots & & & \\ 0 & 0 & \cdots & 1 \end{bmatrix}.$$

1.6. In $P_2(\mathrm{R})$ let an inner product be given by

$$(p \mid q) = \int_0^1 p(x)\, q(x)\, dx.$$

The following calculations yield its matrix relative to the basis $(1, x, x^2)$:

$$(1 \mid 1) = \int_0^1 1 \quad dx = 1 \qquad (1 \mid x) = \int_0^1 x\, dx = \tfrac{1}{2}$$

$$(1 \mid x^2) = \int_0^1 x^2\, dx = \tfrac{1}{3} \qquad (x \mid x) = \int_0^1 x^2\, dx = \tfrac{1}{3}$$

$$(x \mid x^2) = \int_0^1 x^3\, dx = \tfrac{1}{4} \qquad (x^2 \mid x^2) = \int_0^1 x^4\, dx = \tfrac{1}{5}.$$

The matrix is

$$\begin{bmatrix} 1 & \frac{1}{2} & \frac{1}{3} \\ \frac{1}{2} & \frac{1}{3} & \frac{1}{4} \\ \frac{1}{3} & \frac{1}{4} & \frac{1}{5} \end{bmatrix}.$$

1.7. For vectors $\alpha = (x_1, y_1)$ and $\beta = (x_2, y_2)$ in R^2, define

$$(\alpha \mid \beta) = x_1 x_2 + 5x_1 y_2 + 5x_2 y_1 + 26 y_1 y_2 .$$

The function so defined for all α, β in R^2 is an inner product in R^2. Using the formulas in (6), we find the matrix relative to the ordered basis ((1, 1), (−1, 1)) to be

$$\begin{bmatrix} 37 & 25 \\ 25 & 17 \end{bmatrix}.$$

The matrix of this inner product relative to the standard ordered basis is

$$\begin{bmatrix} 1 & 5 \\ 5 & 26 \end{bmatrix},$$

whereas its matrix relative to the basis ((1, 0), (−5, 1)) is

$$\begin{bmatrix} 1 & 0 \\ 0 & 1 \end{bmatrix}.$$

EXERCISES

1.1. Let V be a euclidean space with inner product $(\,|\,)$. Show the following:
(a) $(0\,|\,\alpha) = 0$ for all α in V,
(b) If α_0 is an element of V such that $(\alpha_0\,|\,\beta) = 0$ for all β in V, then $\alpha_0 = 0$.
(c) If $\mathscr{B}$ is a basis of V and α_0 an element of V such that $(\alpha_0\,|\,\beta) = 0$ for all β in $\mathscr{B}$, then $\alpha_0 = 0$.

1.2. Let V be an n-dimensional euclidean space with inner product $(\,|\,)$. Show that if $\mathscr{B} = \{\alpha_1, \ldots, \alpha_n\}$ is a set of n elements of V such that $(\alpha_i\,|\,\alpha_j) = \delta_{ij}$ for $1 \leq i, j \leq n$ then
(a) $\mathscr{B}$ is a basis of V,
(b) for each α in V, $\alpha = \sum_1^n (\alpha\,|\,\alpha_i)\alpha_i$.

1.3. Determine which of the following functions f on $R^3 \times R^3$ into R are inner products in R^3.
(a) $f(\alpha, \beta) = (x_1 - x_2)^2 + (y_1 - y_2)^2 + (z_1 - z_2)^2$, where $\alpha = (x_1, y_1, z_1)$ and $\beta = (x_2, y_2, z_2)$.
(b) $f(\alpha, \beta) = 4x_1x_2 + 3y_1y_2 + z_1z_2 + 2x_1y_2 + 2x_2y_1 - y_1z_2 - z_1y_2$, where $\alpha = (x_1, y_1, z_1)$ and $\beta = (x_2, y_2, z_2)$.

1.4. The function $(\,|\,)$ on $R^2 \times R^2$ into R, where

$$(\alpha\,|\,\beta) = x_1x_2 - x_1y_2 - x_2y_1 + 3y_1y_2$$

and $\alpha = (x_1, y_1)$, $\beta = (x_2, y_2)$, is an inner product in R^2.
(a) Find its matrix relative to the standard basis of R^2.
(b) Find an ordered basis (α_1, α_2) of R^2 such that

$$(\alpha_i\,|\,\alpha_j) = \delta_{ij}, \qquad i, j = 1, 2.$$

1.5. In R^2 define a real-valued function $(\,|\,)$ on ordered pairs (α, β) of vectors as follows: if $\alpha = (x_1, y_1)$ and $\beta = (x_2, y_2)$, then

$$(\alpha \mid \beta) = (x_1, y_1)\begin{bmatrix} 1 & 1 \\ 1 & 1 \end{bmatrix}(x_2, y_2)^t.$$

Determine whether this is an inner product in R^2.

1.6. Find the matrix of the standard inner product in R^3 relative to the basis $((1, 0, 1), (0, -2, 1), (-2, 1, 3))$.

1.7. Let $V = P_3(\mathsf{R})$. An inner product is defined by

$$(p \mid q) = \int_{-1}^{1} p(x)\, q(x)\, dx.$$

(a) Find the matrix of this inner product relative to the basis $(1, x, x^2, x^3)$ of V.
(b) Find a basis (f_0, f_1, f_2, f_3) of V such that $(f_i \mid f_j) = \delta_{ij}$ for $0 \le i, j \le 3$.

1.8. Let V be a euclidean space with inner product $(\,|\,)$. Show that if S is a nonempty subset of V then
(a) $S^\perp = \{\alpha \in V \mid (\alpha \mid \beta) = 0 \text{ for all } \beta \in S\}$ is a subspace of V,
(b) $(S^\perp)^\perp$ is a subspace of V that includes $[S]$.

1.9. Let V be the set of all real solutions of the differential equation

$$\frac{d^3y}{dx^3} - 6\frac{d^2y}{dx^2} + 11\frac{dy}{dx} - 6y = 0.$$

Then V together with the familiar operation of addition of functions and multiplication of a function by a scalar is a vector space. Show that
(a) $\dim V = 3$,
(b) the function $(\,|\,)$ on $V \times V$ into R such that

$$(u \mid v) = \int_{-\infty}^{0} u(x)\, v(x)\, dx$$

is an inner product in V.

2. ORTHONORMAL BASES

In Example 1.4 we observed that if V is an n-dimensional vector space over R with $\mathscr{B} = (\alpha_1, \ldots, \alpha_n)$ as a basis then the function $(\,|\,)$ defined on $V \times V$ into R such that

$$(\alpha \mid \beta) = \sum_{i=1}^{n} x_i\, y_i,$$

where

$$(x_1, \ldots, x_n) = [\alpha]_{\mathscr{B}}, \qquad (y_1, \ldots, y_n) = [\beta]_{\mathscr{B}},$$

is an inner product in V. In particular, $(\alpha_i \mid \alpha_j) = \delta_{ij}$, and so the matrix of this inner product relative to the $\mathscr{B}$-basis is the $n \times n$ matrix (δ_{ij}). In brief, we *defined* an inner product in V whose matrix relative to a *given* basis is (δ_{ij}). We now raise the question: if we are *given* an inner product in V, can we *determine* a basis $\mathscr{B}' = (\alpha_1', \ldots, \alpha_n')$ of V in relation to which the matrix of the inner product is (δ_{ij})? The question deserves consideration because of the ease with which inner products (that is, values of an inner-product function) can be computed in relation to such a basis: if $(\alpha_i' \mid \alpha_j') = \delta_{ij}$ and $\alpha = \sum_1^n x_i \alpha_i'$, $\beta = \sum_1^n y_i \alpha_1'$, then $(\alpha \mid \beta) = \sum_1^n x_i y_i$. A set S of vectors of a euclidean space which has distinctive features like those we have attributed to $\mathscr{B}'$ deserves a name (if such sets exist!). So let us introduce some definitions.

☐ **Definition.** In a euclidean space V a vector α is called *orthogonal to a vector* β if and only if $(\alpha \mid \beta) = 0$. The vector α is called *orthogonal to a nonempty subset* S of V if and only if it is orthogonal to each element of S.

Orthogonality is a symmetric relation; that is, $(\alpha \mid \beta) = 0$ if and only if $(\beta \mid \alpha) = 0$. As such, it is permissible to speak simply of the orthogonality of two vectors. The zero vector is orthogonal to every vector of a euclidean space. This follows from the identity

$$(\alpha \mid 0) = (\alpha \mid 0 + 0) = (\alpha \mid 0) + (\alpha \mid 0).$$

We note further that if a vector α is orthogonal to each member of a subset S of a euclidean space then α is orthogonal to $[S]$.

☐ **Definition.** A subset S of a euclidean space V is called an *orthogonal set* if and only if each of its members is orthogonal to every other member. An orthogonal set such that for each of its members α, $(\alpha \mid \alpha) = 1$, is called an *orthonormal set*. An orthonormal set of V which is a basis of V is called an *orthonormal basis* of V.

An orthogonal set of nonzero vectors can be converted into an orthonormal set by multiplying each element α by the scalar $1/\sqrt{(\alpha \mid \alpha)}$. For proof we note that

$$\left(\frac{1}{\sqrt{(\alpha \mid \alpha)}} \alpha \,\middle|\, \frac{1}{\sqrt{(\alpha \mid \alpha)}} \alpha \right) = \frac{1}{(\alpha \mid \alpha)} (\alpha \mid \alpha) = 1,$$

and if $(\alpha \mid \beta) = 0$ then $(a\alpha \mid b\beta) = 0$ for all scalars a and b.

EXAMPLES

2.1. In the euclidean space R^3 with standard inner product $\{(1, 0, 1), (-1, 0, 1), (0, 1, 0)\}$ is an orthogonal set of vectors and

$$S = \left\{\frac{1}{\sqrt{2}}(1, 0, 1), \frac{1}{\sqrt{2}}(-1, 0, 1), (0, 1, 0)\right\}$$

is an orthonormal set. As a linearly independent set of three vectors, S is an orthonormal basis of R^3.

2.2. In the space $C[-\pi, \pi]$ with standard inner product

$$S = \{1, \sin x, \cos x, \ldots, \sin nx, \cos nx, \ldots\}$$

is an orthogonal set of vectors since

$$\int_{-\pi}^{\pi} \sin mx \cdot \sin nx \, dx = \int_{-\pi}^{\pi} \cos mx \cdot \cos nx \, dx = 0, \; m \neq n,$$

and whether or not m and n are distinct

$$\int_{-\pi}^{\pi} \cos mx \sin nx \, dx = 0.$$

If each element f of S is multiplied by the scalar $1/\sqrt{(f, f)}$, an orthonormal set results.

☐ **Theorem 2.1.** An orthogonal set of nonzero vectors of a euclidean space is linearly independent.

PROOF. Let S be an orthogonal set of nonzero vectors. Let $\{\alpha_1, \ldots, \alpha_p\}$ be a finite subset of S such that $\sum_1^p a_i \alpha_i = 0$. Then, for $1 \leq j \leq p$,

$$0 = (0 \mid \alpha_j) = \left(\sum_{i=1}^{p} a_i \alpha_i \mid \alpha_j\right) = a_j(\alpha_j \mid \alpha_j).$$

Since $(\alpha_j \mid \alpha_j) \neq 0$, it follows that $a_j = 0$. Hence $\{\alpha_1, \ldots, \alpha_p\}$, and in turn S, is linearly independent. ◇

By definition, an orthonormal set S in a euclidean space V has two properties: if $\alpha \in S$, then $(\alpha \mid \alpha) = 1$, and if α, β are distinct elements of S, then $(\alpha \mid \beta) = 0$. Therefore the question raised in the first paragraph of this section can be phrased: can we determine an orthonormal basis of a finite-dimensional euclidean space? The answer is in the affirmative and appears as a corollary to the following theorem.

□ **Theorem 2.2.** If $\{\alpha_1, \ldots, \alpha_n\}$, $n \geq 1$, is a linearly independent subset of a euclidean space V, then vectors $\gamma_1, \ldots, \gamma_n$ can be constructed such that

(i) $\{\gamma_1, \ldots, \gamma_n\}$ is an orthonormal set and

(ii) γ_i is a linear combination of $\alpha_1, \ldots, \alpha_i$ for $1 \leq i \leq n$ (hence the γ's generate the same space as the α's).

REMARKS. Before we present a proof we call the reader's attention to several facts. First, notice that the statement of the theorem is a modification of the converse of Theorem 3.1. Second, the proof of the theorem will provide an algorithm for constructing the γ's, one by one, as linear combinations of the α's. The algorithm is called the *Gram-Schmidt orthogonalization process*. Finally, in the proof we shall supply more details than are necessary for a proof by induction. This is done to assist the reader in computational problems.

PROOF. Consider the given linearly independent set as the ordered set $(\alpha_1, \ldots, \alpha_n)$. We shall first construct an orthogonal set $(\beta_1, \ldots, \beta_n)$ and then "normalize" each β to obtain the orthonormal set that is required. We begin by defining β_1 as α_1. If $n > 1$, we continue by determining a β_2 orthogonal to β_1 and of the form

$$\beta_2 = a\beta_1 + b\alpha_2.$$

Since b may not equal 0, we may assume that

$$\beta_2 = \alpha_2 + a\beta_1$$

for some scalar a. Since

$$(\beta_1 \mid \beta_2) = (\beta_1 \mid \alpha_2) + a(\beta_1 \mid \beta_1),$$

we obtain the desired orthogonality of β_1 and β_2 by setting

$$a = -\frac{(\beta_1 \mid \alpha_2)}{(\beta_1 \mid \beta_1)}.$$

Thus

$$\beta_2 = \alpha_2 - \frac{(\beta_1 \mid \alpha_2)}{(\beta_1 \mid \beta_1)}\beta_1 = \alpha_2 - \frac{(\beta_1 \mid \alpha_2)}{(\beta_1 \mid \beta_1)}\alpha_1,$$

and, since $\{\alpha_1, \alpha_2\}$ is a linearly independent set, $\beta_2 \neq 0$. Next, we determine a β_3 that is orthogonal to $\{\beta_1, \beta_2\}$ and is a linear combination of

$\{\alpha_3, \beta_1, \beta_2\}$, hence a linear combination of $\{\alpha_1, \alpha_2, \alpha_3\}$. Suppose that

$$\beta_3 = \alpha_3 + a\beta_1 + b\beta_2 .$$

Setting each of the inner products $(\beta_1 \mid \beta_3)$ and $(\beta_2 \mid \beta_3)$ equal to 0 leads to the following values for a and b:

$$a = -\frac{(\beta_1 \mid \alpha_3)}{(\beta_1 \mid \beta_1)}, \qquad b = -\frac{(\beta_2 \mid \alpha_3)}{(\beta_2 \mid \beta_2)}.$$

Thus

$$\beta_3 = \alpha_3 - \sum_{i=1}^{2} \frac{(\beta_i \mid \alpha_3)}{(\beta_i \mid \beta_i)} \beta_i$$

meets the requirements and is different from 0.

Assume that $\beta_1, \ldots, \beta_k$ have been defined in such a way that $\{\beta_1, \ldots, \beta_k\}$ is an orthogonal set of nonzero vectors and each β_i is a linear combination of $\{\alpha_1, \ldots, \alpha_i\}$. Let

$$\beta_{k+1} = \alpha_{k+1} - \sum_{i=1}^{k} \frac{(\beta_i \mid \alpha_{k+1})}{(\beta_i \mid \beta_i)} \beta_i .$$

Then $\beta_{k+1} \neq 0$, since otherwise α_{k+1} would be expressible as a linear combination of $\{\alpha_1, \ldots, \alpha_k\}$, contrary to the linear independence of $\{\alpha_1, \ldots, \alpha_{k+1}\}$. Further, by our induction assumption β_{k+1} is a linear combination of $\{\alpha_1, \ldots, \alpha_{k+1}\}$ and

$$(\beta_j \mid \beta_{k+1}) = \left(\beta_j \,\middle|\, \alpha_{k+1} - \sum_{i=1}^{k} \frac{(\beta_i \mid \alpha_{k+1})}{(\beta_i \mid \beta_i)} \beta_i \right)$$

$$= (\beta_j \mid \alpha_{k+1}) - \frac{(\beta_j \mid \alpha_{k+1})}{(\beta_j \mid \beta_j)} (\beta_j \mid \beta_j) = 0, \qquad 1 \le j \le k.$$

Hence $\{\beta_1, \ldots, \beta_{k+1}\}$ is an orthogonal set such that each β_i is a linear combination of $\{\alpha_1, \ldots, \alpha_i\}$, $1 \le i \le k+1$. It follows by the principle of mathematical induction that for each n we can construct an orthogonal set $\{\beta_1, \ldots, \beta_n\}$ such that each β_i is a linear combination of $\{\alpha_1, \ldots, \alpha_i\}$.

Finally, set

$$\gamma_i = \frac{1}{\sqrt{(\beta_i \mid \beta_i)}} \beta_i , \qquad 1 \le i \le n.$$

Then $\{\gamma_1, \ldots, \gamma_n\}$ meets the requirements stated in the theorem. ◇

□ **Corollary.** A finite-dimensional euclidean space has an orthonormal basis.

PROOF. Suppose that $\mathscr{B} = (\alpha_1, \ldots, \alpha_n)$ is a basis of V. Let $\mathscr{C} = (\gamma_1, \ldots, \gamma_n)$ be an orthonormal set of vectors constructed from the α's as described in the theorem. Then $\mathscr{C}$ is a linearly independent set of n vectors and, hence, an orthonormal basis of V. ◇

EXAMPLES

2.3. Consider R^2 with the inner product defined in Example 1.2: if $\alpha = (x_1, y_1)$ and $\beta = (x_2, y_2)$, then

$$(\alpha \mid \beta) = x_1x_2 + 2x_1y_2 + 2y_1x_2 + 5y_1y_2 .$$

Let us derive an orthonormal basis from the standard basis (ϵ_1, ϵ_2). Following the proof of Theorem 2.2, we begin by setting $\beta_1 = \epsilon_1 = (1, 0)$. Then we define β_2 to be

$$\epsilon_2 - \frac{(\beta_1 \mid \epsilon_2)}{(\beta_1 \mid \beta_1)} \epsilon_1 = (0, 1) - \frac{2}{1}(1, 0) = (-2, 1).$$

It is a coincidence that both β_1 and β_2 are normalized, that is, $(\beta_1 \mid \beta_2) = (\beta_2 \mid \beta_2) = 1$. Since this is the case, (β_1, β_2) is an orthonormal basis.

2.4. In $P_3(\mathsf{R})$ with inner product

$$(f \mid g) = \int_0^1 f(x)\, g(x)\, dx$$

let us derive an orthonormal basis from the basis $(1, x, x^2, x^3)$. The β's are computed as follows:

$$\beta_1 = 1,$$

$$\beta_2 = x - \frac{(\beta_1 \mid x)}{(\beta_1 \mid \beta_1)} \beta_1 = x - \tfrac{1}{2},$$

$$\beta_3 = x^2 - \frac{(\beta_2 \mid x^2)}{(\beta_2 \mid \beta_2)} \beta_2 - \frac{(\beta_1 \mid x^2)}{(\beta_1 \mid \beta_1)} \beta_1 = x^2 - x + \tfrac{1}{6},$$

$$\beta_4 = x^3 - \frac{(\beta_3 \mid x^3)}{(\beta_3 \mid \beta_3)} \beta_3 - \frac{(\beta_2 \mid x^3)}{(\beta_2 \mid \beta_2)} \beta_2 - \frac{(\beta_1 \mid x^3)}{(\beta_1 \mid \beta_1)} \beta_1$$

$$= x^3 - (\tfrac{3}{2})\, x^2 + (\tfrac{3}{5})\, x - \tfrac{1}{20}.$$

This orthogonal set can be normalized to obtain a basis of the required kind.

The proof of the next theorem is left as an exercise.

□ **Theorem 2.3.** If $(\alpha_1, \ldots, \alpha_n)$ is an orthonormal basis of a euclidean space V, then for all α, β in V

$$\alpha = \sum_{i=1}^{n} (\alpha \mid \alpha_i)\alpha_i \quad \text{and} \quad (\alpha \mid \beta) = \sum_{i=1}^{n} (\alpha \mid \alpha_i)(\beta \mid \alpha_i).$$

We turn now to a somewhat more profound matter. Suppose that we partially order by inclusion the collection of all orthonormal sets of vectors in a euclidean space V. A maximal element of this set is called a *maximal orthonormal set*. If V is finite-dimensional and $V \neq \{0\}$, it is clear that maximal elements exist—indeed, in view of the Gram-Schmidt process, that the maximal elements are precisely the orthonormal bases of V. For an infinite-dimensional space the situation may be different, as we shall now show. Our illustration employs a theorem from analysis that may not be familiar to the reader. It is usually proved in connection with a study of the approximation of continuous functions by polynomial functions and reads as follows: if f is a continuous function on the closed interval $[a, b]$ and $\int_a^b f(x)x^k\,dx = 0$ $(k = 0, 1, 2, \ldots)$, then f is the zero function on $[a, b]$.†

Now for the example. In the euclidean space $V = C[a, b]$ of all continuous functions on $[a, b]$ with the inner product

$$(f \mid g) = \int_a^b f(x)\,g(x)\,dx,$$

$\mathscr{P} = \{1, x, x^2, \ldots\}$ is a linearly independent set but certainly is not a basis. Indeed, $\mathscr{P}$ is a basis for the proper subspace $P_\infty(\mathsf{R})$ of V. Using the Gram-Schmidt process, we can derive from $\mathscr{P}$ an orthonormal set $\mathscr{P}' = \{p_0, p_1, p_2, \ldots\}$, where p_n is a polynomial of degree n, which is a basis of the same space, $P_\infty(\mathsf{R})$. We now prove that $\mathscr{P}'$ is a maximal orthonormal set in V. Assume to the contrary that $\mathscr{P}'$ is a proper subset of some orthonormal set. Then there exists a function f, different from the zero function

† A proof is given by R. P. Boas, Jr., in *A Primer of Real Functions*, The Mathematical Association of America (1960).

on $[a, b]$ which is orthogonal to each element of $\mathscr{P}'$ and, hence, is orthogonal to $P_\infty(\mathsf{R})$, the space spanned by $\mathscr{P}'$. Therefore, in particular, $(f \mid x^k) = 0$ $(k = 0, 1, 2, \ldots)$. Using the theorem stated above, we find that f is the zero function on $[a, b]$. In summary we may say that in infinite-dimensional euclidean spaces there can exist maximal orthonormal sets of vectors that are not bases.

EXERCISES

2.1. Show that the set $\{(1, 2, 2), (2, 1, -2), (2, -2, 1)\}$ of vectors in R^3 with the standard inner product is orthogonal. Normalize it.

2.2. Apply the Gram-Schmidt process to $\{(1, 0, 1), (1, 0, -1), (1, 3, 4)\}$ to obtain an orthonormal basis for R^3 with the standard inner product.

2.3. Consider R^4 with the standard inner product. Let W be the subspace consisting of all vectors that are orthogonal to both $(1, 0, -1, 1)$ and $(2, 3, -1, 2)$. Find an orthonormal basis of W and extend it to an orthonormal basis of R^4.

2.4. Let $(\alpha_1, \ldots, \alpha_n)$ be a linearly independent set of vectors in a euclidean space V. Suppose that several of the initial members of this set, for instance, α_1, α_2, and α_3, are orthogonal to one another. Show that the Gram-Schmidt process will produce them as the initial members of an orthogonal set that spans $[\alpha_1, \ldots, \alpha_n]$.

2.5. Suppose that in applying the Gram-Schmidt process to a set of vectors the zero vector is obtained. What can be said about the given set of vectors?

2.6. Prove Theorem 2.3.

2.7. Let V be a euclidean space and $\{\alpha_1, \ldots, \alpha_n\}$, a subset of V such that $(\alpha_i \mid \alpha_i) = 1$ and $\sum_1^n |(\beta \mid \alpha_i)|^2 = (\beta \mid \beta)$ for $1 \leq i \leq n$ and all β in V. Prove that the α's form a basis of V.

2.8. Let V be a euclidean space, W, a subspace, and α_0, an element of V such that

$$(\alpha_0 \mid \beta) + (\beta \mid \alpha_0) \leq (\beta \mid \beta)$$

for all β in W. Prove that $(\alpha_0 \mid \beta) = 0$ for all β in W.

2.9. Given a vector of $\alpha_0 \neq 0$ in a finite-dimensional euclidean space, determine the set of all vectors β such that $\beta - \alpha_0$ is orthogonal to $\beta + \alpha_0$.

2.10 Let V be a euclidean space, let W be the subspace having the orthonormal basis $\{\alpha_1, \ldots, \alpha_n\}$, and let β be an element of V. Show that the vector

$$\beta - \sum_{i=1}^{n} (\beta \mid \alpha_i)\alpha_i$$

is orthogonal to every vector in W.

2.11. Let S be a subspace of the finite-dimensional euclidean space V. Show that every vector α in V can be written in the form $\alpha = \beta + \gamma$, where $\beta \in S$ and γ is orthogonal to S, and that

$$(\alpha \mid \alpha) = (\beta \mid \beta) + (\gamma \mid \gamma).$$

2.12. Let $\{\alpha_1, \ldots, \alpha_n\}$ be a basis of a euclidean space V with inner-product $(\mid)$. If $a_1, \ldots, a_n$ are n given scalars, show that there exists exactly one vector β in V such that $(\beta \mid \alpha_i) = a_i$, $1 \leq i \leq n$.

2.13. Let V be a finite-dimensional vector space over R. If $(\mid)_1$ and $(\mid)_2$ are two inner products in V, show that there exists a scalar c such that $(\alpha \mid \alpha)_1 \leq c(\alpha \mid \alpha)_2$ for all $\alpha \in V$.

2.14. In the euclidean space $C[0, 1]$ with standard inner product the function h such that $h(x) = x^2 + ax + b$ is orthogonal to 1 and x. What are the values of a and b?

2.15. Do maximal orthonormal sets always exist in an inner-product space? Prove or disprove: Any two maximal orthonormal sets in an inner-product space can be put into 1-1 correspondence.

3. DISTANCES AND NORMS

Let V be a finite-dimensional euclidean space with inner product $(\mid)$ and $\mathscr{B} = (\alpha_1, \ldots, \alpha_n)$ be an ordered orthonormal basis of V. If $\alpha, \beta \in V$, then $\alpha = \sum_1^n x_i \alpha_i$ and $\beta = \sum_1^n y_i \alpha_i$, where $x_i = (\alpha \mid \alpha_i)$ and $y_i = (\beta \mid \alpha_i)$, $1 \leq i \leq n$ (see Theorem 2.3). It follows that $\alpha - \beta = \sum_1^n (x_i - y_i)\alpha_i$ and so (see Theorem 2.3)

$$(\alpha - \beta \mid \alpha - \beta) = \sum_{i=1}^{n} (x_i - y_i)^2.$$

If V is R^3, $(\mid)$ is the standard inner product in R^3 and $\mathscr{B}$ is the standard (orthonormal) basis $(\epsilon_1, \epsilon_2, \epsilon_3)$, the square root of the right-hand side of the above equation is $\sqrt{\sum_1^3 (x_i - y_i)^2}$ for $\alpha = (x_1, x_2, x_3)$ and $\beta = (y_1, y_2, y_3)$. Interpreting α and β as points of E_3, this expression is the familiar one for the distance between points with coordinates x_1, x_2, x_3 and y_1, y_2, y_3 relative to some rectangular coordinate system. This observation suggests the following definition for arbitrary euclidean spaces.

☐ **Definition.** The *distance* $d(\alpha, \beta)$ between vectors α and β of a euclidean space with inner product $(\mid)$ is

$$d(\alpha, \beta) = \sqrt{(\alpha - \beta \mid \alpha - \beta)}.$$

Further, by the *norm* (or *magnitude*) $\|\alpha\|$ of a vector is meant $d(\alpha, 0)$; that is

$$\|\alpha\| = \sqrt{(\alpha \mid \alpha)}.$$

A vector with norm 1 is called a *unit vector*.

EXAMPLES

3.1. In R^n with standard inner product, if $\alpha = (x_1, \ldots, x_n)$ and $\beta = (y_1, \ldots, y_n)$, then

$$d(\alpha, \beta) = \sqrt{\sum_1^n (x_i - y_i)^2}$$

and

$$\|\alpha\| = \sqrt{\sum_1^n x_i^2}.$$

3.2. In $C[a, b]$ with standard inner product

$$d(f, g) = \left[\int_a^b [f(x) - g(x)]^2 \, dx\right]^{1/2}$$

and

$$\|f\| = \left[\int_a^b f^2(x) \, dx\right]^{1/2}$$

In order to establish basic properties of the two functions just defined, we need a preliminary result.

☐ **Theorem 3.1** (Schwarz's inequality). For all vectors α and β in a euclidean space with inner product $(\mid)$

$$(\alpha \mid \beta)^2 \leq (\alpha \mid \alpha)(\beta \mid \beta)$$

or, what amounts to the same,

$$|(\alpha \mid \beta)| \leq \|\alpha\| \, \|\beta\|.$$

The equality sign holds if and only if $\{\alpha, \beta\}$ is linearly dependent.

PROOF. Observe that if f is a real polynomial such that

$$f(x) = ax^2 - 2bx + c, \qquad a > 0$$

then, since

$$f(x) = a\left(x - \frac{b}{a}\right)^2 + c - \frac{b^2}{a},$$

f has a minimum value of $c - b^2/a$. Hence, if it is the case that $f(x) \geq 0$ for all x, then $c - b^2/a \geq 0$ or $b^2 \leq ac$.

Now let α and β be vectors of a euclidean space. If $\beta = 0$, the assertion is true trivially. So assume that $\beta \neq 0$. Let x be a scalar and consider

$$(\alpha - x\beta \mid \alpha - x\beta) = (\beta \mid \beta)x^2 - 2(\alpha \mid \beta)x + (\alpha \mid \alpha).$$

This number is nonnegative for each x and $(\beta \mid \beta) > 0$. Hence we may infer from the above preliminary result that

$$(\alpha \mid \beta)^2 \leq (\alpha \mid \alpha)(\beta \mid \beta).$$

The proof of the remaining statement in the theorem is left as an exercise. ◇

We are now able to prove that our definitions of norm and distance are satisfactory in the sense that they enjoy the basic properties associated with these notions in familiar contexts.

☐ **Theorem 3.2.** In a euclidean space the norm function has the following properties:

(a) $\|a\alpha\| = |a| \, \|\alpha\|$, where $|a|$ is the absolute value of the scalar a.
(b) $\|\alpha\| \geq 0$; $\|\alpha\| = 0$ if and only if $\alpha = 0$.
(c) $\|\alpha + \beta\| \leq \|\alpha\| + \|\beta\|$.

Further, distance has the following properties:

(d) $d(\alpha, \beta) = d(\beta, \alpha)$
(e) $d(\alpha, \beta) \geq 0$; $d(\alpha, \beta) = 0$ if and only if $\alpha = \beta$.
(f) $d(\alpha, \beta) \leq d(\alpha, \gamma) + d(\gamma, \beta)$.
(g) $d(\alpha + \gamma, \beta + \gamma) = d(\alpha, \beta)$.

PROOF. Properties (a), (b), (d), (e), and (g) follow directly from the definitions in view of properties assigned to an inner product. The proof of (c), which is below, uses Schwarz's inequality.

$$\begin{aligned}\|\alpha+\beta\|^2 &= (\alpha+\beta \mid \alpha+\beta)\\ &= (\alpha \mid \alpha) + 2(\alpha \mid \beta) + (\beta \mid \beta)\\ &\le (\alpha \mid \alpha) + 2[(\alpha \mid \alpha)(\beta \mid \beta)]^{1/2} + (\beta \mid \beta)\\ &= \|\alpha\|^2 + 2\|\alpha\|\,\|\beta\| + \|\beta\|^2\\ &= (\|\alpha\| + \|\beta\|)^2.\end{aligned}$$

The proof of (f) now follows easily:

$$\|\alpha-\beta\| = \|\alpha-\gamma+\gamma-\beta\| \le \|\alpha-\gamma\| + \|\gamma-\beta\|;$$

that is,

$$d(\alpha, \beta) \le d(\alpha, \gamma) + d(\gamma, \beta). \ \diamond$$

Schwarz's inequality has another application of a geometric nature: for nonzero vectors α and β of a euclidean space we deduce from $|(\alpha \mid \beta)| \le \|\alpha\|\ \|\beta\|$ that

$$-1 \le \frac{(\alpha \mid \beta)}{\|\alpha\|\ \|\beta\|} \le 1.$$

Recalling the behavior of the cosine function of $[0, \pi]$, we may conclude that there exists one number θ in $[0, \pi]$ such that

$$\cos\theta = \frac{(\alpha \mid \beta)}{\|\alpha\|\ \|\beta\|}.$$

This number θ we call the (measure of the) *angle* between the nonzero vectors α and β. To make the definition complete we define θ as 0 if either α or β is 0. The reader should compare the foregoing with the introductory paragraph of this section. Further, he should note that the definition of orthogonality, which is suggested by the above (namely, α is orthogonal to β if and only if $\theta = \pi/2$, hence $\cos\theta = 0$), is in agreement with the definition that we gave in Section 2.

EXERCISES

3.1. Prove that $|a\cos nx + b\sin nx|^2 \le a^2 + b^2$ for each positive integer n, using results in this section.

3.2. Establish the following instances of Schwarz's inequality.

(a) For real numbers $a_1, \ldots, a_n$ and $b_1, \ldots, b_n$

$$\left(\sum_{i=1}^{n} a_i b_i\right)^2 \le \left(\sum_{i=1}^{n} a_i^2\right)\left(\sum_{i=1}^{n} b_i^2\right).$$

(b) For functions f and g, which are continuous on $[a, b]$,

$$\left(\int_a^b f(t)\,g(t)\,dt\right)^2 \leq \left(\int_a^b f^2(t)\,dt\right)\left(\int_a^b g^2(t)\,dt\right).$$

3.3. Prove that equality holds in Theorem 3.1, if and only if $\|\alpha\|\,\beta = \pm\|\beta\|\,\alpha$.

3.4. Prove that two vectors α and β in a euclidean space are orthogonal if and only if

$$\|\alpha + \beta\|^2 = \|\alpha\|^2 + \|\beta\|^2.$$

3.5. Prove that if α and β are vectors in a euclidean space and $\|\alpha\| = \|\beta\|$ then $\alpha - \beta$ and $\alpha + \beta$ are orthogonal. Give an interpretation of this result in E_2.

3.6. Prove that if α and β are vectors in a euclidean space then

$$\|\alpha + \beta\|^2 + \|\alpha - \beta\|^2 = 2(\|\alpha\|^2 + \|\beta\|^2).$$

Give an interpretation of this result in E_2.

3.7. Show that if α and β are orthogonal vectors in a euclidean space, then $d(\alpha, \beta) = (\|\alpha\|^2 + \|\beta\|^2)^{1/2}$.

3.8. Suppose that $S = (\alpha_1, \ldots, \alpha_n, \ldots)$ is a sequence of vectors in a euclidean space V and that S is an orthonormal set.

(a) Show that if $\alpha \in V$ and n is any positive integer,

$$\sum_{i=1}^{n} (\alpha \mid \alpha_i)^2 \leq \|\alpha\|^2 \qquad \textit{(Bessel's inequality)}.$$

(*Hint.* Consider the inner product of the vector $\alpha - \sum_1^n (\alpha \mid \alpha_i)\alpha_i$ with itself.)

(b) Investigate the convergence of the infinite series $\sum_1^\infty (\alpha \mid \alpha_i)^2$ for $\alpha \in V$.

3.9. Let α, β, and γ be three nonzero vectors in $\mathbf{R}^2$ with the standard inner product. Prove that the equation

$$\|\alpha - \beta\|\,\|\gamma\| = \|\beta - \gamma\|\,\|\alpha\| + \|\gamma - \alpha\|\,\|\beta\|$$

holds if and only if α, β, γ, and 0 lie on a circle such that the pairs α, β and γ, 0 separate each other.

4. ORTHOGONAL COMPLEMENTS AND ORTHOGONAL PROJECTIONS

Some interesting properties of euclidean spaces are concerned with the notion that was introduced in Exercise 1.8. We start from the beginning by repeating the definition given there.

Let V be a euclidean space with inner-product $(\ \mid\ )$ and S a nonempty subset of V. Define

$$S^\perp = \{\alpha \in V \mid (\alpha \mid \delta) = 0, \text{ all } \delta \text{ in } S\}.$$

If S is any nonempty subset of V, then $S^{\perp}$ (read: S perp) is a subspace of V, since, if $\alpha, \beta \in S^{\perp}$, then $(a\alpha + b\beta \mid \delta) = a(\alpha \mid \delta) + b(\beta \mid \delta) = 0 + 0 = 0$ for scalars a and b and $\delta \in S$. Also, $[S] \cap S^{\perp} = \{0\}$, since $\delta \in [S] \cap S^{\perp}$ implies that $(\delta \mid \delta) = 0$, hence $\delta = 0$. Moreover, it is clear that $(S^{\perp})^{\perp} \supseteq S$, hence that $(S^{\perp})^{\perp} \supseteq [S]$. Since $S^{\perp} = [S]^{\perp}$, we shall restrict our attention to cases in which S is a subspace.

For a suggestion regarding the relationship between a subspace S and the accompanying subspace $S^{\perp}$ we consider an example. Let S be the one-dimensional subspace of R^3 with standard inner product generated by $(1, -1, 2)$. Then

$$S^{\perp} = \{(x, y, z) \in \mathsf{R}^3 \mid x - y + 2z = 0\}$$

is a two-dimensional subspace. Interpreting elements of R^3 as points of E_3, the image of S is a line L through the origin and the point $(1, -1, 2)$ and the image of $S^{\perp}$ is a plane p through the origin and perpendicular to this line. Reflecting for a moment on this picture suggests the plausibility of the following statements: $S = (S^{\perp})^{\perp}$ and, $S^{\perp}$ is a complement of S. These statements are true not merely for our example but for any finite-dimensional subspace of a euclidean space. They do not, however, extend to infinite-dimensional subspaces as the reader is asked to show in Exercise 4.3. Before we continue with this matter still another definition is in order.

☐ **Definition.** Let S be a subspace of a euclidean space V. The subspace $S^{\perp}$ of V is called the *orthogonal complement of* S if and only if $S^{\perp}$ is a complement of S.

Since for a subspace S it is always true that $S \cap S^{\perp} = \{0\}$, the subspace $S^{\perp}$ is the orthogonal complement of S if and only if $V = S + S^{\perp}$.

☐ **Theorem 4.1.** If S is a finite-dimensional subspace of a euclidean space V, then $S^{\perp}$ is the orthogonal complement of S.

PROOF. Let S be a finite-dimensional subspace of V. The fact that $S \cap S^{\perp} = \{0\}$ has already been mentioned. It remains to show that $S + S^{\perp} = V$. Let $\{\alpha_1, \ldots, \alpha_n\}$ be an orthonormal basis of S (the existence of which is assured by the Gram-Schmidt process) and α, an element of V. Then the vector

$$\alpha_S = (\alpha \mid \alpha_1)\alpha_1 + \cdots + (\alpha \mid \alpha_n)\alpha_n$$

is in S. Defining β by

$$\beta = \alpha - \alpha_S,$$

we observe that

$$\begin{aligned}(\beta \mid \alpha_i) &= (\alpha - \alpha_S \mid \alpha_i) = (\alpha \mid \alpha_i) - (\alpha_S \mid \alpha_i) \\ &= (\alpha \mid \alpha_i) - \left(\sum_1^n (\alpha \mid \alpha_k)\alpha_k \mid \alpha_i\right) \\ &= (\alpha \mid \alpha_i) - (\alpha \mid \alpha_i) = 0, \qquad 1 \le i \le n.\end{aligned}$$

Therefore, since β is orthogonal to every α_i, it is orthogonal to S. Hence $\beta \in S^\perp$. This means that $S + S^\perp = V$ and the proof is complete. ◇

EXAMPLES

4.1. In R^3 the orthogonal complement of $S = \{(x, y, z) \in \mathsf{R}^3 \mid ax + by + cz = 0\}$ is $S^\perp = [(a, b, c)]$.

4.2. In the space $V = C[-a, a]$ with standard inner product let

$$S = \{f \in V \mid f(-x) = -f(x)\}.$$

Thus S is the set (a subspace of V) of all *odd functions* in V. By definition,

$$S^\perp = \left\{g \in V \,\middle|\, \int_{-a}^{a} f(x)\, g(x)\, dx = 0, \text{ all } f \in S\right\}.$$

If g is any element of V, then

$$g(x) = \frac{g(x) + g(-x)}{2} + \frac{g(x) - g(-x)}{2}.$$

Let $E(x) = (g(x) + g(-x))/2$ and $O(x) = g(x) - g(-x))/2$. Then E is an even function, that is, $E(-x) = E(x)$, whereas O is odd.

Now assume that $g \in S^\perp$. Defining E and O as above, we conclude that

$$0 = \int_{-a}^{a} O(x)\, g(x)\, dx$$

by virtue of the definition of $S^\perp$. Replacing g with its equal $E + O$

in this integral, we conclude that

$$0 = \int_{-a}^{a} O(x)\,[E(x) + O(x)]\,dx$$
$$= \int_{-a}^{a} O(x)\,E(x)\,dx + \int_{-a}^{a} O^2(x)\,dx$$

Since OE is an odd function, the first integral on the right-hand side has value 0. Thus

$$\int_{-a}^{a} O^2(x)\,dx = 0.$$

Hence, in turn, $O^2(x)$, $O(x)$, and $g(x) - g(-x)$ are each equal to 0 for all x in $[-a, a]$; that is, g is an even function. Conversely, it is clear that if g is an even function then $g \in S^\perp$. Hence $S^\perp$ is the set of all even functions in V. It follows that $V = S + S^\perp$ and so $S^\perp$ is the orthogonal complement of S.

☐ **Theorem 4.2.** Let V be a euclidean space, S be a finite-dimensional subspace of V, and $S^\perp$ be the orthogonal complement of S. The representation of a vector α of V in the form

$$\alpha = \alpha_S + \alpha_{S^\perp}$$

where α_S (the so-called *orthogonal projection* of α on S) is in S and $\alpha_{S^\perp}$ is in $S^\perp$ (hence orthogonal to S), has the following properties:

(a) It is unique.
(b) The Pythagorean theorem holds: $\|\alpha\|^2 = \|\alpha_S\|^2 + \|\alpha_{S^\perp}\|^2$.
(c) The vector α_S is the best approximation in S to α in the sense that $d(\alpha, \gamma)$, with $\gamma \in S$, is minimized when $\gamma = \alpha_S$.

PROOF. The uniqueness of the representation of α in the stated form is a consequence of the fact that $S^\perp$ is a complement of S. For (b) we offer the computation

$$\|\alpha\|^2 = (\alpha_S + \alpha_{S^\perp} \mid \alpha_S + \alpha_{S^\perp}) = \|\alpha_S\|^2 + 2(\alpha_S \mid \alpha_{S^\perp}) + \|\alpha_{S^\perp}\|^2$$
$$= \|\alpha_S\|^2 + \|\alpha_{S^\perp}\|^2.$$

To prove (c) it is sufficient to show that $\|\alpha - \gamma\|^2$, with $\gamma \in S$, is minimized when $\gamma = \alpha_S$. Now

$$\alpha - \gamma = (\alpha_S + \alpha_{S^\perp}) - \gamma = (\alpha_S - \gamma) + \alpha_{S^\perp},$$

from which follows the desired result immediately. ◇

EXAMPLE

4.3. In R^3 with standard inner product let $S = \{(x, y, z) \mid x + y + z = 0\}$. Then $\{\alpha_1, \alpha_2\}$, where

$$\alpha_1 = \frac{1}{\sqrt{2}}(1, -1, 0), \qquad \alpha_2 = \frac{1}{\sqrt{6}}(1, 1, -2)$$

is found to be an orthonormal basis of S. Now consider a vector $\alpha = (x_0, y_0, z_0)$ in R^3. According to the proof of Theorem 4.1, $\alpha_S = (\alpha \mid \alpha_1)\alpha_1 + (\alpha \mid \alpha_2)\alpha_2$. Evaluating the expression on the right-hand side of this equation, we find that

$$\alpha_S = \tfrac{1}{3}(2x_0 - y_0 - z_0, -x_0 + 2y_0 - z_0, -x_0 - y_0 + 2z_0).$$

Another computation gives

$$d(\alpha, \alpha_S) = \|\alpha - \alpha_S\| = \frac{1}{\sqrt{3}}|x_0 + y_0 + z_0|.$$

That α_S enjoys the property stated in part (c) of Theorem 4.2 may be checked by the reader with methods of analytic geometry.

The reader will have recognized at this point that the notion of the orthogonal projection of a vector of a euclidean space on a finite-dimensional subspace is a generalization of the familiar notion of projections of arrows in E_3. How closely the abstract version parallels the geometric version deserves mentioning. In the proof of Theorem 4.1 we found that the projection of α on S is the vector $\alpha_S = (\alpha \mid \alpha_1)\alpha_1 + \cdots + (\alpha \mid \alpha_n)\alpha_n$, where $\{\alpha_1, \ldots, \alpha_n\}$ is an orthonormal basis of S. Recalling the formula for the angle between two vectors (Section 3), we have for the angle θ_i between α and α_i

$$\cos \theta_i = \frac{(\alpha \mid \alpha_i)}{\|\alpha\|}.$$

Hence

$$(\alpha \mid \alpha_i)\alpha_i = (\|\alpha\| \cos \theta_i)\alpha_i,$$

which may be interpreted geometrically as the projection of α on α_i. Thus α_S may be interpreted as the sum of the projections of α on the basis $\{\alpha_1, \ldots, \alpha_n\}$ of S. In addition, we note that the square of its magnitude is

$$\|\alpha_S\|^2 = \sum_1^n (\alpha \mid \alpha_i)^2,$$

hence

$$\begin{aligned} d(\alpha, \alpha_S) = \|\alpha - \alpha_S\| = \|\alpha_{S^\perp}\| &= \sqrt{\|\alpha\|^2 - \|\alpha_S\|^2} \\ &= \sqrt{\|\alpha\|^2 - \sum_1^n (\alpha \mid \alpha_i)^2}. \end{aligned}$$

It is of interest to interpret the Gram-Schmidt process in terms of projections. Writing the defining equation for β_{k+1} given in the proof of Theorem 2.2 in the form

$$\beta_{k+1} = \alpha_{k+1} - \sum_{i=1}^{k} \left(\alpha_{k+1} \,\middle|\, \frac{\beta_i}{\|\beta_i\|}\right) \frac{\beta_i}{\|\beta_i\|},$$

we see that β_{k+1} is obtained by subtracting from α_{k+1} its projection on the space having the orthonormal basis $(\beta_1/\|\beta_1\|, \ldots, \beta_k/\|\beta_k\|)$.

It is possible to extend the definition of the orthogonal projection of a vector on a subspace to the case of a linear manifold. Let

$$M = \beta + S$$

be a finite-dimensional linear manifold of a euclidean space V. If

$$\beta = \beta_S + \beta_{S^\perp},$$

where $\beta_S \in S$ and $\beta_{S^\perp} \in S^\perp$, then

$$M = \beta_{S^\perp} + S.$$

If α is any vector in V and α_S is its orthogonal projection on S, the vector

$$\alpha_M = \beta_{S^\perp} + \alpha_S$$

in M is called the *orthogonal projection* of α on M. In the sense of distance α_M is the best approximation in M to α. To prove this statement, consider any vector μ in M. Suppose that $\mu = \beta_{S^\perp} + \delta$, where $\delta \in S$. Then

$$\begin{aligned} \alpha - \mu &= (\alpha_S + \alpha_{S^\perp}) - (\beta_S + \delta) \\ &= (\alpha_S - \delta) + (\alpha_{S^\perp} - \beta_{S^\perp}). \end{aligned}$$

The pythagorean theorem may be applied to yield

$$\|\alpha - \mu\|^2 = \|\alpha_S - \delta\|^2 + \|\alpha_{S^\perp} - \beta_{S^\perp}\|^2.$$

Hence $\|\alpha - \mu\|$ is minimized when $\delta = \alpha_S$, that is, when $\mu = \beta_{S^\perp} + \alpha_S = \alpha_M$. We shall call $\|\alpha - \alpha_M\|$ the *distance* from α to M. If $\{\alpha_1, \ldots, \alpha_n\}$ is an orthonormal basis of S, then

$$\|\alpha - \alpha_M\| = \|\alpha_{S^\perp} - \beta_{S^\perp}\|,$$

where

$$\alpha_{S^\perp} = \alpha - \sum_{i=1}^{n} (\alpha \mid \alpha_i)\alpha_i, \qquad \beta_{S^\perp} = \beta - \sum_{i=1}^{n} (\beta \mid \alpha_i)\alpha_i,$$

EXERCISES

4.1. In R^4 with standard inner product let $S = [(1, 0, 1, 1), (1, 1, 2, 0)]$.
(a) Find an orthonormal basis of $S^\perp$.
(b) Find the distance from $\alpha = (0, 1, -1, 2)$ to the subspace S.

4.2. Let S be a subspace of a euclidean space V. Show that if $S^\perp$ is a complement of S, then $(S^\perp)^\perp = S$. What about the converse of this statement?

4.3. In the space V of all continuous functions on R into R with inner product $(\,|\,)$ defined by

$$(f \mid g) = \int_0^1 f(x)\, g(x)\, dx,$$

let $S = P_\infty(\mathsf{R})$. Show that $S^\perp = \{0\}$, hence that $(S^\perp)^\perp = V$. Observe that $(S^\perp)^\perp \neq S$.

4.4. For subspaces S_1 and S_2 of a finite-dimensional euclidean space, show that

$$(S + T)^\perp = S^\perp \cap T^\perp \text{ and } (S \cap T)^\perp = S^\perp + T^\perp.$$

4.5. Find a polynomial p in $P_2(\mathsf{R})$ such that $\int_0^1 (p(x) - e^x)^2\, dx$ is minimized.

4.6. Given a continuous function f on $[a, b]$ and a positive integer n, describe how one can construct a polynomial $p \in P_n(\mathsf{R})$ such that $\int_a^b (p)(x) - f(x))^2\, dx$ is minimized.

4.7. In R^3 with standard inner product let $P = (a, b, c)$ and $S = \{(x, y, z) \mid Ax + By + Cz = 0\}$. Find the distance from P to S.

4.8. In R^3 with standard inner product let $M = \{(x, y, z) \mid 3x + y - z = 2\}$.
(a) Find a subspace S of R^3 and a vector $\beta \in S^\perp$ such that $M = \beta + S$.
(b) Find the orthogonal projection of $(1, 1, 1)$ on M and the distance of this vector from M.

5. UNITARY SPACES

Let V be a vector space over C, the field of complex numbers. If we attempt to introduce functions on $V \times V$ into C for the purpose of defining distances between vectors and magnitudes of vectors such that the properties stated in Theorem 3.2 hold, the earlier requirements for an inner-product function must be modified. To substantiate this assertion recall the first two properties of norms given in Theorem 3.2: (a) $\|c\alpha\| = |c|\ \|\alpha\|$ and (b) $\|\alpha\| \geq 0$. The second is an immediate consequence of the property

(p_1) $\qquad (\alpha \mid \alpha) \geq 0, (\alpha \mid \alpha) = 0 \quad$ if and only if $\alpha = 0$, for all α,

and the first follows from the property

(p_2) $\qquad (c\alpha \mid \beta) = c(\alpha \mid \beta)$, $\quad$ for all scalars c and all vectors α, β,

in conjunction with the symmetric property of an inner-product function. If we desire an inner-product function $(\ |\ )$ on a vector space V over C to have properties (p_1) and (p_2), then we must sacrifice the symmetric property, for if it holds we can derive a contradiction as follows: in (p_2) choose α to be any nonzero vector, set $c = i$ $(i^2 = -1)$, and set $\beta = i\alpha$. Then

$$(i\alpha \mid i\alpha) = i(\alpha \mid i\alpha) = i(i\alpha \mid \alpha) = i^2(\alpha \mid \alpha) = -(\alpha \mid \alpha),$$

which, together with (p_1), is a contradiction. Fortunately a simple modification of the symmetric property restores law and order. It is incorporated in the next definition.

☐ **Definition.** An *inner product* in a vector space V over C is a function $(\ |\ )$ on $V \times V$ into C such that
(1) $(\alpha \mid \beta) = \overline{(\beta \mid \alpha)}$ (complex conjugation) $\quad$ for all α, β in V.
(2) $(a\alpha + b\beta \mid \gamma) = a(\alpha \mid \gamma) + b(\beta \mid \gamma)$ for all α, β, γ in V and all a, b in C,
(3) $(\alpha \mid \alpha) > 0$† for all $\alpha \neq 0$ in V.
A *unitary space* is a vector space over C with a specified inner product in the space.

To obtain an example of a unitary space consider the space C^n over C. For $\alpha = (x_1, \ldots, x_n)$ and $\beta = (y_1, \ldots, y_n)$ define

$$(\alpha \mid \beta) = \sum_{i=1}^{n} x_i \bar{y}_i .$$

† Observe that by virtue of (1) the inner product of a vector with itself is a real number. Hence the requirement that $(\alpha \mid \alpha)$ be positive is meaningful.

The function so defined is an inner product in $\mathbf{C}^n$, as the reader can verify. We call it the *standard inner product* in $\mathbf{C}^n$.

At this point the reader should begin to work his way through Sections 1 to 4 to discover those results that carry over unchanged from euclidean spaces to unitary spaces and to make the necessary modifications in the other cases. Below is a summary of the conclusions that he should reach.

(4) $$(\alpha \mid a\beta) = \bar{a}(\alpha, \beta).$$

(5) $$\left(\sum_{i=1}^{n} a_i \alpha_i \mid \sum_{i=1}^{m} b_j \beta_j\right) = \sum_{i=1}^{n} \sum_{j=1}^{m} a_i(\alpha_i \mid \beta_j)\bar{b}_j.$$

(6) If V is a finite-dimensional unitary space and $\mathscr{B} = (\alpha_1, \ldots, \alpha_n)$ is an ordered basis of V, then the matrix (c_{ij}) of the inner product relative to $\mathscr{B}$ has the following properties:

$$c_{ii} \geq 0 \quad \text{and} \quad c_{ij} = \bar{c}_{ji},$$

where $c_{ij} = (\alpha_i \mid \alpha_j)$.

(7) With the definitions of orthogonality and orthonormal sets of vectors unchanged, the Gram-Schmidt process can be applied exactly as in the case of a euclidean space to convert a finite linearly independent set of vectors into an orthonormal set. In particular, an orthonormal basis can be constructed for a finite-dimensional unitary space.

(8) If V is a finite-dimensional unitary space, $\mathscr{B} = (\gamma_1, \ldots, \gamma_n)$ is an orthonormal basis of V, and $\alpha, \beta \in V$, then

$$\alpha = \sum_{i=1}^{n} (\alpha \mid \gamma_i)\gamma_i, \qquad \beta = \sum_{i=1}^{n} (\beta \mid \gamma_i)\gamma_i,$$

and

$$(\alpha \mid \beta) = \sum_{i=1}^{n} (\alpha \mid \gamma_i)\overline{(\beta \mid \gamma_i)}.$$

(9) Schwarz's inequality can be stated for vectors of a unitary space in the following form:

$$(\alpha \mid \beta)\overline{(\alpha \mid \beta)} \leq (\alpha \mid \alpha)(\beta \mid \beta).$$

Again the proof is trivial if $\beta = 0$. If $\beta \neq 0$, let

$$\gamma = \frac{\beta}{\sqrt{(\beta \mid \beta)}},$$

so that $(\gamma \mid \gamma) = 1$. Then

$$
\begin{aligned}
0 \le (\alpha - (\alpha \mid \gamma)\gamma \mid \alpha - (\alpha \mid \gamma)\gamma) \\
&= (\alpha \mid \alpha) - (\overline{\alpha \mid \gamma})(\alpha \mid \gamma) - (\alpha \mid \gamma)(\gamma \mid \alpha) + (\alpha \mid \gamma)(\overline{\alpha \mid \gamma}) \\
&= (\alpha \mid \alpha) - (\alpha \mid \gamma)(\alpha \mid \gamma) \\
&= (\alpha \mid \alpha) - \frac{(\alpha \mid \beta)(\overline{\alpha \mid \beta})}{(\beta \mid \beta)}.
\end{aligned}
$$

This gives the desired inequality immediately.

(10) With the definitions of distance between two vectors and the magnitude of a vector unchanged, Theorem 3.2 can be proved. In the identity $\|\alpha c\| = |a| \, \|\alpha\|$ the number $|a|$ is $\sqrt{a\bar{a}}$.

Incidentally, Schwarz's inequality may be stated in terms of norms as

$$|(\alpha \mid \beta)| \le \|\alpha\| \, \|\beta\|.$$

Finally, the results of Section 4 carry over to unitary spaces with minor modifications (see Exercise 5.1).

We conclude the chapter with a definition of the spaces mentioned in its title.

□ **Definition.** An *inner-product space* V over the field F is either a euclidean space, in which case $F = \mathsf{R}$, or a unitary space, in which case $F = \mathsf{C}$.

EXERCISES

5.1. Let S be a finite-dimensional subspace of a unitary space. Show that the distance from a vector α to S is given by

$$\sqrt{\|\alpha\|^2 - \textstyle\sum_1^n (\alpha \mid \alpha_i)(\overline{\alpha \mid \alpha_i})} = \sqrt{\|\alpha\|^2 - \textstyle\sum_1^n |(\alpha \mid \alpha_i)|^2}$$

if $\{\alpha_1, \ldots, \alpha_n\}$ is an orthonormal basis of S.

5.2. Prove Bessel's inequality for unitary spaces (see Exercise 3.7).

5.3. If V is a finite-dimensional inner-product space, show that for any sequence $\{\alpha_1, \ldots, \alpha_n, \ldots\}$ of vectors $\alpha_1, \ldots, \alpha_n, \ldots,$ in V such that

$$\lim_{n,m\to\infty} \|\alpha_n - \alpha_m\| = 0$$

there exists a vector α_0 in V such that $\lim_{n\to\infty} \|\alpha_0 - \alpha_n\| = 0$. (*Note.* This exercise presupposes familiarity with Cauchy sequences.)

5.4. If $z = a + ib \in \mathbf{C}$, we shall denote the real part of z by $\text{Re}(z)$; thus $\text{Re}(z) = a$. Now let $(\ |\ )$ be an inner product in a unitary space. Prove that if $\text{Re}[(\alpha \mid \beta)]$ is known for all α, β, then $(\alpha \mid \beta)$ is known for all α, β.

5.5. Let V be a unitary space with inner product $(\ |\ )$. A function h on V into V such that for all $\alpha, \beta \in V$ and $c \in \mathbf{C}$

$$h(\alpha + \beta) = h(\alpha) + h(\beta)$$

$$h(c\alpha) = \bar{c}h(\alpha)$$

$$h(h(\alpha)) = \alpha$$

$$(h(\alpha) \mid h(\beta)) = (\alpha \mid \beta)$$

is called a conjugation in V.

(a) Show that if V is finite-dimensional, then a conjugation in V always exists.

(b) If h is a conjugation in V, a vector $\alpha \in V$ is called real (relative to h) if and only if $h(\alpha) = \alpha$. Show that the set of all real vectors form a vector space over $\mathbf{R}$. If $\dim V = n$, show that the space of all real vectors also has dimension n.

(c) Show further that each $\alpha \in V$ has a unique representation in the form $\alpha = x + iy$, where x and y are real vectors.

4 Linear Transformations

In this chapter we return to a consideration of vector spaces over an arbitrary field and concentrate on the notion which in Section 2.2 was labeled a homomorphism of V into W (where both V and W are vector spaces over the same field F). A more common name for this is a linear transformation of V into W. Thus in linear algebra, as in everyday life, objects and concepts acquire various names.

The study of linear transformations of vector spaces occupies a central position in the remainder of this book. This fact simply indicates their importance within the theory known as linear algebra. Linear transformations occur not only in many areas within mathematics but in the mathematical idealizations of many classes of problems in the natural and the social sciences (see Chapter 9).

1. DEFINITION OF A LINEAR TRANSFORMATION

For convenience we repeat the earlier definition of a homomorphism for vector spaces and assign this concept its new name.

□ **Definition.** If V and W are vector spaces over the same field F, a mapping A of V into W is called a *linear transformation of V into W* if and only if for all α, β in V and all a in F

$$(\alpha + \beta)A = \alpha A + \beta A$$

and

$$(a\alpha)A = a(\alpha A)$$

or, what is equivalent,

$$(a\alpha + b\beta)A = a(\alpha A) + b(\beta A)$$

for all α, β in V and all a, b in F. A linear transformation of V into V is called a *linear transformation on* V.

EXAMPLES

1.1. Recall that a linear transformation A of V into W maps the zero vector of V onto the zero vector of W and $-\alpha$ onto $(-\alpha A)$ for all α in V.

1.2. Let V be an n-dimensional vector space over F, with $\mathscr{B} = (\alpha_1, \ldots, \alpha_n)$ as an ordered basis, and let W be an m-dimensional space over F with $\mathscr{B}' = (\beta_1, \ldots, \beta_m)$ as an ordered basis. Corresponding to each $n \times m$ matrix (a_{ij}) over F there is a linear transformation A of V into W defined in the following way:

$$\text{(1) If } \alpha = \sum_{i=1}^{n} x_i \alpha_i, \quad \text{then} \quad \alpha A = \left(\sum_{i=1}^{n} x_i a_{i1}\right)\beta_1 + \cdots + \left(\sum_{i=1}^{n} x_i a_{im}\right)\beta_m .$$

To prove that the mapping A is a linear transformation is a routine exercise that we leave for the reader. We call A the *linear transformation induced by the matrix* (a_{ij}) relative to the pair $(\mathscr{B}, \mathscr{B}')$ of bases $\mathscr{B}$ of V and $\mathscr{B}'$ of W.

The definition of A can be recast in a form that is easily remembered. If $[\alpha A]_{\mathscr{B}'} = (x_1', \ldots, x_m')$, then (1) means that

$$x_j' = \sum_{i=1}^{n} x_i a_{ij}, \qquad 1 \leq j \leq m.$$

These defining equations for $x_1', \ldots, x_m'$ may be written as the single matrix equation

$$(x_1, \ldots, x_n)\begin{bmatrix} a_{11} & \cdots & a_{1m} \\ \vdots & & \\ a_{n1} & \cdots & a_{nm} \end{bmatrix} = (x_1', \ldots, x_m')$$

or, in the more compact form,

$$[\alpha]_{\mathscr{B}}(a_{ij}) = [\alpha A]_{\mathscr{B}'}.$$

Turning matters around, suppose that A is a given linear transformation of V into W. Relative to the ordered bases $\mathscr{B}$ and $\mathscr{B}'$, we define a matrix in the following way. If

$$\begin{aligned} \alpha_1 A &= a_{11}\beta_1 + \cdots + a_{1m}\beta_m \\ &\vdots \\ \alpha_n A &= a_{n1}\beta_1 + \cdots + a_{nm}\beta_m, \end{aligned} \tag{2}$$

then by the matrix of A relative to the pair $(\mathscr{B}, \mathscr{B}')$ of bases $\mathscr{B}$ and $\mathscr{B}'$ we mean

$$(a_{ij}) = \begin{bmatrix} a_{11} \cdots a_{1m} \\ \vdots \\ a_{n1} \cdots a_{nm} \end{bmatrix}.$$

It is left to the reader to convince himself fully of the following facts: starting with an $n \times m$ matrix (a_{ij}), if rule (1) is applied to determine a linear transformation A and then rule (2) is applied to A to determine a matrix, the matrix (a_{ij}) is obtained. Conversely, starting with a linear transformation A, if rule (2) is applied to determine a matrix (a_{ij}) and then rule (1) is applied to (a_{ij}) to determine a linear transformation, the linear transformation A is the result. Of course, it is assumed that the rules are applied relative to a fixed pair of bases. It follows that by means of any one pair $(\mathscr{B}, \mathscr{B}')$ of bases of V and W, respectively, there is determined a 1-1 correspondence between the set of all linear transformations of V into W and the set of all $n \times m$ matrices over F (the field of scalars for both spaces).

1.3. Let V be an inner-product space. For a fixed vector α_0 in V define a mapping A on V into F by

$$\alpha A = (\alpha \mid \alpha_0), \qquad \text{all } \alpha \text{ in } V.$$

Now a field F may be classified as a vector space over itself. Although this may suggest an abstract instance of incest, it is mathematically correct. Indeed, if the definition of a vector space is recalled and the product $ab (a, b \in F)$ is interpreted as the product of the scalar a and the vector b, our assertion becomes evident. The dimension of the space F over the field F is 1; any nonzero element of F is a basis of F over F. Returning to the mapping A defined above, we may conclude, therefore, that it maps a vector space over F into a vector space over F. It is an easy exercise to show that A is a linear transformation.

1.4. Let V be a euclidean space and let S be a finite-dimensional subspace. In view of part (a) of Theorem 3.4.2, a function, A, let us say, on V into S is defined by the following rule:

$$\alpha A = \alpha_S \text{ (the projection of } \alpha \text{ onto } S).$$

This function is a linear transformation on V onto S.

1.5. Let V, W, and X be vector spaces over F and suppose that both

$$A: V \to W \quad \text{and} \quad B: W \to X$$

are linear transformations. Define the function $AB: V \to X$ by

$$\alpha(AB) = (\alpha A)B, \quad \text{for all } \alpha \text{ in } V.$$

Then it is a routine exercise to show that AB is a linear transformation of V into X. We shall call AB the *product* of A and B.

Suppose that each of these spaces is finite-dimensional and that

$$\mathscr{B} = (\alpha_1, \ldots, \alpha_n), \qquad \mathscr{B}' = (\beta_1, \ldots, \beta_m), \qquad \mathscr{B}'' = (\gamma_1, \ldots, \gamma_p)$$

are ordered bases of V, W, and X, respectively. Let (a_{ik}) be the matrix of A relative to $(\mathscr{B}, \mathscr{B}')$ and let (b_{rj}) be the matrix of B relative to $(\mathscr{B}', \mathscr{B}'')$. We shall prove that $(a_{ik})(b_{rj})$ is the matrix of AB relative to $(\mathscr{B}, \mathscr{B}'')$. First observe that (a_{ik}) is $n \times m$ and (b_{rj}) is $m \times p$, so that the indicated product is defined and is an $n \times p$ matrix. Thus it has the correct number of rows and columns. Now for the details. Let the $n \times p$ matrix (c_{ij}) be the matrix of AB relative to $(\mathscr{B}, \mathscr{B}'')$. By the definition of the matrix of a linear transformation relative to a pair of bases c_{ij} is the jth coordinate of $\alpha_i(AB)$. By our definitions of (a_{ik}) and (b_{rj}) we have

$$\alpha_i A = \sum_{k=1}^{m} a_{ik}\beta_k, \qquad 1 \le i \le n$$

and

$$\beta_k B = \sum_{j=1}^{p} b_{kj}\gamma_j, \qquad 1 \le k \le m.$$

Hence

$$
\begin{aligned}
\alpha_i(AB) &= (\alpha_i A)B \\
&= \left(\sum_{k=1}^{m} a_{ik}\beta_k\right)B \\
&= \sum_{k=1}^{m} a_{ik}(\beta_k B) \\
&= \sum_{k=1}^{m} a_{ik}\left(\sum_{j=1}^{p} b_{kj}\gamma_j\right) \\
&= \sum_{j=1}^{p}\left(\sum_{k=1}^{m} a_{ik}b_{kj}\right)\gamma_j, \qquad 1 \le i \le n.
\end{aligned}
$$

Thus

$$
c_{ij} = \sum_{k=1}^{m} a_{ik}b_{kj}, \qquad 1 \le i \le n, \quad 1 \le j \le p.
$$

It follows that $(c_{ij}) = (a_{ik})(b_{kj})$ by the definition of matrix multiplication.

EXERCISES

1.1. Show that the mapping defined in Example 1.3 [viz., if V is an inner-product space over F and α_0 is a fixed vector of V, then $A: V \to F$, where $\alpha A = (\alpha \mid \alpha_0)$] is a linear transformation.

1.2. Try to find a mapping f of R^3 into R^3 such that $(\alpha + \beta)f = \alpha f + \beta f$ for all α, β in R^3 but is not a linear transformation.

1.3. Find a mapping g of R^3 into R^3 such that $(a\alpha)g = a(\alpha g)$ for a in R and all α in R^3 but is not a linear transformation.

1.4. Which of the following functions A of R^2 into R^2 are linear transformations?

(a) $(x, y)A = (1 + x, y)$ (b) $(x, y)A = (y, x)$
(c) $(x, y)A = (x^2, y)$ (d) $(x, y)A = (\sin x, y)$
(e) $(x, y)A = (x - y, 0)$ (f) $(x, y)A = (2x - 3y, x - y)$.

1.5. Show that each of the following mappings is a linear transformation of R^2 into R^3. Find the matrices relative to the pair of standard bases of each space, ordered in the natural way.

(a) $(x, y)A = (x + y, x - y, y)$
(b) $(x, y)A = (x + y, y - x, -x)$.

1.6. Find the matrices of the linear transformations defined in Exercise 1.5 relative to the pair of bases $\mathscr{B} = ((1, 1), (1, 2))$ and $\mathscr{B}' = ((1, 0, 1), (0, 1, 1), (0, 1, 0))$ of R^2 and R^3, respectively.

1.7. Find $(x, y, z)A$, where A is the linear transformation induced by the matrix

$$\begin{bmatrix} -2 & 1 \\ 4 & 0 \\ 3 & 4 \end{bmatrix}$$

relative to the pair $(\mathscr{B}, \mathscr{B}')$, where $\mathscr{B} = (\epsilon_1, \epsilon_2, \epsilon_3)$ is the standard basis of R^3 and $\mathscr{B}' = ((1, 1), (2, 3))$ is a basis of R^2.

1.8. Let $V = P_n(\mathsf{R})$ and let D be the derivative operator. Then D is a linear transformation of V into V. Find
(a) the matrix of D relative to the pair $(\mathscr{B}, \mathscr{B})$, where $\mathscr{B} = (1, x, \ldots, x^n)$;
(b) the matrix of D relative to the pair $(\mathscr{B}, \mathscr{B}')$, where $\mathscr{B} = (1, x, \ldots, x^n)$ and $\mathscr{B}' = (1, (x - a), \ldots, (x - a)^n)$ for some nonzero a.

1.9. Assuming that the inner-product space V of Example 1.3 is finite-dimensional, find the matrix of A relative to a pair of bases consisting of some orthonormal basis of V and the basis $\{1\}$ of F.

1.10 Let V be an n-dimensional vector space over a field F and let W be any vector space over F. Let $(\alpha_1, \ldots, \alpha_n)$ be an ordered basis of V and let $(\beta_1, \ldots, \beta_n)$ be an arbitrary n-tuple of vectors in W. Show that there exists exactly one linear transformation A of V into W such that $\alpha_i A = \beta_i$, $1 \leq i \leq n$.

1.11 Let V and W be vector spaces over a field F. Let $\{\alpha_1, \ldots, \alpha_n\}$ be a linearly independent set of vectors in V and $\{\beta_1, \ldots, \beta_n\}$ be a collection of n vectors (not necessarily distinct) in W. Show that there exists a linear transformation A of V into W such that $\alpha_i A = p_i$, $1 \leq i \leq n$.

2. RANGE, NULL SPACE, RANK, AND NULLITY

When a homomorphism of a vector space V into a vector space W is renamed a linear transformation, some of the other terminology introduced in Section 2.2 is changed. We describe these alterations now. Let A be a linear transformation of V into W. The set that was called the kernel of A earlier we rename the *null space* of A and symbolize it N_A. Thus

$$N_A = \{\alpha \in V \mid \alpha A = 0\}.$$

In harmony with this notation we symbolize the range of A by R_A. Thus

$$R_A = \{\beta \in W \mid \beta = \alpha A \text{ for some } \alpha \text{ in } V\}.$$

We recall that the null space and range are subspaces of V and W, respectively. If R_A is finite-dimensional, then dim R_A is called the *rank* of

A. If N_A is finite-dimensional, then dim N_A is called the *nullity* of A. According to Theorem 2.2.1, A is 1-1 if and only if $N_A = \{0\}$ and, by definition, A is onto W if and only if $R_A = W$. Incidentally, a 1-1 linear transformation is also called a *regular* or a *nonsingular linear transformation.*

If the linear transformation A of V into W is both 1-1 and onto W (that is, A is an isomorphism of V onto W), then the inverse A^{-1} of A is defined and

$$\beta A^{-1} = \alpha \text{ if and only if } \alpha A = \beta, \qquad \text{all } \beta \text{ in } W.$$

Thus

$$(\alpha A)A^{-1} = \alpha \quad \text{and} \quad \beta(A^{-1}A) = \beta$$

for all α in V and all β in W. The inverse of A is a 1-1 linear transformation of W onto V, which we call the *inverse transformation* of A.

We prove next a fundamental identity relating the rank and nullity of a linear transformation.

□ **Theorem 2.1.** Let V be a finite-dimensional vector space and let A be a linear transformation of V into some vector space W. Then

$$\dim V = (\text{rank of } A) + (\text{nullity of } A).$$

PROOF. Let S be a complementary subspace of N_A; then $V = N_A \oplus S$. Now the restriction of A to S is 1-1, for suppose that $\alpha A = \beta A$, where $\alpha, \beta \in S$. Then $(\alpha - \beta)A = 0$, hence $\alpha - \beta \in N_A$. Thus $\alpha - \beta \in S \cap N_A$ and consequently $\alpha - \beta = 0$ or $\alpha = \beta$. Further $(S)A = R_A$. Thus the restriction of A to S is an isomorphism between R_A and S. Since S is finite-dimensional, it follows that $\dim S = \dim R_A$. Thus

$$\dim V = \dim N_A + \dim S = \dim N_A + \dim R_A. \; \diamond$$

□ **Corollary.** If A is a linear transformation on a finite-dimensional vector space V, then the following statements are equivalent:

(a) A has an inverse.
(b) A is nonsingular (that is, 1-1).
(c) A is onto V (that is, $R_A = V$).

PROOF. That (a) implies (b) is obvious. That (b) implies (c) follows from the theorem, since the assumption implies that $N_A = \{0\}$. Finally, (c) implies (a), since from the assumption dim $N_A = 0$; hence, in turn, $N_A = \{0\}$, A is 1-1. Thus, if (c) holds, A is both 1-1 and onto, so A has an inverse. ◇

EXAMPLES

2.1. Let A be the linear transformation on R^4 into R^3 induced by the matrix

$$\begin{bmatrix} 1 & -1 & 0 \\ 2 & 1 & 1 \\ 0 & -1 & 1 \\ -2 & 0 & 1 \end{bmatrix}$$

relative to the pair of standard bases of R^4 and R^3, respectively. What are the null space and range of A? Since

$$(x_1, x_2, x_3, x_4)A = (x_1 + 2x_2 - 2x_4, -x_1 + x_2 - x_3, x_2 + x_3 + x_4),$$

the null space N_A of A is the solution space of the system of homogeneous equations

$$\begin{aligned} x_1 + 2x_2 \qquad\quad - 2x_4 &= 0 \\ -x_1 + \ x_2 - x_3 \qquad\quad &= 0 \\ x_2 + x_3 + \ x_4 &= 0. \end{aligned}$$

The matrix of this system can be reduced to the following echelon matrix:

$$\begin{bmatrix} 1 & 0 & 0 & -\frac{3}{2} \\ 0 & 1 & 0 & -\frac{1}{4} \\ 0 & 0 & 1 & \frac{1}{4} \end{bmatrix}.$$

We infer that $N_A = [(6, 1, -1, 4)]$, that the nullity of A is 1, and that the rank of A is $4 - 1 = 3$. If we were to follow the method of proof of Theorem 2.1 to find a basis of R_A, we would extend $\{(6, 1, -1, 4)\}$ to a basis $\{(6, 1, -1, 4), \alpha_2, \alpha_3, \alpha_4\}$ of R^4. Then $\{\alpha_2, \alpha_3, \alpha_4\}$ is a basis of a complement of N_A and $\{\alpha_2 A, \alpha_3 A, \alpha_4 A\}$ is a basis of R_A. A shorter method is to compute the image of the basis $\{\epsilon_1, \epsilon_2, \epsilon_3, \epsilon_4\}$ of R^4 under A. This set spans R_A, and a basis of R_A can be found by the method given in Section 2.3. Notice that the alternative we propose amounts to merely reducing the matrix that defines A to echelon form.

2.2. The derivative operator D is a linear transformation on $P_\infty(\mathsf{R})$. We observe that N_D is the subspace of constant polynomials and R_D is $P_\infty(\mathsf{R})$.

2.3. Let V be the space of all continuous functions of R into R. Define A by

$$(fA)(x) = \int_0^x f(t)\,dt.$$

Then A is a linear transformation on V. We notice that $N_A = \{0\}$ and R_A is the set of all g in V such that g has a continuous first derivative and $g(0) = 0$.

EXERCISES

2.1. A mapping A on R^3 into R^3 is defined by

$$(x, y, z)A = (x - y + 2z,\ 2x + y,\ -x - 2y + 2z).$$

(a) Show that A is a linear transformation.
(b) State necessary and sufficient conditions on a, b, and c for (a, b, c) to be in R_A. What is the rank of A?
(c) State similar conditions on a, b, and c for (a, b, c) to be in N_A. What is the nullity of A?
(d) Find the matrix of A relative to the pair $(\mathscr{B}, \mathscr{B}')$ of bases where

$$\mathscr{B} = (\epsilon_1, \epsilon_2, \epsilon_3) \text{ and } \mathscr{B}' = ((1, 0, 1), (-1, -1, 0), (0, 2, 3)).$$

2.2. Describe explicitly a linear transformation on R^3 which has $[(1, 1, 0), (-1, 2, 3)]$ as its range.

2.3. Let A be the linear transformation of R^4 into R^3 induced by the matrix

$$\begin{bmatrix} -1 & 0 & 1 \\ 2 & 1 & 1 \\ 0 & -1 & 2 \\ 1 & 3 & -1 \end{bmatrix}$$

relative to the standard bases of R^4 and R^3, respectively. Find a basis of R_A and a basis of N_A.

2.4. Let A be the linear transformation on R^3 defined by

$$(x, y, z)A = (x + y + z,\ x + z,\ y + 2z).$$

(a) Show that A is nonsingular.
(b) Find A^{-1}.

(c) Find the matrices of A and A^{-1} relative to the pair $(\mathscr{B}, \mathscr{B})$, where $\mathscr{B} = (\epsilon_1, \epsilon_2, \epsilon_3)$.

2.5. Let S be a subspace of a vector space V.
(a) Describe as explicitly as you can a linear transformation A on V such that $N_A = S$.
(b) Describe as explicitly as you can a linear transformation B on V such that $\alpha B = \alpha$ if and only if $\alpha \in S$.

2.6. Let V be the vector space of all continuous functions on the interval $[a, b]$ into R. Show that the mapping A on V into V such that $(fA)(x) = xf(x)$ is a linear transformation. Find N_A and R_A. Is the function $\sin x$ in R_A?

2.7. Let V and W be vector spaces over a field F, let P be a linear transformation of V into V, Q be a linear transformation of W into W, and A be a linear transformation of V into W. Then PA and AQ are linear transformations of V into W. Show that
(a) $N_{AQ} \supseteq N_A$,
(b) $R_{PA} \subseteq R_A$.
Further, show that if P and Q have inverses all inclusion signs may be replaced by equality signs.

2.8. Let (a_{ij}) be an $n \times m$ matrix over a field F. Then the set S of rows of (a_{ij}) is a subset of F^m. Show that if A is the linear transformation induced by (a_{ij}), relative to the pair $(\mathscr{B}, \mathscr{B}')$ of bases of F^n and F^m, respectively, the rank of A is equal to $\dim [S]$.

2.9. Let V be a euclidean space with inner product $(\,|\,)$. Let A be a linear transformation of V into V. Show that if $(\alpha A \mid \beta A) = (\alpha \mid \beta)$ for all $\alpha, \beta \in V$, then $N_A = \{0\}$.

2.10 Let V be an n-dimensional euclidean space with the orthonormal basis $\mathscr{B}$. Let (a_{ij}) be an $n \times n$ matrix over R such that $a_{ij} = a_{ji}$, $1 \le i, j \le n$. Show that the linear transformation A induced by (a_{ij}) relative to $(\mathscr{B}, \mathscr{B})$ has the following property:

$$(\alpha A \mid \beta) = (\alpha \mid \beta A) \quad \text{all } \alpha, \beta \in V.$$

Show, conversely, that if a linear transformation $A: V \to V$ satisfies the above property the matrix (a_{ij}) of A, relative to $(\mathscr{B}, \mathscr{B})$, where $\mathscr{B}$ is an orthonormal basis of V, has the property that $a_{ij} = a_{ji}$, $1 \le i, j \le n$.

3. THE VECTOR SPACES $\mathscr{L}(V, W)$ AND $\mathscr{L}(V, V)$

Suppose that V and W are vector spaces over the field F. Let $\mathscr{L}(V, W)$ denote the set of all linear transformations of V into W. We introduce an operation of addition in this set and a multiplication of an element by a scalar in such a way that a vector space over F results.

☐ **Definition.** If A and B are elements of $\mathscr{L}(V, W)$, then their sum $A + B$ is the mapping of V into W such that

$$\alpha(A + B) = \alpha A + \beta B, \quad \text{for all } \alpha \text{ in } V.$$

For a in F the *scalar multiple* aA of A is the mapping on V into W such that

$$\alpha(aA) = (a\alpha)A, \quad \text{for all } \alpha \text{ in } V.$$

Since

$$\begin{aligned}(a\alpha + b\beta)(A + B) &= (a\alpha + b\beta)A + (a\alpha + b\beta)B \\ &= a(\alpha A) + b(\beta A) + a(\alpha B) + b(\beta B) \\ &= a(\alpha A + \alpha B) + b(\beta A + \beta B) \\ &= a(\alpha(A + B)) + b(\beta(A + B))\end{aligned}$$

and

$$\begin{aligned}(a\alpha + b\beta)(cA) &= (c(a\alpha + b\beta))A \\ &= ((ca)\alpha + (cb)\beta)A \\ &= (ac)(\alpha A) + (bc)(\beta A) \\ &= a(\alpha(cA)) + b(\beta(cA)),\end{aligned}$$

we may conclude that $A + B$ and cA are linear transformations of V into W, hence in $\mathscr{L}(V, W)$.

The function on V into W which maps each element of V onto the zero vector of W is in $\mathscr{L}(V, W)$; we denote this member of $\mathscr{L}(V, W)$ by 0 and call it the *zero transformation*. Further, if $A \in \mathscr{L}(V, W)$ then $-A$, which we define by the rule

$$\alpha(-A) = -(\alpha A) \quad \text{all } \alpha \in V$$

is in $\mathscr{L}(V, W)$. It is a routine exercise to show that $(\mathscr{L}(V, W), +, 0, F, \cdot)$ is a vector space.

In passing we note that the zero transformation is equal to $0A$, the scalar product of the zero scalar and any A in $\mathscr{L}(V, W)$, since

$$\alpha(0_F A) = (0_F \alpha)A = 0_V A = 0_W, \qquad \text{all } \alpha \in V.$$

Also $-A = (-1)A$ for each A in $\mathscr{L}(V, W)$, as the reader can show.

Suppose now that V and W are finite-dimensional, say $\dim V = n$ and $\dim W = m$. Let $\mathscr{B} = (\alpha_1, \ldots, \alpha_n)$ and $\mathscr{B}' = (\beta_1, \ldots, \beta_m)$ be ordered bases of V and W, respectively. Suppose that (a_{ij}) and (b_{ij}) are the matrices of

A and B, respectively, relative to the pair $(\mathscr{B}, \mathscr{B}')$. Then the matrix of $A + B$ relative to $(\mathscr{B}, \mathscr{B}')$ is easily seen to be (c_{ij}), where $c_{ij} = a_{ij} + b_{ij}$ for all i and j. Because of this, we define the *sum* $(a_{ij}) + (b_{ij})$ of two $n \times m$ matrices by

$$(a_{ij}) + (b_{ij}) = (c_{ij})$$

where $c_{ij} = a_{ij} + b_{ij}$.

Thus the sum of two $n \times m$ matrices is an $n \times m$ matrix. If d is a scalar, then the matrix of the linear transformation dA relative to $(\mathscr{B}, \mathscr{B}')$ is (c_{ij}), where $c_{ij} = da_{ij}$ for all i and j. On account of this we define the product $d(a_{ij})$ of an $n \times m$ matrix (a_{ij}) by a scalar d as follows:

$$d(a_{ij}) = (c_{ij}),$$

where $c_{ij} = da_{ij}$. Observe that $d(a_{ij})$ is an $n \times m$ matrix.

□ **Theorem 3.1.** The set $M_{n \times m}(F)$ of all $n \times m$ matrices over F, together with the operations of addition and scalar multiplication defined above, is a vector space of dimension nm over F.

PROOF. No computations are necessary to prove that a vector space is at hand. Let V be an n-dimensional vector space over F and let W be an m-dimensional space over F. (For instance, we may choose V to be F^n and W to be F^m.) Relative to a pair $(\mathscr{B}, \mathscr{B}')$ of bases of V and W, respectively, there is determined a 1-1 correspondence f between $\mathscr{L}(V, W)$ and $M_{n \times m}(F)$, as explained in Example 1.2. The mapping f has the following properties: $(A + B)f = Af + Bf$ and $(aA)f = a(Af)$ for $A, B \in \mathscr{L}(V, W)$ and $a, b \in F$. Indeed, our definitions of operations for $M_{n \times m}(F)$ were chosen so that this would be true! Hence, by virtue of the result in Exercise 2.2.10, $M_{n \times m}(F)$, when equipped with the operations introduced above, is a vector space.

The reader can easily show that the dimension of this space is nm by proving that the nm matrices

$$E_{ij} = i\begin{bmatrix} & \overset{j}{\vdots} & \\ \cdots & 1 & \cdots \\ & \vdots & \end{bmatrix}, \qquad 1 \leq i \leq n,\ 1 \leq j \leq m,$$

with 1 as their (i, j)th entry and 0 as all other entries, is a basis. ◇

Note that as a consequence of the theorem we have such results (whose proofs are trivial) as the addition of matrices is a commutative and an associative operation and the sum of an $n \times m$ matrix (a_{ij}) and the $n \times m$ *zero matrix*

$$O = \begin{bmatrix} 0 \cdots 0 \\ \vdots \\ 0 \cdots 0 \end{bmatrix},$$

each of whose entries is 0, is equal to (a_{ij}).

Now let us consider the space $\mathscr{L}(V, V)$ of all linear transformations of a vector space V over F into itself. Its structure is more interesting because of the fact that, besides the operation of addition, an operation of multiplication is provided by the definition given in Example 1.5. Before summarizing the basic properties of this type of mathematical system we introduce some abbreviations. In $\mathscr{L}(V, V)$ are those mappings A that have the form

$$\alpha A = a\alpha$$

for some scalar a and all α in V. Clearly, these are in 1-1 correspondence with F. We call them *scalar transformations* and denote them by the corresponding scalar. (Our earlier agreement to denote the zero transformation by 0 is consistent with this convention.) The linear transformation denoted by 1 is the identity map on V; we call it the *identity transformation*. Observe that the term

$$aA$$

where $a \in F$ and $A \in \mathscr{L}(V, V)$ has two interpretations: on the one hand it is the scalar product of A and the scalar a and on the other hand it is the product of the linear transformations a and A. This causes no confusion, since both interpretations define the same linear transformation.

☐ **Theorem 3.2.** A system consisting of the set $\mathscr{L}(V, V)$ of all linear transformations of some vector space V over a field F, together with the operations of addition, scalar multiplication, and multiplication, as they have been defined, has the following properties.

(a) $(\mathscr{L}(V, V), +, 0, F, \cdot)$ is a vector space.
(b) For all A, B, and C in $\mathscr{L}(V, V)$,

$$A(BC) = (AB)C.$$

(c) For all A, B, and C in $\mathscr{L}(V, V)$,

$$A(B + C) = AB + AC \quad \text{and} \quad (B + C)A = BA + CA.$$

(d) For all A, B, in $\mathscr{L}(V, V)$ and all a in F,

$$a(AB) = (aA)B = A(aB).$$

(e) $1A = A1 = A$.

(f) If A in $\mathscr{L}(V, V)$ has an inverse (that is, A is an isomorphism of V onto V), then

$$AA^{-1} = A^{-1}A = 1.$$

(g) If A and B in $\mathscr{L}(V, V)$ have inverses, then so does AB, and $(AB)^{-1} = B^{-1}A^{-1}$.

PROOF. The reader is asked to decide which statements require further proof and then to supply such proofs. ◇

For our next theorem about $\mathscr{L}(V, V)$ we supply some background remarks. Suppose that V and W are vector spaces over a field F, such that $\dim V = n$, and $\dim W = m$. Implicit in the proof of Theorem 3.1 is the fact that the 1-1 correspondence between $\mathscr{L}(V, W)$ and $M_{n \times m}(F)$, which is defined when a pair $(\mathscr{B}, \mathscr{B}')$ of ordered bases is selected for V and W, respectively, is an isomorphism between these spaces. In the event that $W = V$, it is natural to choose $\mathscr{B}' = \mathscr{B}$ and we shall do this. The matrix of a linear transformation A in $\mathscr{L}(V, V)$, relative to the pair $(\mathscr{B}, \mathscr{B})$, is called simply the matrix of A relative to $\mathscr{B}$ and symbolized

$$[A]_{\mathscr{B}}.\dagger$$

In terms of our new notation, the relationship between $\mathscr{L}(V, V)$ and $M_{n \times n}(F)$ may be described as the map $f: \mathscr{L}(V, V) \to M_{n \times n}(F)$ such that $Af = [A]_{\mathscr{B}}$ is a 1-1 correspondence such that $(A + B)f = [A]_{\mathscr{B}} + [B]_{\mathscr{B}}$

† Thus, if $\mathscr{B} = (\alpha_1, \ldots, \alpha_n)$ and

$$\begin{aligned} \alpha_1 A &= a_{11}\alpha_1 + \cdots + a_{1n}\alpha_n \\ &\vdots \\ \alpha_n A &= a_{n1}\alpha_1 + \cdots + a_{nn}\alpha_n, \end{aligned}$$

then

$$[A]_{\mathscr{B}} = \begin{bmatrix} a_{11} & \cdots & a_{1n} \\ \vdots & & \\ a_{n1} & \cdots & a_{nn} \end{bmatrix}$$

and $(aA)f = a[A]_{\mathscr{B}}$. In addition (this is new!), $(AB)f = [A]_{\mathscr{B}}[B]_{\mathscr{B}}$. This result is a by-product of the calculation made in Example 1.5. For reference we summarize our findings as a theorem.

□ **Theorem 3.3.** Let V be an n-dimensional vector space over a field F and let $\mathscr{B}$ be a basis of V. The mapping $f : \mathscr{L}(V, V) \to M_{n\times n}(F)$ such that $Af = [A]_{\mathscr{B}}$ is a 1-1 correspondence with the following properties:

$$[A + B]_{\mathscr{B}} = [A]_{\mathscr{B}} + [B]_{\mathscr{B}},$$

$$[AB]_{\mathscr{B}} = [A]_{\mathscr{B}}[B]_{\mathscr{B}},$$

$$[aA]_{\mathscr{B}} = a[A]_{\mathscr{B}}.$$

□ **Corollary.** Let (a_{ij}), (b_{ij}), and (c_{ij}) be $n \times n$ matrices over F. Then

$$(a_{ij})((b_{ij})(c_{ij})) = ((a_{ij})(b_{ij}))(c_{ij})$$

$$(a_{ij})((b_{ij}) + (c_{ij})) = (a_{ij})(b_{ij}) + (a_{ij})(c_{ij}),$$

$$((a_{ij}) + (b_{ij}))(c_{ij}) = (a_{ij})(c_{ij}) + (b_{ij})(c_{ij})$$

PROOF. These results follow immediately from Theorem 3.3 and parts (b), (c) of Theorem 3.2. For example, to prove the *associativity of matrix multiplication*, that is, the first of the above properties, choose an n-dimensional vector space V over F and a basis $\mathscr{B}$ of V. Let A, B, and C be the linear transformations induced by (a_{ij}), (b_{ij}), and (c_{ij}), respectively, relative to $\mathscr{B}$. If f is the isomorphism between $\mathscr{L}(V, V)$ and $M_{n\times m}(F)$ determined by B, then $Af = (a_{ij})$, $Bf = (b_{ij})$, and $Cf = (c_{ij})$. According to Theorem 3.3,

$$(A(BC))f = (Af)((BC)f) = (a_{ij})((b_{ij})(c_{ij}))$$

and

$$((AB)C)f = ((AB)f)(Cf) = ((a_{ij})(b_{ij}))(c_{ij}).$$

Since $A(BC) = (AB)C$, the associativity of matrix multiplication follows. ◇

Although Theorem 3.2 may be made the basis for proofs of the identities given the corollary only in the case of $n \times n$ matrices, that is, so-called *square matrices*, they extend to all matrices such that the operations involved are defined. This is easily seen on considering appropriate $\mathscr{L}(V, W)$'s.

That the foregoing properties of matrix operations are nontrivial will be appreciated readily by the reader if he attempts detailed, direct proofs. This is our first example of a powerful method for proving algebraic properties of matrices; namely, to prove the corresponding properties for associated linear transformations and then to rely on an isomorphism of a space of linear transformations and that of a space of matrices. Such an isomorphism is a two-way street and there is traffic in the other direction; for example, the (easily proved) fact that $M_{n\times m}(F)$ has dimension nm implies that $\mathscr{L}(V, W)$, where V and W are vector spaces over F such that $\dim V = n$ and $\dim W = m$, has dimension nm.

We turn now to a discussion of several properties of multiplication of linear transformations and of square matrices. Multiplication of linear transformations is not a commutative operation, nor does it have the property that if $AB = 0$ for linear transformations A and B then either A or B is the zero transformation. As an illustration, with $\mathscr{B}$ the standard basis R^2, define A and B in $\mathscr{L}(\mathsf{R}^2, \mathsf{R}^2)$ by

$$[A]_{\mathscr{B}} = \begin{bmatrix} 1 & 1 \\ 0 & 0 \end{bmatrix}$$

and

$$[B]_{\mathscr{B}} = \begin{bmatrix} 1 & 0 \\ -1 & 0 \end{bmatrix}$$

Then, if $(x, y) \in \mathsf{R}^2$,

$$(x, y)A = (x, y)\begin{bmatrix} 1 & 1 \\ 0 & 0 \end{bmatrix} = (x, x),$$

$$(x, y)B = (x, y)\begin{bmatrix} 1 & 0 \\ -1 & 0 \end{bmatrix} = (x - y, 0),$$

$$(x, y)AB = (x, x)B = (0, 0),$$

$$(x, y)BA = (x - y, 0)A = (x - y, x - y).$$

From the last equation it is clear that $BA \neq 0$. The next to the last equation implies that $AB = 0$. Thus $AB \neq BA$. Further, although neither A nor B is the zero transformation, the product AB is.

The failure of multiplication to be a commutative operation suggests various questions. They may be posed in terms of the notion of commuting transformations. Linear transformations A and B are said to *commute* if and only if $AB = BA$. Then one question may be stated as which linear transformations of $\mathscr{L}(V, V)$ commute with all members of $\mathscr{L}(V, V)$? Scalar transformations have this property. Indeed, if A is the scalar transformation corresponding to the scalar a (thus $\alpha A = a\alpha$ for all α), then for any B in $\mathscr{L}(V, V)$ and any α in V

$$\alpha(AB) = (a\alpha)B = a(\alpha B) = (\alpha B)A = \alpha(BA).$$

Thus $AB = BA$ for all B. Actually, the scalar transformations exhaust the set of linear transformations that commute with all members of $\mathscr{L}(V, V)$. We shall deduce this as a consequence of the following important result.

□ **Theorem 3.4.** If $\{\alpha_1, \ldots, \alpha_k\}$ is a linearly independent set of vectors in a vector space V over F and if $\{\beta_1, \ldots, \beta_k\}$ is an arbitrary set of (not necessarily distinct) k vectors in a vector space W over F, then there exists an A in $\mathscr{L}(V, W)$ such that

$$\alpha_i A = \beta_i, \qquad 1 \le i \le k.$$

Proof. Let $S = [\alpha_1, \ldots, \alpha_k]$ and let S' be a complement of S. Then each α in V can be written uniquely as

$$\alpha = \gamma + \gamma', \quad \text{for} \quad \gamma \in S \quad \text{and} \quad \gamma' \in S'.$$

In turn, γ can be written uniquely as

$$\gamma = \sum_{i=1}^{k} a_i \alpha_i .$$

Thus a mapping A of V into W is specified on defining

$$\alpha A = \sum_{i=1}^{k} a_i \beta_i$$

for each α in V. It is a routine exercise to show that $A \in \mathscr{L}(V, W)$. Moreover, it is clear that $\alpha_i A = \beta_i$ for $1 \le i \le k$. ◇

□ **Theorem 3.5.** An element of $\mathscr{L}(V, V)$ commutes with all elements of $\mathscr{L}(V, V)$ if and only if it is a scalar transformation.

Proof. It remains to prove that if A in $\mathscr{L}(V, V)$ commutes with all elements of $\mathscr{L}(V, V)$, then A is a scalar transformation. Assume that A commutes with all elements of $\mathscr{L}(V, V)$. Then, for each α in V, $\{\alpha, \alpha A\}$ is a linearly dependent set. Otherwise, by Theorem 3.4 there exists a B in

$\mathscr{L}(V, V)$ such that $\alpha B = 0$ and $(\alpha A)B \neq 0$. But then

$$(\alpha A)B = \alpha(AB) = \alpha(BA) = (\alpha B)A = 0A = 0$$

contrary to our choice of B. This contradiction establishes our assertion.

Now let α_0 be a nonzero vector in V. Since $\{\alpha_0, \alpha_0 A\}$ is linearly dependent, $\alpha_0 A = a\alpha_0$ for some scalar a. We shall prove that $\beta A = a\beta$ for every β in V. Let β be given. By Theorem 3.4 there exists a B in $\mathscr{L}(V, V)$ such that $\alpha_0 B = \beta$. Since $\{\beta, \beta A\}$ is linearly dependent, $\beta A = b\beta$ for some scalar b. If $\beta = 0$, then we may choose $b = a$. If $\beta \neq 0$, we have

$$b\beta = \beta A = (\alpha_0 B)A = (\alpha_0 A)B = a(\alpha_0 B) = a\beta;$$

hence

$$(b - a)\beta = 0 \quad \text{and} \quad b = a.$$

Thus $\beta A = a\beta$ for all β in V, so that A is a scalar transformation. ◇

In order to state the matrix version of Theorem 3.5 some definitions are required. In any isomorphism of $\mathscr{L}(V, V)$ and $M_{n \times n}(F)$ the image of the identity transformation 1 is the matrix

$$\begin{bmatrix} 1 & 0 & \cdots & 0 \\ 0 & 1 & \cdots & 0 \\ \vdots & & & \\ 0 & 0 & \cdots & 1 \end{bmatrix}.$$

We shall call this the $n \times n$ *identity matrix* and symbolize it by I. A scalar multiple of the identity is called a *scalar matrix*. The scalar matrix aI is the image of the scalar transformation a under any isomorphism of $\mathscr{L}(V, V)$ and $M_{n \times n}(F)$. The matrix analogue of Theorem 3.5 is as follows:

□ **Corollary.** An element of $M_{n \times n}(F)$ commutes with all elements of $M_{n \times n}(F)$ if and only if it is a scalar matrix.

Another question that we might raise concerning commuting transformations is this: for a given element A of $\mathscr{L}(V, V)$ can the set of all B in $\mathscr{L}(V, V)$ that commute with A be characterized? The answer is in the affirmative, but it is too difficult a problem for us to tackle. We must be content with a partial answer. If $A \in \mathscr{L}(V, V)$, we define A^n, n a nonnegative integer, by induction as follows:

$$A^0 = 1$$
$$A^{n+1} = A^n A.$$

A linear transformation of the form

$$p(A) = a_k A^k + \cdots + a_1 A + a_0,$$

where k is a nonnegative integer and $a_0, a_1, \ldots, a_k$ are scalars, is called a *polynomial in* A. Then it is an easy matter to prove that *each polynomial in* A *commutes with* A. The details are left as an exercise.

We conclude this section with a corollary to Theorem 3.4 which is of sufficient importance to be classified as a theorem.

□ **Theorem 3.6.** Let V be a finite-dimensional vector space over a field F and let $\{\alpha_1, \ldots, \alpha_n\}$ be a basis of V. If $\{\beta_1, \ldots, \beta_n\}$ is an arbitrary set of n vectors (not necessarily distinct) in a vector space W over F, then there exists exactly one linear transformation A on V into W such that

$$\alpha_i A = \beta_i, \qquad 1 \leq i \leq n.$$

PROOF. The existence of such a linear transformation is insured by Theorem 3.4. Reviewing the proof of that theorem we note that in the current circumstances the linear transformation described is the mapping A on V into W such that if $\alpha = \sum_1^n a_i \alpha_i$, then $\alpha A = \sum_1^n a_i \beta_i$.

To establish the uniqueness of A, suppose that B is also a linear transformation on V into W such that $\alpha_i \beta = \beta_i$ for $1 \leq i \leq n$. Then, if $\alpha = \sum_1^n a_i \alpha_i$,

$$\alpha B = \left(\sum_1^n a_i \alpha_i\right) B = \sum_1^n a_i(\alpha_i B) = \sum_1^n a_i \beta_i;$$

that is, $B = A$. ◇

EXERCISES

3.1. Compute $(a_{ij})(b_{ij})$ if

(a) $(a_{ij}) = \begin{bmatrix} 1 & -3 \\ 2 & 1 \\ 1 & 2 \end{bmatrix}$ and $(b_{ij}) = \begin{bmatrix} 1 & 2 & 3 & 0 \\ 4 & 3 & -2 & 1 \end{bmatrix}$,

(b) $(a_{ij}) = \begin{bmatrix} -2 & 4 & 1 & 2 \\ 3 & 1 & 1 & -1 \\ 4 & -2 & 0 & 1 \end{bmatrix}$ and $(b_{ij}) = \begin{bmatrix} 2 & 0 \\ 1 & 1 \\ -1 & 1 \\ 1 & -1 \end{bmatrix}$.

3.2. Verify the associative law for matrix multiplication if

$$(a_{ij}) = (1, -1, 1), \qquad (b_{ij}) = \begin{bmatrix} 2 & 1 & -5 \\ 0 & 3 & 0 \\ 1 & 2 & 3 \end{bmatrix}, \qquad (c_{ij}) = \begin{bmatrix} -2 \\ 1 \\ 2 \end{bmatrix}.$$

3.3. The linear transformations A and B in $\mathscr{L}(\mathsf{R}^3, \mathsf{R}^3)$ are defined by

$$(x, y, z)A = (x + y, y - 2z, x - y + z),$$

$$(x, y, z)B = (x + z, y - x, x + y + z).$$

Find the matrices of A, B, $A + B$, and AB relative to the basis $\mathscr{B} = ((1, 1, 0), (0, 1, 0), (-1, 0, 1))$ of R^3.

3.4. If (a_{ij}) and (b_{ij}) are matrices over F such that both $(a_{ij})(b_{ij})$ and $(b_{ij})(a_{ij})$ are defined, must (a_{ij}) and (b_{ij}) be square matrices?

3.5. Let (a_{ij}) and (b_{ij}) be $n \times n$ matrices over a field F and let V be some n-dimensional vector space over F. Show that (a_{ij}) and (b_{ij}) commute if and only if the same is true of the linear transformations of V into V induced by (a_{ij}) and (b_{ij}) relative to any basis of V.

3.6. Find all 2×2 matrices over R that commute with

$$\begin{bmatrix} 1 & 1 \\ 2 & 1 \end{bmatrix}.$$

3.7. Let V be a two-dimensional vector space over a field F. Show that if $A \in \mathscr{L}(V, V)$ and the matrix of A relative to some basis of V is

$$\begin{bmatrix} a & b \\ c & d \end{bmatrix}.$$

then $A^2 - (a + d)A + (ad - bc)$ is the zero transformation.

3.8. Let A be the linear transformation on R^3 induced by the matrix

$$\begin{bmatrix} 1 & 0 & -1 \\ 2 & 1 & 0 \\ 0 & 1 & 1 \end{bmatrix}.$$

relative to the standard basis.

(a) Show that A has an inverse,

(b) Find the matrix of A^{-1} relative to the same basis,

(c) Compute the product of it and the given matrix.

3.9. If

$$(a_{ij}) = \begin{bmatrix} 2 & 2 & -10 \\ 1 & 2 & -7 \\ 0 & 2 & -4 \end{bmatrix}$$

(a) show that $(a_{ij})^3 = O$ but $(a_{ij})^2 \neq O$;
(b) find a basis $(\alpha_1, \alpha_2, \alpha_3)$ of $\mathbf{R}^3$ such that the linear transformation A induced by (a_{ij}) relative to the standard basis of $\mathbf{R}^3$ has the following properties:

$$\alpha_1 A = \alpha_2, \qquad \alpha_2 A = \alpha_3, \qquad \alpha_3 A = 0.$$

(c) Determine the matrix of A relative to the basis $(\alpha_1, \alpha_2, \alpha_3)$.

3.10. Let V be a vector space whose dimension is greater than 1. Prove that there exist elements A and B of $\mathscr{L}(V, V)$ such that $AB \neq BA$.

3.11. Let V be a finite-dimensional vector space and $A \in \mathscr{L}(V, V)$. Show that A is singular if and only if there exists a nonzero transformation B in $\mathscr{L}(V, V)$ such that $AB = 0$.

3.12. Let V be a vector space and suppose that A is a member of $\mathscr{L}(V, V)$ such that $A^2 = A$. Show that R_A and N_A are complementary subspaces. If V is finite-dimensional, say dim $V = n$, find a basis $\{\alpha_1, \ldots, \alpha_n\}$ of V such that

$$\alpha_i A = \alpha_i, \qquad 1 \leq i \leq k \quad (0 \leq k \leq n)$$

and

$$\alpha_i A = 0, \qquad k < i \leq n.$$

What is the matrix of A relative to this basis?

3.13. Let V and W be vector spaces over F, let S be a subspace of V, and let A be a member of $\mathscr{L}(S, W)$. Show that there exists an $\bar{A}$ in $\mathscr{L}(V, W)$ such that $\alpha \bar{A} = \alpha A$ for all α in S.

3.14. If A and B are isomorphisms of a vector space V onto itself, show that AB is also an isomorphism of V onto V and that $(AB)^{-1} = B^{-1}A^{-1}$.

3.15. Let V be a vector space and A be an element of $\mathscr{L}(V, V)$. Show that there exists a B in $\mathscr{L}(V, V)$ such that $ABA = A$ [from which it follows that $(AB)^2 = AB$ and $(BA)^2 = BA$].

3.16. Let V be a vector space and let A be an element of $\mathscr{L}(V, V)$. Show that if there exists a unique B in $\mathscr{L}(V, V)$ such that $AB = 1$, then $BA = 1$; that is, A is an isomorphism of V onto V. Further show that if V is finite-dimensional the uniqueness of B follows from its existence.

3.17. With the same assumptions as in Exercise 3.16, show that if there exist distinct B and C in $\mathscr{L}(V, V)$ such that $AB = AC = I$ then there exist infinitely many D's in $\mathscr{L}(V, V)$ such that $AD = I$.

3.18. Prove that if $A, B \in \mathscr{L}(V, V)$ and rank $B \leq$ rank A, then there exist $P, Q \in \mathscr{L}(V, V)$ such that $B = PAQ$. Show also that if rank $B =$ rank A then P and Q are invertible.

3.19. Let V be a vector space and let $S = \{A \in \mathscr{L}(V, V) \mid R_A$ is finite-dimensional$\}$. Show that
(a) if $A, B \in S$, then $A - B \in S$;
(b) if $A \in S$, then for any C in $\mathscr{L}(V, V)$, AC and CA are in S;
(c) if $\{\alpha_1, \ldots, \alpha_k\}$ is a linearly independent subset of V and $\{\beta_1, \ldots, \beta_k\}$ is any set of k vectors in V, then there exists an A in S such that $\alpha_i A = \beta_i$, $1 \leq i \leq k$.

3.20. Let V be a vector space and S be a subset of $\mathscr{L}(V, V)$. Call S *dense* in $\mathscr{L}(V, V)$ if and only if for any finite linearly independent set $\{\alpha_1, \ldots, \alpha_n\}$ of vectors in V and any collection $\{\beta_1, \ldots, \beta_p\}$ of vectors in V, there exists an A in S such that $\alpha_i A = \beta_i$, $1 \leq i \leq p$.
(a) Show that if V is finite-dimensional, then $\mathscr{L}(V, V)$ is the only dense set in $\mathscr{L}(V, V)$.
(b) Show that $S = \{A \in \mathscr{L}(V, V) \mid R_A$ is finite-dimensional$\}$ is a smallest dense subset of $\mathscr{L}(V, V)$.
(c) Suppose that S is a dense subset of $\mathscr{L}(V, V)$ and that $A \in \mathscr{L}(V, V)$. Show that if A commutes with every element in S, then A is a scalar transformation.

4. LINEAR FUNCTIONALS AND DUAL SPACES

Linear transformations of the type defined in Example 1.3, namely, on a vector space into its field of scalars, are of sufficient importance to deserve a special name.

☐ **Definition.** A linear transformation of a vector space V over F into the vector space F over F is called a *linear functional* on V. The space $\mathscr{L}(V, F)$ of all linear functionals on V is called the *dual space* of V and is denoted by V^*.

EXAMPLES

4.1. Let V be the space over R of all differentiable functions on $[a, b]$. Each of the following mappings of V into R is a linear functional on V.
(a) $A: V \to \mathsf{R}$ such that $fA = f'(c)$, the derivative of f evaluated at a fixed point c in $[a, b]$.
(b) $A: V \to \mathsf{R}$ such that $fA = \int_a^b f(t)\, dt$.

(c) $A: V \to \mathsf{R}$ such that $fA = \int_a^b f(t)\, g(t)\, dt$, where g is a fixed element of V.

4.2. Let F be a field and α be an element of the space F^n over F. Let $(a_1, \ldots, a_n)$ be an n-tuple of n (not necessarily distinct) scalars. If $\alpha = (x_1, \ldots, x_n)$, define a mapping f of F^n into F by

$$\alpha f = \sum_{i=1}^{n} a_i\, x_i .$$

Then f is a linear functional on F^n which we refer to as the linear functional *induced by* $(a_1, \ldots, a_n)$.

4.3. Let $a_1, \ldots, a_n, t_1, \ldots, t_n$ be fixed real numbers. Let $f: P_\infty(\mathsf{R}) \to \mathsf{R}$ such that $pf = a_1 p(t_1) + \cdots + a_n p(t_n)$. Then f is a linear functional on $P_\infty(\mathsf{R})$.

4.4. Let F be a field. The mapping f of $M_{n\times n}(F)$ into F such that

$$(a_{ij})f = a_{11} + \cdots + a_{nn}$$

is a linear functional on the vector space $M_{n\times n}(F)$. The quantity $(a_{ij})f$ is called the *trace* of (a_{ij}).

4.5. Let V be an inner-product space over F and A an element of $\mathscr{L}(V, V)$. Define $f: V \to F$ such that

$$\alpha f = (\alpha A \mid \alpha_0),$$

where α_0 is a fixed element of V. This mapping is a linear functional on V.

As in Examples 4.2 to 4.5 we usually denote linear functionals by lower-case latin letters. We turn our attention now to dual spaces. They have many interesting properties and important applications.

□ **Theorem 4.1.** Let V be a vector space over a field F and V^* be its dual space. If $\{\alpha_1, \ldots, \alpha_n\}$ is a linearly independent set of vectors in V, then there exist linear functionals $f_1, \ldots, f_n$ in V^* such that

$$\alpha_i f_j = \delta_{ij}, \qquad 1 \le i, j \le n.$$

The set $\{f_1, \ldots, f_n\}$ is linearly independent and if $\{\alpha_1, \ldots, \alpha_n\}$ is a basis of V then $\{f_1, \ldots, f_n\}$ is a basis of V^*. In this event the f's are uniquely determined and $\{f_1, \ldots, f_n\}$ is called the *dual basis* of $\{\alpha_1, \ldots, \alpha_n\}$.

PROOF. Assume that $\{\alpha_1, \ldots, \alpha_n\}$ is a linearly independent set of vectors in V. According to Theorem 3.4 (taking the space W to be F), there exists an element f_1 of $\mathscr{L}(V, F)$, that is, a linear functional on V, such that

$$\alpha_1 f_1 = 1, \; \alpha_j f_1 = 0, \qquad 2 \leq j \leq n.$$

The existence of the remaining f's is established similarly.

To prove that any set $\{f_1, \ldots, f_n\}$ which meets the requirements that $\alpha_i f_j = \delta_{ij}$ is linearly independent, suppose that

$$\sum_{i=1}^{n} a_i f_i = 0.$$

(The 0 on the right-hand side of this equation is the zero element of V^*, which under the circumstances, we shall call the *zero functional* on V). Then

$$0 = \alpha_i(a_1 f_1 + \cdots + a_n f_n) = a_i, \qquad 1 \leq i \leq n.$$

This implies that $\{f_1, \ldots, f_n\}$ is linearly independent.

Now assume that $\{\alpha_1, \ldots, \alpha_n\}$ is a basis of V. Let $f \in V^*$ and suppose that

$$\alpha_i f = b_i, \qquad 1 \leq i \leq n.$$

Then

$$\alpha_i(b_1 f_1 + \cdots + b_n f_n) = b_i = \alpha_i f, \qquad 1 \leq i \leq n.$$

This means that the linear functionals $\sum_1^n b_i f_i$ and f coincide on a basis of V. Hence

$$f = \sum_{i=1}^{n} b_i f_i.$$

Thus $\{f_1, \ldots, f_n\}$ is a basis of V^*. The uniqueness of the f's follows from Theorem 3.6. ◇

□ **Corollary 1.** If α is a nonzero vector in V, then there exists a linear functional f in V^* such that $\alpha f \neq 0$.

□ **Corollary 2.** If $\dim V = n$, then $\dim V^* = n$.

Observe that the proof that the f's form a basis of V^* when the α's form a basis of V follows directly from their linear independence, since

$\dim \mathscr{L}(V, F) = \dim V \cdot \dim F = n \cdot 1 = n$. We chose to establish directly the fact that they span V^* to exhibit the form of an arbitrary linear functional f, relative to such a basis:

(1) $$\text{If } f \in V^*, \text{ then } f = \sum_{i=1}^{n} (\alpha_i f) f_i.$$

That is, f is known in terms of the dual basis $\mathscr{B}^* = (f_1, ,\ldots, f_n)$ of the basis $\mathscr{B} = (\alpha_1, \ldots, \alpha_n)$ of V as soon as its values at members of $\mathscr{B}$ are known.

Continuing with the finite-dimensional case, if

$$\alpha = \sum_{1}^{n} x_i \alpha_i,$$

then

$$\alpha f_j = \sum_{1}^{n} x_i (\alpha_i f_j) = x_j, \quad 1 \leq j \leq n.$$

that is,

(2) $$\text{if } \alpha \in V, \text{ then } \alpha = \sum_{i=1}^{n} (\alpha f_i) \alpha_i.$$

So α is known in terms of the $\mathscr{B}$-basis as soon as its images at the members of the dual basis $\mathscr{B}^*$ are known. Further, if $\alpha \in V, f \in V^*$ and

$$[\alpha]_{\mathscr{B}} = (x_1, \ldots, x_n), \qquad [f]_{\mathscr{B}^*} = (a_1, \ldots, a_n),$$

then

(3) $$\alpha f = \sum_{i=1}^{n} a_i x_i;$$

that is, if vectors of V are designated by their coordinates $(x_1, \ldots, x_n)$ relative to an ordered basis of V and a linear functional f is designated by its coordinates $(a_1, \ldots, a_n)$ relative to the ordered dual basis, then values of f are computed in accordance with (3). In particular, we note that every linear functional f on the space F^n has the form

$$(x_1, \ldots, x_n) f = \sum_{i=1}^{n} a_i x_i$$

for scalars $a_1, \ldots, a_n$. The converse of this statement appears in Example 4.2.

EXAMPLES

4.6. The vectors

$$\alpha_1 = (1, 0, 1), \qquad \alpha_2 = (-1, 1, 0), \qquad \alpha_3 = (0, 1, 2)$$

form a basis $\mathscr{B}$ of R_3. Let us find its dual basis $\{f_1, f_2, f_3\}$. The representation of a vector (x, y, z) relative to this basis is

$$(x, y, z) = (2x + 2y - z)\alpha_1 + (x + 2y - z)\alpha_2 + (-x - y + z)\alpha_3 .$$

Then, according to (2),

$$(x, y, z)f_1 = 2x + 2y - z,$$

$$(x, y, z)f_2 = x + 2y - z,$$

$$(x, y, z)f_3 = -x - y + z,$$

and we have determined the f's explicitly.

Next let us determine the representation of a linear functional f on R^3 relative to this basis. We have shown that such a function is given by a recipe of the form

$$(x, y, z)f = ax + by + cz,$$

where, according to (1), $a = \alpha_1 f$, $b = \alpha_2 f$, and $c = \alpha_3 f$. Hence

$$f = (a + c)f_1 + (-a + b)f_2 + (b + 2c)f_3 .$$

4.7. Consider next the converse to the foregoing type of problem. Suppose that $\mathscr{B}^* = \{g_1, g_2, g_3\}$ is a given basis of $(R^3)^*$—to find the basis of R^3 that has $\mathscr{B}^*$ as its dual. As a linear functional on R^3, we know that

$$(x, y, z)g_i = a_i x + b_i y + c_i z, \ 1 \leq i \leq 3,$$

for suitable scalars a_i, b_i, and c_i. Now let $\{f_1, f_2, f_3\}$ be the dual basis of the standard basis $\{\epsilon_1, \epsilon_2, \epsilon_3\}$ of R^3. Then, from (1),

$$\begin{aligned} g_i &= (\epsilon_1 g_i)f_1 + (\epsilon_2 g_i)f_2 + (\epsilon_3 g_i)f_3 \\ &= a_i f_1 + b_i f_2 + c_i f_3, \qquad 1 \leq i \leq 3. \end{aligned}$$

Under the isomorphism between R^3 and $(R^3)^*$ determined by the respective bases $(\epsilon_1, \epsilon_2, \epsilon_3)$ and (f_1, f_2, f_3), the image of (a_i, b_i, c_i) is $a_i f_1 + b_i f_2 + c_i f_3$, $1 \leq i \leq 3$, and hence the linear independence of $\{g_1, g_2, g_3\}$ implies the linear independence of $\{(a_1, b_1, c_1), (a_2, b_2, c_2), (a_3, b_3, c_3)\}$. Therefore, for any choice of the scalars

d_1, d_2, and d_3, the system of equations

$$a_i x + b_i y + c_i z = d_i, \qquad 1 \le i \le 3,$$

has a unique solution. If we choose for (d_1, d_2, d_3), in turn, ϵ_1, ϵ_2, and ϵ_3, we will obtain as solutions three vectors α_1, α_2, and α_3, which form a basis whose dual is $\mathscr{B}^*$.

We return to the case in which the vector space V is not necessarily finite-dimensional and consider a class of linear functionals on V^*. With α_0 a fixed element of V and f ranging over V^*, $\alpha_0 f$ defines a function on V^* into F. Let us denote this function by h_{α_0}; in other words,

$$f h_{\alpha_0} = \alpha_0 f, \qquad f \in V^*.$$

Now h_{α_0} is a linear functional on V^* because

$$(af + bg)h_{\alpha_0} = \alpha_0(af + bg) = a(\alpha_0 f) + b(\alpha_0)g = a(f h_{\alpha_0}) + b(g h_{\alpha_0}).$$

Thus h_{α_0} is in the dual space of V^*. We denote this space as V^{**} and call it the *second dual space* of V. Since, in addition,

$$f h_{a\alpha + b\beta} = (a\alpha + b\beta)f = a(\alpha f) + b(\beta f) = a(f h_\alpha) + b(f h_\beta),$$

we may conclude that

$$h_{a\alpha + b\beta} = a h_\alpha + b h_\beta .$$

This means that the mapping φ on V into V^{**} given by $\alpha\varphi = h_\alpha$ is a linear transformation (or homomorphism) of V into V^{**}. Further, φ is 1-1, for if α, β are distinct elements of V then $\alpha - \beta \ne 0$ and there exists a linear functional f such that $(\alpha + \beta)f = \alpha f - \beta f \ne 0$ (corollary, Theorem 4.1). Hence $\alpha f \ne \beta f$ and, in turn, $h_\alpha \ne h_\beta$. Our results so far are summarized in the following theorem.

☐ **Theorem 4.2.** Let V be a vector space over the field F. For α in V define the linear functional h_α on V^* by

$$f h_\alpha = \alpha f, \quad \text{all } f \text{ in } V^*.$$

Then the mapping φ of V into V^{**} such that

$$\alpha\varphi = h_\alpha$$

is a nonsingular linear transformation of V into V^{**}.

□ **Corollary.** If V is finite-dimensional, then φ is an isomorphism of V onto V^{**}.

PROOF. This is left as an exercise. ◇

Ordinarily we do not identify two vector spaces simply because they are isomorphic. The case of V and V^{**} (when V is finite-dimensional), however, is distinctive because the isomorphism φ is independent of a basis of V; that is, there is a "natural" pairing of the elements of V^{**} with those of V. This being the case, we identify V and V^{**}, keeping in mind that this identification is made with the isomorphism φ.

EXERCISES

4.1. In $\mathbf{R}^3$ let $\alpha_1 = (1, -1, 0)$, $\alpha_2 = (0, 1, 2)$, and $\alpha_3 = (-1, 0, 1)$.

(a) If f is the linear functional on $\mathbf{R}^3$ such that

$$\alpha_1 f = 2, \qquad \alpha_2 f = 1, \qquad \alpha_3 f = -1,$$

find $(x, y, z)f$.

(b) Determine a linear functional g on $\mathbf{R}^3$ such that

$$\alpha_1 g \neq 0 \quad \text{and} \quad \alpha_2 g = \alpha_3 g = 0.$$

4.2. The set $\mathscr{B} = \{\alpha_1, \alpha_2, \alpha_3\}$ of Exercise 4.1 is a basis of $\mathbf{R}^3$. Find the dual basis of $\mathscr{B}$.

4.3. Linear functionals f_1, f_2, and f_3 of $\mathbf{R}^3$ are defined as follows:

$$(x, y, z)f_1 = x + 2y + z,$$

$$(x, y, z)f_2 = y - z,$$

$$(x, y, z)f_3 = x - y + 2z.$$

Show that they form a basis of the dual space of $\mathbf{R}^3$ and find a basis $\mathscr{B}$ of $\mathbf{R}^3$ such that $\mathscr{B}^* = (f_1, f_2, f_3)$.

4.4. Let t_1, t_2, t_3 be distinct real numbers. Define linear functionals f_1, f_2, f_3 in $V = P_2(\mathbf{R})$ as follows: $pf_i = p(t_i)$, $i = 1, 2, 3$. Show that $\mathscr{B}^* = \{f_1, f_2, f_3\}$ is a basis of V^*. Determine a basis $\mathscr{B}$ of V such that $\mathscr{B}^*$ is the dual basis of $\mathscr{B}$.

4.5. Let f and g be linear functionals on an n-dimensional vector space V. Let N_f and N_g be the null spaces of f and g, respectively. Assume that $f \neq g$, that neither is the zero functional, and that $N_f \neq N_g$. Find the dimensions of each of the following spaces: N_f, N_g, $N_f \cap N_g$ and $N_f + N_g$.

4.6. Let V be a vector space over a field F. Let f be a nonzero linear functional on V and let N_f be its null space. Let α_0 be a fixed element of $V - N_f$. Prove that for each α in V there is a uniquely determined scalar a and a uniquely determined vector β in N_f such that $\alpha = a\alpha_0 + \beta$.

4.7. Let V be a finite-dimensional vector space over F having a basis $\mathscr{B}$. Let f be a linear functional on V. Determine the matrix of F relative to the pair $(\mathscr{B}, \{1\})$.

4.8. Prove the corollary to Theorem 4.2 by using a dimension argument.

4.9. Show that if V is a finite-dimensional vector space and V^* is its dual space then each basis of V^* is the dual basis of some basis of V.

4.10 Let V be a vector space and let V^* be its dual space. Show that if $f, g \in V^*$ have the property that whenever $\alpha f = 0$ then $\alpha g = 0$, there exists a scalar a such that $g = af$.

4.11 Let V be a finite-dimensional inner-product space with inner-product $(\mid)$. Show that for each f in V^* there is a unique α_0 in V such that $\alpha f = (\alpha \mid \alpha_0)$ for all α in V.

5. ANNIHILATORS

In this section we present an introductory account of a relationship that exists between the subspaces of a vector space and the subspaces of its dual space and has applications in projective geometry. This material is optional, but we urge the reader to examine at least the last part of the section in which we complete our study of systems of linear equations by proving two theorems (using techniques developed in this chapter) that appear as exercises in Section 2.3 (and so can be proved by more elementary means). Also, the result appearing in Exercise 5.7 should be understood.

□ **Definition.** If S is a nonempty subset of a vector space V, the *annihilator* S^a of S is the set of all linear functionals on V whose restriction to S is the zero functional on S; that is,

$$S^a = \{f \in V^* \mid \alpha f = 0, \quad \text{all } \alpha \in S\}.$$

The reader can show easily that S^a is a subspace of V^*, that $\{0\}^a = V^*$, and that $V^a = \{0\}$. In studying annihilators of subsets of a vector space V, we might as well restrict our attention to subspaces of V, since if S is a subset of V then $S^a = [S]^a$ (proof?).

What are the properties of the mapping $S \to S^a$ that assigns to each subspace S of V its annihilator S^a in V? To investigate this matter in the finite-dimensional case we need the following result.

□ **Theorem 5.1.** If V is a finite-dimensional vector space and S is a subspace of V, then

$$\dim S + \dim S^a = \dim V.$$

PROOF. Let $(\alpha_1, \ldots, \alpha_k)$ be a basis of S and the extension $\mathscr{B} = (\alpha_1, \ldots, \alpha_k, \alpha_{k+1}, \ldots, \alpha_n)$ be a basis of V. Let $(f_1, \ldots, f_k, f_{k+1}, \ldots, f_n)$ be the dual basis of $\mathscr{B}$ and, finally, let $N = [f_{k+1}, \ldots, f_n]$. To establish the theorem it is sufficient to show that $N = S^a$.

An element f of N has the form $f = \sum_{k+1}^n a_j f_j$. Since $\alpha_i f_j = \delta_{ij}$, it is clear that $\alpha_i f = 0$ for $1 \le i \le k$. Hence $\alpha f = 0$ for each α in S. Thus $f \in S^a$ and we have proved that $N \subseteq S^a$.

To prove the converse consider an element g in S^a. Then $\alpha_i g = 0$ for $1 \le i \le k$, but if $g = \sum_1^n a_i f_i$ then a_i is given by the formula $a_i = \alpha_i g$, $1 \le i \le n$. Hence $a_1 = \cdots = a_k = 0$ and g is a linear combination of $\{f_{k+1}, \ldots, f_n\}$. Hence $g \in N$ and $S^a \subseteq N$. It follows that $S^a = N$ and the proof is complete. ◇

Consider now a subspace S of a finite-dimensional vector space V, its annihilator S^a, and, in turn, the annihilator of S^a. The annihilator of S^a, which we denote by S^{aa}, is a subspace of V^{**}. Earlier we agreed to identify V and V^{**} via the isomorphism φ which pairs α in V with h_α in V^{**}. As a consequence, S^{aa} is a subspace of V. We wish to prove that $S^{aa} = S$. We show first that $S \subseteq S^{aa}$. By definition,

$$S^{aa} = \{h \in V^{**} \mid fh = 0, \quad \text{all } f \in S^a\}.$$

Now let T be the subspace of V such that $T\varphi = S^{aa}$. By definition of φ

$$T = \{\alpha \in V \mid \alpha\varphi = h_\alpha, \quad \text{where } fh_\alpha = 0, \quad \text{all } f \in S^a\}.$$

Since, if $\alpha\varphi = h_\alpha$, then $fh_\alpha = \alpha f$ for all $f \in V^*$, it follows that

$$T = \{\alpha \in V \mid \alpha f = 0, \quad \text{all } f \in S^a\}.$$

With V and V^{**} identified, T and S^{aa} are identified, and clearly $S \subseteq S^{aa}$ by virtue of the definition of S^a. The equality of S and S^{aa} will follow if we can show that they have the same dimension. By Theorem 5.1 we see that

$$\dim S + \dim S^a = \dim V$$

and

$$\dim S^a + \dim S^{aa} = \dim V^*.$$

Since $\dim V = \dim V^*$, it follows that $\dim S = \dim S^{aa}$. This completes the proof. We state our result as the following theorem.

□ **Theorem 5.2.** If S is a subspace of a finite-dimensional vector space V, then $S = S^{aa}$.

□ **Corollary 1.** If S_1 and S_2 are subspaces of V such that $S_1{}^a = S_2{}^a$, then $S_1 = S_2$.

PROOF. If $S_1{}^a = S_2{}^a$, then $S_1{}^{aa} = S_2{}^{aa}$ or, in other words, $S_1 = S_2$. ◇

□ **Corollary 2.** If U is a subspace of V^*, then $U = S^a$ for some subspace S of V.

PROOF. Let U be a given subspace of V^*. With V^{**} identified with V, $U^a = S$ is a subspace of V. It follows that $U^{aa} = S^a$ or, in other words, $U - S^a$. ◇

Together, the corollaries mean that the mapping ψ which assigns to each subspace S of V the subspace S^a of V^* is a 1-1 correspondence between the collection of all subspaces of V and those of V^*. It is left as an exercise to prove that ψ has the additional features stated in the next theorem.

□ **Theorem 5.3.** Let V be a finite-dimensional vector space. Let ψ be the 1-1 correspondence $S \to S^a$ between the subspaces of V and those of V^*. Then for all subspaces S and S_2 of V $(S_1 \cap S_2)\psi = (S_1)\psi + (S_2)\psi$ and $(S_1 + S_2)\psi = (S_1)\psi \cap (S_2)\psi$.

The remainder of this section is concerned primarily with applications of some of our results to systems of linear equations. As a preliminary, we review a result derived in Section 4. In Example 4.2 we mentioned that an n-tuple $(x_1, \ldots, x_n)$ of scalars of a field F induces a linear functional f on F^n defined as follows. If $(a_1, \ldots, a_n) \in F^n$, then

$$(a_1, \ldots, a_n)f = \sum_{i=1}^{n} a_i x_i .$$

Conversely, given $f \in (F^n)^*$, let $\epsilon_i f = x_i$, $1 \le i \le n$, where $\{\epsilon_1, \ldots, \epsilon_n\}$ is the standard basis of F^n. If $\alpha = (a_1, \ldots, a_n) \in F^n$, then $\alpha f = (a_1 \epsilon_1 + \cdots + a_n \epsilon_n)f = \sum_1^n a_i x_i$; that is, f is the linear functional induced by the n-tuple $(x_1, \ldots, x_n)$ of elements of F. Thus every linear functional of F^n is induced by an ordered set of n scalars.

Consider now a system of m homogeneous linear equations in n unknowns over a field F:

$$\sum_{j=1}^{n} a_{ij} x_j = 0, \qquad 1 \le i \le m.$$

Let S denote the row space of the matrix (a_{ij}) of the system. Thus

$$S = [\alpha_1, \ldots, \alpha_m], \qquad \alpha_i = (a_{i1}, \ldots, a_{in}), \qquad 1 \le i \le m.$$

The dimension of S is called the *rank* of the system of equations. We observe that an n-tuple $(x_1, \ldots, x_n)$ is a solution of the system if and only if the linear functional that it induces is in S^a. Thus the solution space of the system has the same dimension as S^a. According to Theorem 5.1, dim $S^a = n - \dim S$. Hence

$$\dim \text{(solution space)} = \text{(number of unknowns)} - \text{(rank of the system)}.$$

(The reader may recall that he was asked to prove this result in Exercise 2.3.10.) In particular, a system of homogeneous linear equations having fewer equations than unknowns has a nontrivial solution (for its solution space has dimension greater than 0).

With systems of linear equations still on our minds, we prove next a generalization of Theorem 3.4 in the case in which $V = F^n$, which has an application to systems of nonhomogeneuos equations.

□ **Theorem 5.4.** Let $M = \{\alpha_1, \ldots, \alpha_m\}$ be a set of m vectors in F^n and $(b_1, \ldots, b_m)$ be an m-tuple of elements of the field F. Suppose that

$$\alpha_i = (a_{i1}, \ldots, a_{in}), \qquad 1 \le i \le m$$

and define

$$\alpha_i' = (a_{i1}, \ldots, a_{in}, b_i), \qquad 1 \le i \le m.$$

Then there exists a linear functional f on F^n such that

$$\alpha_i f = b_i, \qquad 1 \le i \le m$$

if and only if dim $[M]$ = dim $[M']$, where $M' = \{\alpha_1', \ldots, \alpha_m'\} \subseteq F^{n+1}$.

PROOF. If every α_i is the zero n-tuple, then, clearly, an f with the required property exists if and only if every $b_i = 0$, and so the theorem is true in this trivial case.

Suppose that some α_i is different from $(0, \ldots, 0)$. Then we can choose a maximal linearly independent subset of M. Without loss of generality, we may assume that $\{\alpha_1, \ldots, \alpha_k\}$ is such a set. Then dim $[M] = k$ and, in the event that $k < m$, there exist scalars c_{ij} such that

$$\alpha_j = c_{j1}\alpha_1 + \cdots + c_{jk}\alpha_k, \qquad k < j \le m.$$

Further, $\{\alpha_1', \ldots, \alpha_k'\}$ is a linearly independent subset of M', since if $\sum_1^k d_i\alpha_i' = 0$ (the zero vector in F^{n+1}), then $\sum_1^k d_i\alpha_i = 0$ (the zero vector in F^n).

With these preliminaries out of the way, we commence the proof. Assume that there exists a linear functional f such that $\alpha_i f = b_i, 1 \le i \le m$. Then, if $k < m$,

$$\begin{aligned}\alpha_j f = b_j &= c_{j1}(\alpha_1 f) + \cdots + c_{jk}(\alpha_k f)\\ &= c_{j1}b_1 + \cdots + c_{jk}b_k, \qquad k < j \le m,\end{aligned}$$

hence

$$\alpha_j' = c_{j1}\alpha_1' + \cdots + c_{jk}\alpha_k', \qquad k < j \le m.$$

It follows that $\{\alpha_1', \ldots, \alpha_k'\}$ is a maximal linearly independent subset of M'. Thus $\dim [M] = \dim [M']$. Of course, the same is true if $k = m$.

For the converse assume that $\dim [M] = \dim [M'] = k$. Since $\{\alpha_1', \ldots, \alpha_k'\}$ is a linearly independent set of k elements of M', it is a maximal linearly independent set. Hence, if $k < m$, there exist scalars c_{ij} such that

$$\alpha_j' = c_{j1}\alpha_1' + \cdots + c_{jk}\alpha_k', \qquad k < j \le m.$$

It follows that

$$b_j = c_{j1}b_1 + \cdots + c_{jk}b_k$$

and

$$\alpha_j = c_{j1}\alpha_1 + \cdots + c_{jk}\alpha_k, \qquad k < j \le m.$$

By Theorem 3.4 there exists a linear functional f on F^n such that $\alpha_i f = b_i$, $1 \le i \le k$. This linear functional has the property stated in the theorem because, in addition (if $k < m$),

$$\begin{aligned}\alpha_j f &= c_{j1}(\alpha_1 f) + \cdots + c_{jk}(\alpha_k f)\\ &= c_{j1}b_1 + \cdots + c_{jk}b_k = b_j, \qquad k < j \le m.\end{aligned}$$

If $k = m$, the existence of f follows from Theorem 3.4. ◇

Consider now a system of nonhomogeneous linear equations in $x_1, \ldots, x_n$ over a field F:

$$\sum_{j=1}^{n} a_{ij}x_j = b_i, \qquad 1 \le i \le m.$$

Let $M = \{\alpha_1, \ldots, \alpha_m\}$, where $\alpha_i = (a_{i1}, \ldots, a_{in})$ for $1 \le i \le m$. Then an n-tuple $(x_1, \ldots, x_n)$ is a solution if and only if the linear functional f which it induces has the property that

$$\alpha_i f = b_i, \qquad 1 \le i \le m.$$

According to Theorem 5.4, there exists such a functional and consequently there exists a solution of the system of equations if and only if the row spaces of the matrices

$$\begin{bmatrix} a_{11} & \cdots & a_{1n} \\ \vdots & & \\ a_{m1} & \cdots & a_{mn} \end{bmatrix} \quad \text{and} \quad \begin{bmatrix} a_{11} & \cdots & a_{1n} & b_1 \\ \vdots & & & \\ a_{m1} & \cdots & a_{mn} & b_m \end{bmatrix}$$

have equal dimension.

Calling (i) the dimension of the row space of a matrix its *row rank*, (ii) the matrix on the left the *coefficient matrix* of the given system of equations and, (iii) the matrix on the right the *augmented matrix* of the system, we have proved the result which is commonly stated in the following form.

□ **Theorem 5.5.** A system of nonhomogeneous linear equations has a solution if and only if its coefficient matrix and its augmented matrix have the same row rank.

□ **Corollary.** A system of n linear equations in n unknowns has a unique solution if and only if the coefficient matrix of the system has row rank n.

PROOF. The reader is asked to deduce this from Theorem 3.6. ◇

We complete our study of systems of linear equations with the following remark. The foregoing result is not a useful way to determine whether a system of linear equations has solutions. In practice, the reader will proceed in accordance with the result he obtained in answering Exercise 2.3.8, namely, by reducing the augmented matrix of a given system to echelon form to conclude that no solutions exist if and only if the last nonzero row has the form

$$(0, \ldots, 0, 1).$$

EXERCISES

5.1. Complete the proof of Theorem 5.3. *Hint*. First prove that if S_1 and S_2 are subspaces of V such that $S_1 \subseteq S_2$, then $S_1{}^a \supseteq S_2{}^a$.

5.2. (a) Let S be the subspace of R^3 spanned by $\{(1, 3, 1), (1, 4, 1)\}$. Determine S^a.
(b) Let S be the subspace of R^4 spanned by $\{(1, 2, -3, 4), (1, 3, -1, 0), (6, 1, 0, 4)\}$. Determine S^a.

5.3. Give a proof of the corollary of Theorem 5.5.

5.4. Find the rank of each of the following systems of homogeneous linear equations and determine the solution space.

(a)
$$\begin{aligned} x_1 + 2x_2 - 3x_3 + 4x_4 &= 0,\\ x_1 + 3x_2 - x_3 \qquad\quad &= 0,\\ 6x_1 + 4x_2 \qquad\quad + 2x_4 &= 0, \end{aligned}$$

(b)
$$\begin{aligned} x_1 + x_2 + x_3 + x_4 + x_5 &= 0,\\ x_1 + 2x_2 \qquad\qquad\qquad &= 0,\\ 4x_1 + 7x_2 + x_3 + x_4 + x_5 &= 0. \end{aligned}$$

5.5. Determine all solutions of each of the following systems of equations:

(a)
$$\begin{aligned} x + 2y + u + v &= 6,\\ x \qquad + u - v &= 0,\\ 2y + u + v &= 5,\\ -2x + y + u \qquad &= 0, \end{aligned}$$

(b)
$$\begin{aligned} 2x + 3y + u + v &= 1,\\ x - y - u + v &= 1,\\ 3x + y + u + 2v &= 0. \end{aligned}$$

5.6. Determine those values of k and m for which the following system of equations is *consistent* (that is, has a solution).

$$\begin{aligned} 3x + 2y + 3u + 4v &= 0,\\ 5x + 3y + 2u + v &= l,\\ 11x + 7y + 12u + 9v &= k,\\ 4x + 3y + 13u + 11v &= m. \end{aligned}$$

5.7. Let V and W be vector spaces over a field F and V^* and let W^* be the respective dual spaces. If $A \in \mathscr{L}(V, W)$, define the mapping $A^t\colon W^* \to V^*$ as follows. For $g \in W^*$, gA^t is the mapping on V such that

$$\alpha(gA^t) = (\alpha A)g, \quad \text{all} \in V.$$

Show that
(a) $gA^t \in V^*$, for each $g \in W^*$,
(b) $A^t \in \mathscr{L}(W^*, V^*)$,
(c) $N_{A^t} = R_A{}^a$, the annihilator of R_A in W (which implies that the rank of A is equal to the rank of A^t if V and W are finite-dimensional).
(d) If $\dim V = m$, $\dim W = n$, and (a_{ij}) is the matrix of A relative to the pair $(\mathscr{B}_1, \mathscr{B}_2)$ of ordered bases $\mathscr{B}_1$ and $\mathscr{B}_2$ of V and W, respectively,

show that the matrix (c_{ij}) of A^t relative to the pair $(\mathscr{B}_2^*, \mathscr{B}_1^*)$ of dual bases of $\mathscr{B}_2$ and $\mathscr{B}_1$, respectively, is given by

$$c_{ij} = a_{ji}, \qquad 1 \le i \le m, 1 \le j \le n.$$

REMARK. The matrix (c_{ij}) is called the *transpose* of (a_{ij}) and is denoted by

$$(a_{ij})^t.$$

5.8. Imitating the definitions of row space and row rank of a matrix, give definitions of the *column space* and *column rank* of a matrix (a_{ij}) over a field F. Prove that the row rank and column rank of (a_{ij}) are equal.

REMARK. Henceforth we shall simply speak of the *rank* of a matrix.

6. ADJOINTS

We have observed that every linear functional f on the space F^n is the linear functional induced by some n-tuple of scalars. Indeed, if $\epsilon_i f = a_i$, $1 \le i \le n$, then for $\alpha = (x_1, \ldots, x_n) \in F^n$, $\alpha f = \sum_1^n a_i x_i$. If $F = \mathsf{R}$ and we assign to R^n the standard inner product, then $\alpha f = (\alpha \mid \beta_0)$ if $\beta_0 = (a_1, \ldots, a_n)$. This is an instance of the following characterization of linear functionals on finite-dimensional inner-product spaces.

□ **Theorem 6.1.** For each linear functional f on a finite-dimensional inner-product space V with inner product $(\mid)$, there exists a unique vector β_0 in V such that

$$\alpha f = (\alpha \mid \beta_0) \quad \text{all} \quad \alpha \text{ in } V.$$

PROOF. Let $\{\alpha_1, \ldots, \alpha_n\}$ be an orthonormal basis of V and set

$$\beta_0 = \sum_{j=1}^{n} \overline{(\alpha_j f)} \alpha_j .$$

Then

$$(\alpha_i \mid \beta_0) = \left(\alpha_i \middle| \sum_1^n \overline{(\alpha_j f)} \alpha_j\right) = \alpha_i f, \qquad 1 \le i \le n,$$

and so

$$\left(\sum_1^n x_i \alpha_i\right) f = \sum_1^n x_i (\alpha_i f) = \sum_1^n x_i (\alpha_i \mid \beta_0) = \left(\sum_1^n x_i \alpha_i \mid \beta_0\right),$$

or, in other words, $\alpha f = (\alpha \mid \beta_0)$ for all α in V.

Now suppose that $\alpha f = (\alpha \mid \beta_1)$ for all α in V. Then $(\alpha \mid \beta_1 - \beta_0) = 0$ for all α in V. In particular, $(\beta_1 - \beta_0 \mid \beta_1 - \beta_0) = 0$ and this implies that $\beta_1 = \beta_0$. ◇

The technique of defining linear functionals in an inner-product space V by way of inner products with a fixed vector was generalized in Example 4.5. There we observed that if $A \in \mathscr{L}(V, V)$ and α_0 is a fixed element of V then the mapping $f: V \to F$, where $\alpha f = (\alpha A \mid \alpha_0)$, is a linear functional on V. If V is finite-dimensional, it follows from Theorem 6.1 that f determines a unique element β_0 such that

$$\alpha f = (\alpha A \mid \alpha_0) = (\alpha \mid \beta_0) \quad \text{all} \quad \alpha \text{ in } V.$$

With A fixed, let us define A^* to be the composite of the mapping that assigns to each vector α_0 a linear functional and that which assigns to each linear functional a vector β_0 in accordance with the above rules. Thus A^* is a function on V into V such that

$$\alpha_0 A^* = \beta_0 \quad \text{if} \quad \alpha f = (\alpha A \mid \alpha_0) \quad \text{and} \quad (\alpha A \mid \alpha_0) = (\alpha \mid \beta_0), \quad \text{all} \quad \alpha \in V,$$

or, in other words,

$$(\alpha A \mid \gamma) = (\alpha \mid \gamma A^*) \quad \text{all } \alpha, \gamma \text{ in } V.$$

Since, for each α in V,

$$\begin{aligned}(\alpha \mid (a\gamma_1 + b\gamma_2)A^*) &= (\alpha A \mid a\gamma_1 + b\gamma_2)\\ &= \bar{a}(\alpha A \mid \gamma_1) + \bar{b}(\alpha A \mid \gamma_2)\\ &= \bar{a}(\alpha \mid \gamma_1 A^*) + \bar{b}(\alpha \mid \gamma_2 A^*)\\ &= (\alpha \mid a(\gamma_1 A^*) + b(\gamma_2 A^*)),\end{aligned}$$

we conclude that

$$(a\gamma_1 + b\gamma_2)A^* = a(\gamma_1 A^*) + b(\gamma_2 A^*)$$

or, in other words, A^* is a linear transformation on V.

□ **Definition.** If A is a linear transformation on an inner-product space V, a linear transformation A^* on V is called an *adjoint* of A if and only if

$$(\alpha A \mid \beta) = (\alpha \mid \beta A^*) \quad \text{all } \alpha, \beta \text{ in } V.$$

It is left as an exercise to prove that if A has an adjoint then it has exactly one. The computations preceding the definition may now be summarized.

☐ **Theorem 6.2.** The adjoint of a linear transformation on a finite-dimensional inner-product space exists and is unique.

EXAMPLES

6.1. Consider the linear transformation A on R^3 (with standard inner product) such that

$$(x, y, z)A = (x - y, \ -x + y + 2z, \ y + z).$$

We shall determine A^*. If $\beta = (x', y', z') \in \mathsf{R}^3$, then

$$\begin{aligned}(\alpha A \mid \beta) &= (x - y)x' + (-x + y + 2z)y' + (y + z)z' \\ &= x(x' - y') + y(-x' + y' + z') + z(2y' + z') \\ &= (\alpha \mid \beta_0),\end{aligned}$$

where $\beta_0 = (x' - y', \ -x' + y' + z', \ 2y' + z')$. Therefore A^* is the linear transformation on R^3 such that

$$(x, y, z)A^* = (x - y, \ -x + y + z, \ 2y + z).$$

6.2. Let V be the inner-product space of all polynomials with coefficients in C and with the inner product given by

$$(p \mid q) = \int_0^1 p(x)\,\overline{q(x)}\,dx.$$

Let h be a fixed polynomial in V and let A be the linear transformation on V such that

$$pA = ph \quad \text{all } p \text{ in } V.$$

Since

$$(pA \mid q) = \int_0^1 p(x)\,h(x)\,\overline{q(x)}\,dx = \int_0^1 p(x)\,\overline{q(x)}\,h(x)\,dx = (p \mid q\bar{h}),$$

we see that A has an adjoint A^* and

$$pA^* = p\bar{h}.$$

6.3. Let V be the same space as above and let D be the operation of differentiation in V. Then

$$(pD \mid q) = \int_0^1 p'(x)\,\overline{q(x)}\,dx = p(x)\,\overline{q(x)}\Big|_0^1 - \int_0^1 p(x)\,\overline{q'(x)}\,dx$$

$$= p(1)\,\overline{q(1)} - p(0)\,\overline{q(0)} + (p \mid q(-D)).$$

If we restrict our attention to the subspace of V, which consists of polynomials p such that $p(0) = p(1) = 0$, then D has an adjoint, namely, $-D$. It is left as an exercise to prove that D, regarded as a linear transformation of V, has no adjoint.

☐ **Theorem 6.3.** Assume that a linear transformation A on an inner-product space V has an adjoint A^*. Then

(a) $N_{A^*} = R_A^{\perp}$,
(b) if V is finite-dimensional, then A and A^* have equal ranks.

Proof. Assume that A has an adjoint A^*, so that $(\alpha A \mid \beta) = (\alpha \mid \beta A^*)$ for all α, β in V. If $\beta \in N_{A^*}$, then $(\alpha A \mid \beta) = (\alpha \mid \beta A^*) = 0$ for all α in V, hence $\beta \in R_A^{\perp}$. Conversely, if $\beta \in R_A^{\perp}$, then $(\alpha \mid \beta A^*) = 0$ for all α in V. In particular, $(\beta A^* \mid \beta A^*) = 0$ so that in turn $\beta A^* = 0$, $\beta \in N_{A^*}$.

If V has dimension n, then by Theorems 3.4.1 and 2.1, the rank of A is equal to

$$\dim R_A = n - \dim R_A^{\perp} = n - \dim N_{A^*} = \dim R_{A^*},$$

which is the rank of A^*. ◇

The proof of the following theorem is left as an exercise.

☐ **Theorem 6.4.** If V is a finite-dimensional inner-product space, then for all $A, B \in \mathscr{L}(V, V)$ and for all scalars a

(a) $(A + B)^* = A^* + B^*$,
(b) $(AB)^* = B^*A^*$,
(c) $(A^*)^* = A$,
(d) $(aA)^* = \bar{a}A^*$.

Continuing with finite-dimensional inner-product spaces (for which adjoints always exist), we investigate the relationship between the matrix of a linear transformation relative to a basis of V and the matrix of its adjoint relative to the same basis. Let V be an n-dimensional inner-product space and suppose that (a_{ij}) and (b_{ij}) are the matrices of A and

A^*, respectively, relative to the *orthonormal* basis $\mathscr{B} = (\alpha_1, \ldots, \alpha_n)$ of V. Then

$$a_{ij} = (\alpha_i A \mid \alpha_j) = (\alpha_i \mid \alpha_j A^*) = \overline{(\alpha_j A^* \mid \alpha_i)} = \overline{b_{ji}}$$

or

$$b_{ij} = \overline{a_{ji}}, \qquad 1 \leq i, j \leq n.$$

If V is a euclidean space, then $b_{ij} = a_{ji}$ and (b_{ij}) is the transpose $(a_{ij})^t$ of (a_{ij}) (see Exercise 5.7). In the complex case (that is, when V is a unitary space) (b_{ij}) is called the *conjugate transpose* of (a_{ij}). We denote the conjugate transpose of a matrix (a_{ij}) over C by $(a_{ij})^*$. In summary, if (a_{ij}) is the matrix of a linear transformation A on V relative to an orthonormal basis, then its conjugate transpose $(a_{ij})^*$ is the matrix of its adjoint A^* relative to the same basis. Conversely, we can prove that if A is the linear transformation on V induced by an $n \times n$ matrix (a_{ij}) relative to an orthonormal basis $\mathscr{B} = (\alpha_1, \ldots, \alpha_n)$ of V then A^* is the linear transformation induced by its conjugate transpose relative to the same basis. To prove this let B denote the linear transformation on V induced by $(a_{ij})^*$. Then

$$\alpha_i A = a_{i1}\alpha_1 + \cdots + a_{in}\alpha_n,$$

$$\alpha_j B = \bar{a}_{1j}\alpha_1 + \cdots + \bar{a}_{nj}\alpha_n,$$

and so

$$(\alpha_i \mid \alpha_j B) = (\alpha_i \mid \bar{a}_{1j}\alpha_1 + \cdots + \bar{a}_{nj}\alpha_n) = (\alpha_i \mid \bar{a}_{ij}\alpha_i)$$
$$= a_{ij} = (\alpha_i A \mid \alpha_j);$$

but from $(\alpha_i A \mid \alpha_j) = (\alpha_i \mid \alpha_j B)$ for $1 \leq i, j \leq n$ we may infer that $(\alpha A \mid \beta) = (\alpha \mid \beta B)$ for all α, β in V, hence that $B = A^*$.

EXAMPLES

6.4. In R^3 with standard inner product let A be the linear transformation induced by the matrix

$$\begin{bmatrix} -2 & 4 & 6 \\ 0 & 3 & 4 \\ 1 & -2 & 1 \end{bmatrix}$$

relative to the standard (orthonormal) basis. Then the matrix of A^* relative to the same basis is

$$\begin{bmatrix} -2 & 0 & 1 \\ 4 & 3 & -2 \\ 6 & 4 & 1 \end{bmatrix}.$$

6.5. Let $V = P_2(\mathrm{C})$, the space of polynomials of degree at most 2 over C, with the inner product given by

$$(p \mid q) = \int_0^1 p(x)\,\overline{q(x)}\,dx.$$

Let D be the operation of differentiation. The orthonormal basis obtained by the Gram-Schmidt process from the basis $(1, x, x^2)$ is

$$\mathscr{B} = (\alpha_1, \alpha_2, \alpha_3),$$

where

$$\alpha_1 = 1, \qquad \alpha_2 = \sqrt{3}(2x - 1), \qquad \alpha_3 = \sqrt{5}(6x^2 - 6x + 1).$$

The matrix of D relative to $\mathscr{B}$ is

$$\begin{bmatrix} 0 & 0 & 0 \\ 2\sqrt{3} & 0 & 0 \\ 0 & 6\sqrt{5/3} & 0 \end{bmatrix}$$

and so that of D^* is

$$\begin{bmatrix} 0 & 2\sqrt{3} & 0 \\ 0 & 0 & 6\sqrt{5/3} \\ 0 & 0 & 0 \end{bmatrix}$$

relative to the same basis.

☐ **Definition.** A linear transformation A on an inner-product space V is called a *self-adjoint linear transformation* if and only if $A = A^*$, that is,

$$(\alpha A \mid \beta) = (\alpha \mid \beta A) \quad \text{all } \alpha, \beta \text{ in } V.$$

☐ **Definition.** An $n \times n$ matrix (a_{ij}) over a field F is called *symmetric* if and only if

$$(a_{ij}) = (a_{ij})^t.$$

An $n \times n$ matrix (a_{ij}) over C is called *hermitian* if and only if

$$(a_{ij}) = (a_{ij})^*.$$

The computations preceding Example 6.3 imply the following results.

□ **Theorem 6.5.** Let V be a finite-dimensional inner-product space and let $A \in \mathscr{L}(V, V)$. Then A is self-adjoint if and only if its matrix relative to some *orthonormal basis* is hermitian (or symmetric if V is a euclidean space).

EXAMPLE

6.6. The linear transformation A on R^3 (with standard inner product) defined by

$$(x, y, z)A = (2x + 2z, x + z, x + z)$$

is not self-adjoint, since

$$0 = ((0, 1, 0)A \mid (1, 0, 0)) \neq ((0, 1, 0) \mid (1, 0, 0)A) = 1,$$

yet the matrix of A relative to the basis

$$((1, 1, 0), (1, 0, 0), (0, 0, 1))$$

is the symmetric matrix

$$\begin{bmatrix} 1 & 1 & 1 \\ 1 & 1 & 1 \\ 1 & 1 & 1 \end{bmatrix}.$$

Thus the assumption that the basis mentioned in Theorem 6.5 is orthonormal is essential for the conclusion of the theorem.

EXERCISES

6.1. Show that if a linear transformation has an adjoint then it has exactly one.

6.2. Prove the assertion made in the last sentence of Example 6.3.

6.3. Provide a proof of Theorem 6.4.

6.4. Let V be the unitary space consisting of $\mathbf{C}^3$ with the standard inner product. Let A in turn be the linear transformation on V induced by each of the following matrices relative to the standard basis of V. Determine A^* in each case.

$$\text{(a)} \begin{bmatrix} 1 & -1 & 2 \\ 0 & -1 & 1 \\ 1 & 1 & 1 \end{bmatrix}, \qquad \text{(b)} \begin{bmatrix} 1 & i & 2 \\ -i & 0 & 1 \\ 0 & 1 & 3\sqrt{2} \end{bmatrix}.$$

6.5. In the space $\mathbf{C}^2$ with standard inner product let A be the linear transformation defined by

$$(1, 0)A = (1, -2)$$

$$(0, 1)A = (i, -1).$$

Find $(x, y)A^*$ and the matrix of A^* relative to the standard basis of $\mathbf{C}^2$.

6.6. For each of the following matrices (a_{ij}) calculate $(a_{ij})^*$, $(a_{ij})(a_{ij})^*$, and $(a_{ij})^*(a_{ij})$.

$$\text{(a)}\ (a_{ij}) = \begin{bmatrix} 1 & -1 & -1 \\ 1 & 2 & 0 \\ 1 & -1 & 1 \end{bmatrix}, \qquad \text{(b)}\ (a_{ij}) = \begin{bmatrix} a & b \\ -a & b \end{bmatrix}, \qquad a, b \in \mathbf{R}.$$

$$\text{(c)}\ (a_{ij}) = \begin{bmatrix} -\sin^2\varphi + i\cos^2\varphi & (1+i)\sin\varphi\cos\varphi \\ (1+i)\sin\varphi\cos\varphi & -\cos^2\varphi + i\sin^2\varphi \end{bmatrix}, \qquad \varphi \in \mathbf{R}.$$

6.7. Show that if (a_{ij}) is an $n \times n$ matrix over $\mathbf{C}$ then the trace of $(a_{ij})(a_{ij})^*$ is a positive number.

6.8. Show that if (a_{ij}) is an $n \times n$ matrix over $\mathbf{C}$ such that $(a_{ij})(a_{ij})^* = I$ then

$$\sum_{k=1}^{n} a_{ik}\bar{a}_{jk} = \delta_{ij}, \qquad 1 \le i, j \le n.$$

6.9. Let $V = P_2(\mathbf{R})$ with inner product given by

$$(f \mid g) = \int_0^1 f(x)\, g(x)\, dx.$$

(a) If a_0 is a real number, find the polynomial g_0 such that $(f \mid g_0) = f(a_0)$ for all f in V.
(b) Let D be the derivative operator in V. Find D^*.
(c) Find the matrix of D and D^* relative to the basis obtained by normalizing the orthogonal basis $(1, x - \frac{1}{2}, x^2 - x + \frac{1}{6})$.

6.10. Let V be a finite-dimensional inner-product space and $A \in \mathscr{L}(V, V)$. Show that if A has an inverse then so has A^* and

$$(A^*)^{-1} = (A^{-1})^*.$$

6.11. Let V be an inner-product space and γ, δ be fixed vectors of V. Show that the mapping $A: V \to V$ such that $\alpha A = (\alpha \mid \gamma)\delta$ is a linear transformation on V. Show that A has an adjoint and obtain an explicit description of A^*.

6.12. Show that the product of two self-adjoint linear transformations is self-adjoint if and only if they commute. State two theorems about matrices that are immediate consequences of this result.

6.13. If A is a linear transformation on a finite-dimensional inner-product space, show that AA^* and A^*A are self-adjoint transformations.

6.14. A linear transformation A on an inner-product space such that $A = -A^*$ is called *skew symmetric* or *skew hermitian* according as the space is euclidean or unitary. Prove that

(a) every linear transformation A on a finite-dimensional inner-product space can be written as the sum $A = B + C$ of a self-adjoint transformation B and a skew-symmetric transformation C.

(b) If A is a skew-symmetric transformation on a euclidean space, then $(\alpha A \mid \alpha) = 0$ for all vectors α. (Does the converse hold?)

(c) If A is self-adjoint or skew-symmetric and if $\alpha A^2 = 0$, then $\alpha A = 0$.

6.15. Let $V = P_\infty(\mathsf{R})$ with the inner product given by $\int_0^1 f(x)\, g(x)\, dx$. Show that Theorem 6.1 fails for the linear functional F such that $fF = f(c)$, where c is a fixed real number.

6.16. Let V be a unitary space and let A be a self-adjoint linear transformation on V. Show that if there exists a nonzero vector α_0 in V such that $\alpha_0 A = c\alpha_0$ for some scalar c then $c \in \mathsf{R}$.

6.17. Let V be a finite-dimensional unitary space and let A be a linear transformation on V. Show that A is self-adjoint if and only if $(\alpha A \mid \alpha)$ is a real number for every α in V.

6.18. Let V be a finite-dimensional inner-product space and let E be a linear transformation on V such that $E^2 = E$. Prove that E is self-adjoint if and only if $EE^* = E^*E$.

6.19. Let A be a self-adjoint linear transformation on an inner-product space V. Suppose that S is a subspace of V such that $\alpha A \in S$ for all α in S. Show that $S^\perp$ has the same property.

6.20. Show that if A is a linear transformation on an inner-product space and A^* exists then A^{**} exists and is equal to A.

6.21. Let V be a vector space equipped with the inner-product functions f_1 and f_2 having the property that $f_1(\alpha, \beta) = 0$ if and only if $f_2(\alpha, \beta) = 0$. Show that if $A \in \mathscr{L}(V, V)$ has an adjoint A_1^* with respect to f_1 then A has an adjoint A_2^* with respect to f_2 and $A_1^* = A_2^*$.

7. UNITARY AND ORTHOGONAL TRANSFORMATIONS

A mapping T of an inner-product space V into itself is said to *preserve inner products* if

$$(\alpha T \mid \alpha T) = (\alpha \mid \beta) \quad \text{all } \alpha, \beta \text{ in } V.$$

An interest in such transformations of an inner-product space arises in a natural way, since all metric notions are defined in terms of inner products and consequently are preserved by mappings of the kind we have singled out. In more detail we recall that the norm $\|\alpha\|$ of a vector α of an inner product space V is defined by $\|\alpha\| = \sqrt{(\alpha \mid \alpha)}$. If T preserves inner products in V, then

$$\|\alpha T\| = \sqrt{(\alpha T \mid \alpha T)} = \sqrt{(\alpha \mid \alpha)} = \|\alpha\|;$$

hence T preserves norms. Similarly, we find that such a mapping preserves angles and distances between vectors; in particular, it maps an orthonormal set onto an orthonormal set. Of special importance is the fact that an inner-product preserving map is linear. This result appears in our next theorem.

□ **Theorem 7.1.** If T is an inner-product preserving map of an inner-product space V into itself, then

(a) T preserves norms (that is, $\|\alpha T\| = \|\alpha\|$, all α in V) and distances [that is, $d(\alpha T, \beta T) = d(\alpha, \beta)$, all α, β in V];

(b) T is a linear transformation;

(c) T is nonsingular.

PROOF. We have already shown that such a T preserves norms of vectors. To prove that T is linear it is sufficient to show that

$$(a\alpha)T = a(\alpha T) \quad \text{and} \quad (\alpha + \beta)T = \alpha T + \beta T$$

for all scalars a and all α, β in V. To establish the first equality it is sufficient to show that $\|(a\alpha)T - a(\alpha T)\|^2 = 0$. This we now do.

$$\begin{aligned}
\|(a\alpha)T - a(\alpha T)\|^2 &= \|(a\alpha)T\|^2 - \bar{a}((a\alpha)T \mid \alpha T) - a(\alpha T \mid (a\alpha)T) + a\bar{a}\|\alpha T\|^2 \\
&= \|a\alpha\|^2 - \bar{a}(a\alpha \mid \alpha) - a(\alpha \mid a\alpha) + a\bar{a}\,\|\alpha\|^2 \\
&= a\bar{a}\,\|\alpha\|^2 - a\bar{a}\,\|\alpha\|^2 - a\bar{a}\,\|\alpha\|^2 + a\bar{a}\,\|\alpha\|^2 \\
&= 0.
\end{aligned}$$

The proof of the additivity of T is similar and is left as an exercise.

It now follows easily that T preserves distances, for $d(\alpha T, \beta T) = \|\alpha T - \beta T\| = \|(\alpha - \beta)T\| = \|\alpha - \beta\| = d(\alpha, \beta)$.

Finally, T is nonsingular, since if $\alpha T = 0$ then $\|\alpha T\| = \|\alpha\| = 0$, which implies that $\alpha = 0$. ◇

It follows that if T is an inner-product preserving map of a finite-dimensional inner-product space V into itself then T is a linear transformation of V onto V, that is, an isomorphism of V onto V. In contrast, if V is infinite-dimensional, such a T need not map V onto V (Exercise 7.10).

□ **Definition.** Let V be a unitary space and T be a mapping of V onto V which preserves inner products (hence is a linear transformation of V onto V). Then T is called a *unitary transformation* on V. If, instead, V is a euclidean space, then T is called an *orthogonal transformation* on V.

In order to simplify the exposition, we turn our attention to unitary spaces and accompanying unitary transformations. Frequently the method of reasoning used to establish a theorem for the complex case yields a similar conclusion for the real case. Such cases are called to the attention of the reader.

□ **Theorem 7.2.** Let V be a finite-dimensional unitary space. Then the image of an orthonormal basis of V under a unitary transformation is an orthonormal basis. Conversely, if $T \in \mathscr{L}(V, V)$ and T maps some orthonormal basis of V onto an orthonormal basis, then T is a unitary transformation on V.

PROOF. The proof of the first assertion is left as an exercise. For the converse, assume that $\{\alpha_1, \ldots, \alpha_n\}$ is an orthonormal basis of V which is mapped onto an orthonormal basis by T. Then $(\alpha T \mid \beta T) = (\alpha \mid \beta)$ for all α, β in $\{\alpha_1, \ldots, \alpha_n\}$. The linearity of T yields the same conclusion for all α, β in V. Hence T is a unitary transformation. ◇

□ **Theorem 7.3.** A linear transformation T on a unitary space V is unitary if and only if the adjoint T^* of T exists and $TT^* = T^*T = 1$.

PROOF. Assume that T^* exists and $TT^* = T^*T = 1$. Then T has an inverse, namely T^*, hence T is an isomorphism of V onto V. Moreover,

$$(\alpha T \mid \beta T) = (\alpha \mid \beta(TT^*)) = (\alpha \mid \beta)$$

for all α, β, and so T preserves inner products.

Conversely, suppose that T is unitary. Then T has an inverse and

$$(\alpha T \mid \beta) = (\alpha T \mid \beta(T^{-1}T)) = (\alpha T \mid (\beta T^{-1})T) = (\alpha \mid \beta T^{-1})$$

for all α, β in V; hence T^{-1} is the adjoint of T. ◇

□ **Corollary 1.** A linear transformation T on a euclidean space is orthogonal if and only if the adjoint T^* of T exists and $TT^* = T^*T = 1$.

□ **Corollary 2.** Let T be a linear transformation on a finite-dimensional unitary space V and let T^* be its adjoint. Then T is unitary if and only if $TT^* = 1$.

We derive next the characteristic property of matrices of unitary transformations relative to orthonormal bases. Suppose that V is an n-dimensional unitary space, that $\mathscr{B} = (\alpha_1, \ldots, \alpha_n)$ is an orthonormal basis of V, and that (a_{ij}) is the matrix of the linear transformation T on V relative to $\mathscr{B}$. Then we know that $(a_{ij})^*$ is the matrix of T^* relative to $\mathscr{B}$ (see Section 6). In view of Corollary 2, we conclude that T is unitary if and only if

$$(a_{ij})(a_{ij})^* = I.$$

□ **Definition.** An $n \times n$ matrix (a_{ij}) over C is called *unitary* if and only if $(a_{ij})(a_{ij})^* = I$. An $n \times n$ matrix (a_{ij}) over R is called *orthogonal* if $(a_{ij})(a_{ij})^t = I$.

Now we may state our findings as follows.

□ **Theorem 7.4.** Let T be a linear transformation on a finite dimensional unitary space V. Then T is unitary if and only if the matrix of T relative to some orthonormal basis is unitary. Similarly, a linear transformation on a finite-dimensional euclidean space is orthogonal if and only if its matrix relative to some orthonormal basis is orthogonal.

The defining property $(a_{ij})(a_{ij})^* = I$ of a unitary matrix amounts to the condition

$$\sum_{k=1}^{n} a_{ik}\bar{a}_{jk} = \delta_{ij}, \qquad 1 \le i, j \le n,$$

or, in other words, the row vectors of (a_{ij}), regarded as elements of C^n with standard inner product, form an orthonormal set. It follows from the matrix interpretation of the result in Exercise 3.16 that if an $n \times n$ matrix (a_{ij}) has a right inverse (b_{ij}) [that is, $(a_{ij})(b_{ij}) = I$] then $(b_{ij})(a_{ij}) = I$. Thus, if (a_{ij}) is unitary, then $(a_{ij})^*(a_{ij}) = I$. This means that the column vectors of (a_{ij}) also form an orthonormal set.

EXAMPLES

7.1. The matrix of the linear transformation A on R^2 defined by

$$(x, y)A = (x \cos\theta + y \sin\theta,\ -x \sin\theta + y \cos\theta)$$

is

$$\begin{bmatrix} \cos\theta & -\sin\theta \\ \sin\theta & \cos\theta \end{bmatrix}$$

relative to the standard basis. It is obvious that the rows of this matrix form an orthonormal set. Hence A is an orthogonal transformation.

7.2. The matrix

$$\begin{bmatrix} \dfrac{1}{\sqrt{3}} & -\dfrac{1}{\sqrt{6}} & -\dfrac{1}{\sqrt{2}} \\ \dfrac{1}{\sqrt{3}} & \dfrac{2}{\sqrt{6}} & 0 \\ \dfrac{1}{\sqrt{3}} & -\dfrac{1}{\sqrt{6}} & \dfrac{1}{\sqrt{2}} \end{bmatrix}$$

is orthogonal, If V is a three-dimensional euclidean space and $\mathscr{B}$ is an orthonormal basis of V, then the linear transformation A induced by this matrix relative to $\mathscr{B}$ is orthogonal.

7.3. The matrix

$$\frac{1}{\sqrt{12}} \begin{bmatrix} 2 & 1 - \sqrt{3}\,i & -1 - \sqrt{3}\,i \\ -2i & \sqrt{3} - i & \sqrt{3} + i \\ -2 & 2 & -2 \end{bmatrix}$$

is unitary. If V is a three-dimensional unitary space and $\mathscr{B}$ is an orthonormal basis of V, then the linear transformation induced by this matrix relative to $\mathscr{B}$ is unitary.

7.4. Let A be the linear transformation on R^2 (with standard inner product), which is induced by the orthogonal matrix

$$\begin{bmatrix} 0 & 1 \\ -1 & 0 \end{bmatrix}$$

relative to the basis $\mathscr{B} = ((1, 1), (0, 1))$ of R^2. Since $[(x, y)]_{\mathscr{B}} = (x, y - x)$,

$$[(x, y)A]_{\mathscr{B}} = (x, y - x)\begin{bmatrix} 0 & 1 \\ -1 & 0 \end{bmatrix} = (x - y, x).$$

Thus $(x, y)A = (x - y)(1, 1) + x(0, 1) = (x - y, 2x - y)$. Although A is induced by an orthogonal matrix, A is not an orthogonal transformation, since $\mathscr{B}$ is not an orthonormal basis, according to Theorem 7.4. This conclusion is easily verified; for instance, $\|(1, 0)\| = 1$, $(1, 0)A = (1, 2)$, and $\|(1, 2)\| = \sqrt{5}$.

7.5. Let V be the space of all real-valued, integrable functions in $[0, 1]$ with inner product given by

$$(f \mid g) = \int_0^1 f(x)\, g(x)\, dx.$$

One element of V is the function k such that

$$k(x) = \begin{cases} -1, & 0 \le x < \frac{1}{2}, \\ 1, & \frac{1}{2} \le x \le 1. \end{cases}$$

Then the function A on V into V such that $fA = kf$ for all f in V is an orthogonal transformation on V. Indeed, since $(fA \mid g) = (f \mid gA)$, A^* exists and is equal to A. Further, $A^2 = 1$ and so $AA^* = 1 = A^*A$.

With applications to physical problems in mind, it is natural to consider distance-preserving transformations in inner-product spaces. A mapping h of an inner-product space V onto itself that preserves distances between vectors is called a *rigid motion* of V. Here we consider rigid motions of unitary spaces. In Chapter 8 we discuss them for euclidean spaces.

According to Theorem 7.1, a unitary transformation on a unitary space V is a rigid motion. Recalling the definition of a translation of a vector space given in Section 2.5, we note that it is clear that translations of a unitary space are rigid motions. Intuitively, it is also clear that the composite of rigid motions is a rigid motion; hence a unitary transformation followed by a translation is a rigid motion. This result and its converse are proved below.

□ **Theorem 7.5.** A mapping h of a unitary space V onto itself is a rigid motion if and only if $h = AT$, where A is a unitary transformation and T is a translation.

PROOF. If $h = AT$, where A is a unitary transformation on V and T is the translation of V such that $\alpha T = \alpha + \alpha_0$, then

$$\|\alpha h - \beta h\| = \|\alpha A + \alpha_0 - \beta A - \alpha_0\| = \|\alpha A - \beta A\| = \|\alpha - \beta\| .$$

Thus h is a rigid motion.

Conversely, assume that h is a rigid motion of V. Let $\alpha_0 = 0h$ and choose T to be the translation of V such that $\alpha T = \alpha + \alpha_0$. Next, define the mapping A on V into V by

$$\alpha A = \alpha h - \alpha_0 \qquad \text{all } \alpha \text{ in } V.$$

Then $0A = 0h - \alpha_0 = 0$ and $\|\alpha A - \beta A\| = \|\alpha h - \beta h\| = \|\alpha - \beta\|$; so that A preserves inner products (prove this). Since, in addition, A maps V onto V, A is unitary. Finally, since $h = AT$, the proof is complete. ◇

EXERCISES

7.1. Show that each of the following matrices is orthogonal:

$$\frac{1}{3}\begin{bmatrix} 2 & 2 & 1 \\ 1 & -2 & 2 \\ 2 & -1 & -2 \end{bmatrix}, \quad \frac{1}{2}\begin{bmatrix} 1 & 1 & 1 & 1 \\ 1 & 1 & -1 & -1 \\ 1 & -1 & 1 & -1 \\ 1 & -1 & -1 & 1 \end{bmatrix}.$$

7.2. Formulate a definition of the inverse of an $n \times n$ matrix and then find the inverse of each matrix in Exercise 7.1.

7.3. Find the inverse of each of the following matrices:

$$\begin{bmatrix} 1 & 2 & 2 \\ 2 & 1 & -2 \\ -2 & 2 & -1 \end{bmatrix}, \quad \begin{bmatrix} 2 & 1-\sqrt{3}i & -1-\sqrt{3}i \\ -2i & \sqrt{3}-i & \sqrt{3}+i \\ -2 & 2 & 2 \end{bmatrix}.$$

7.4. For each of the matrices (a_{ij}) find an orthogonal matrix (p_{ij}) such that $(p_{ij})(a_{ij}) = (d_{ij})(p_{ij})$, where (d_{ij}) has the form

$$(d_{ij}) = \begin{bmatrix} d_1 & 0 \\ 0 & d_2 \end{bmatrix}.$$

(a) $(a_{ij}) = \begin{bmatrix} 1 & -1 \\ -1 & 2 \end{bmatrix}$, (b) $(a_{ij}) = \begin{bmatrix} 0 & i \\ -i & 0 \end{bmatrix}$

7.5. Show that the matrix

$$\begin{bmatrix} -\sin^2 \varphi + i \cos^2 \varphi & (1+i) \sin \varphi \cos \varphi \\ (1+i) \sin \varphi \cos \varphi & -\cos^2 \varphi + i \sin^2 \varphi \end{bmatrix}$$

is unitary for all φ in R. Suppose that it defines the transformation T on C^2 relative to the standard basis. What is the image of the orthonormal basis $\{(0, i), (-1, 0)\}$ of C^2 under T?

7.6. Show that the $n \times n$ matrix whose (r, s)th entry is $e^{i(s-1)(r-1)\theta}/\sqrt{n}$, where $\theta = 2\pi/n$, is unitary.

7.7. Show that an $n \times n$ matrix (a_{ij}) over R is an orthogonal matrix if and only if it is the matrix of an ordered orthonormal basis $\mathscr{B}$ of an n-dimensional euclidean space over R relative to an ordered orthonormal basis $\mathscr{B}'$ of the same space.

7.8. Let A be the linear transformation on R^3 with the standard inner product induced by the orthogonal matrix

$$\begin{bmatrix} \frac{1}{\sqrt{2}} & 0 & \frac{1}{\sqrt{2}} \\ \frac{-1}{\sqrt{2}} & 0 & \frac{1}{\sqrt{2}} \\ 0 & 1 & 0 \end{bmatrix}$$

relative to the standard basis.
(a) Find a nonzero vector $\alpha_0 \in R^3$ such that $\alpha_0 A = -\alpha_0$.
(b) If $S = [\alpha_0]$, show that for all $\beta \in S^\perp$, $\beta A \in S^\perp$.
(c) Find an orthonormal basis $(\alpha_1, \alpha_2, \alpha_3)$ of R^3, where $\alpha_1 \in S$ and $\alpha_2, \alpha_3 \in S^\perp$, relative to which the matrix of A has the form

$$\begin{bmatrix} -1 & 0 & 0 \\ 0 & \cos\theta & -\sin\theta \\ 0 & \sin\theta & \cos\theta \end{bmatrix}$$

for some real number θ.

7.9. Complete the proof of part (b) of Theorem 7.1.

7.10. Let V be the space of all integrable functions in [0, 1] with the inner product given by

$$(f \mid g) = \int_0^1 f(x)\, g(x)\, dx.$$

The function k such that

$$k(x) = \begin{cases} 1, & x \neq \frac{1}{2} \\ 0, & x = \frac{1}{2} \end{cases}$$

is in V. Prove that the mapping A of V into V such that $fA = kf$ preserves inner products but is not an orthogonal transformation on V (by showing that the range of A is not V).

7.11. Prove that if A and B are unitary transformations of V, then A^{-1} and AB are also unitary transformations of V.

7.12. Let V be an inner-product space and A a mapping of V into V which preserves inner products. Show that if α is a nonzero vector such that $\alpha A = c\alpha$ for some scalar c then $|c| = 1$.

7.13. Find a mapping T of R^2 (with standard inner product) into R^2 such that $\|\alpha T\| = \|\alpha\|$ for all α in R^2 but T is not an orthogonal transformation.

7.14. Show that if a linear transformation A on a unitary space preserves norms of vectors then A preserves inner products.

7.15. Show that if a linear transformation A on a finite-dimensional inner-product space has any two of the following three properties then it has the third: A is self-adjoint, A is unitary (or orthogonal), $A^2 = I$.

7.16. Let S be a subspace of the finite-dimensional euclidean space V. Then each $\alpha \in V$ is expressible uniquely in the form $\alpha = \beta + \gamma$, where $\beta \in S$ and $\gamma \in S^{\perp}$. Define the mapping A on V into V by

$$\alpha A = \beta - \gamma \qquad \text{all } \alpha \in V.$$

(a) Prove that A is both self-adjoint and orthogonal.
(b) Let $V = \mathsf{R}^3$ with the standard inner product and $S = [(1, 1, 1)]$. Find the matrix of A relative to the standard basis.

5 Matrices

For some, matrices have a life of their own, that is, an existence apart from representing linear transformations. This chapter is concerned with properties of matrices that are of importance from at least one of these two possible points of view. In Section 1 we reconcile the use of the term "rank" in connection with linear transformations with that in connection with matrices. In Section 2 we discuss the relationship that exists between linear transformations defined by a given matrix and the relation between matrices of a given linear transformation relative to different bases of underlying space. Consequences of performing elementary row operations on a matrix by multiplying it by suitable matrices is treated in Section 3. Special types of matrices that play an important role in the study of linear transformations are described in Section 4. Finally, in Section 5 we have attempted to give an account of the elementary theory of determinants that is both brief and intelligible.

1. RANK

The term "rank" has been employed in connection with linear transformations (Section 4.2) and matrices (Exercise 4.5.7). In this section we show that (i) all linear transformations induced by a given matrix have the same rank and this common value is the rank of the matrix, and (ii) all matrices of a given linear transformation have the same rank and this common value is the rank of the linear transformation.

To cope with the problem at hand we establish a preliminary result. Consider four finite-dimensional vector spaces V, V', W, and W' over a common field F such that $\dim V = \dim V'$ and $\dim W = \dim W'$. Let $A \in \mathscr{L}(V, W)$ and $A' \in \mathscr{L}(V', W')$ and assume that there exist nonsingular linear transformations P in $\mathscr{L}(V', V)$ and Q in $\mathscr{L}(W, W')$ such that

$$A' = PAQ.$$

(The diagram below may be of assistance in keeping everything in mind).

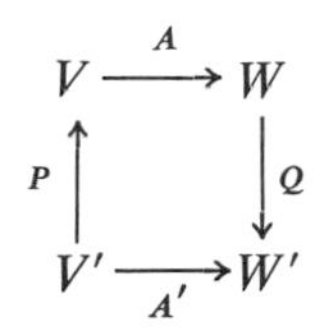

It then follows, as we shall prove, that A and A' have equal ranks. First we show that $N_A = (N_{A'})P$. Suppose that $\alpha' \in N_{A'}$. Then, in turn, $\alpha'A' = ((\alpha'P)A)Q = 0$, $(\alpha'P)A = 0$ (since Q is nonsingular), $\alpha'P \in N_A$, and $(N_{A'})P \subseteq N_A$. Conversely, let $\alpha \in N_A$. Since P is an isomorphism of V' onto V, there exists an α' in V' such that $\alpha'P = \alpha$. Then, in turn, $0 = \alpha A = (\alpha'P)A$, $0 = (\alpha'PA)Q = \alpha'A'$, $\alpha' \in N_{A'}$, $\alpha \in (N_{A'})P$, and $N_A \subseteq (N_{A'})P$. Hence $N_A = (N_{A'})P$, which implies that $\dim N_A = \dim N_{A'}$, since P is nonsingular (see Theorem 2.2.3). Finally, in view of Theorem 4.2.1, we conclude that A and A' have equal ranks.

Now let (a_{ij}) be an $n \times m$ matrix over a field F. Further, let V and W be vector spaces over F of dimensions n and m, respectively, and suppose that $\mathscr{B}_1 = (\alpha_1, \ldots, \alpha_n)$ and $\mathscr{B}_2 = (\beta_1, \ldots, \beta_m)$ are ordered bases of V and W. If A is the linear transformation of V into W induced by (a_{ij}) relative to $(\mathscr{B}_1, \mathscr{B}_2)$, then (see Example 4.1.2)

$$\alpha_i A = \sum_{j=1}^{m} a_{ij}\beta_j, \qquad 1 \le i \le n.$$

We contend that the rank of A is determined solely by (a_{ij}); that is, it is independent of our choice of V and W (within the obvious limitations on their dimensions) and their respective ordered bases. To prove this let V' and W' be vector spaces over F such that $\dim V' = n$ and $\dim W' = m$. Further, let $\mathscr{B}'_1 = (\alpha'_1, \ldots, \alpha'_n)$ and $\mathscr{B}'_2 = (\beta'_1 \ldots, \beta'_m)$ be ordered bases of V' and W', respectively. Let A' be the element of $\mathscr{L}(V', W')$ induced by (a_{ij}) relative to $(\mathscr{B}'_1, \mathscr{B}'_2)$. Now define $P \in \mathscr{L}(V', V)$ and $Q \in \mathscr{L}(W, W')$ by

$$\alpha'_i P = \alpha_i, \qquad 1 \le i \le n \quad \text{and} \quad \beta_j Q = \beta'_j, \qquad 1 \le j \le m.$$

Then P and Q (which exist by Theorem 4.3.6) are nonsingular and, since

$$\alpha_i'(PAQ) = (\alpha_i A)Q = \left(\sum_{j=1}^{m} a_{ij}\beta_j\right)Q = \sum_{j=1}^{m} a_{ij}\beta_j' = \alpha_i' A',$$

we have $A' = PAQ$. Thus we have an instance of the circumstances discussed above and so we may infer that A and A' have equal ranks.

It follows that to determine the rank of a linear transformation induced by (a_{ij}) we may choose the linear transformation A on F^n into F^m induced by (a_{ij}) relative to the pair $(\mathscr{B}_1, \mathscr{B}_2)$, where $\mathscr{B}_1 = (\epsilon_i, \ldots, \epsilon_n)$ and $\mathscr{B}_2 = (\epsilon_1', \ldots, \epsilon_m')$ are the standard bases of F^n and F^m, respectively. This A is the linear transformation such that

$$\epsilon_i A = a_{i1}\epsilon_1' + \cdots + a_{im}\epsilon_m' = (a_{i1}, \ldots, a_{im}), \qquad 1 \le i \le n.$$

Now the rank of A is the dimension of the subspace of F^m spanned by $\{\alpha_1, \ldots, \alpha_n\}$, where $\alpha_i = (a_{i1}, \ldots, a_{im})$, $1 \le i \le n$; but $[\alpha_1, \ldots, \alpha_n]$ is the row space of (a_{ij}) and by definition its dimension is the row rank, that is, the rank, of (a_{ij}). We summarize our findings in the following theorem:

□ **Theorem 1.1.** All linear transformations induced by a matrix have the same rank; the common value is the rank of the matrix.

EXAMPLES

1.1. Suppose that A is a linear transformation induced by the matrix

$$(a_{ij}) = \begin{bmatrix} 1 & 2 & 0 & 1 \\ -1 & 1 & -1 & 0 \\ 2 & 3 & 4 & 5 \end{bmatrix}$$

over R. Applying a few row operations to (a_{ij}), we find that its row rank is 3. Hence the rank of A is 3.

1.2. Let A be the linear transformation R^3 into R^3 such that

$$(x, y, z)A = (x + z, 2x + 3y + 5z, 4y + 4z).$$

The matrix of A relative to the standard basis of R^3 is

$$\begin{bmatrix} 1 & 2 & 0 \\ 0 & 3 & 4 \\ 1 & 5 & 4 \end{bmatrix}$$

and has row rank 2; hence the rank of A is 2.

Turning matters around, suppose that V and W are finite-dimensional vector spaces over a field F and that A is a given linear transformation on V into W. Let (a_{ij}) be the matrix of A relative to some pair of bases of V and W. Then A is the linear transformation induced by (a_{ij}) relative to that pair of bases. According to the foregoing, the rank of (a_{ij}) is equal to that of A. Thus we have proved the following result.

□ **Theorem 1.2.** All matrices of a given linear transformation (on a finite-dimensional vector space into a vector space) have the same rank; the common value is the rank of the transformation.

EXAMPLE

1.3. The matrix of the linear transformation of Example 1.2 relative to the basis ((1, 1, 0), (−2, 0, 1), (0, 1, 1)) is

$$\begin{bmatrix} 1 & 0 & 4 \\ -5 & -2 & 6 \\ -1 & -1 & 9 \end{bmatrix}.$$

Since A has rank 2, so has this matrix.

We conclude this section with several definitions. Suppose that V is an n-dimensional vector space over a field F with an ordered basis $\mathscr{B}$, that $A \in \mathscr{L}(V, V)$ and that $(a_{ij}) = [A]_{\mathscr{B}}$. The statements "$A$ has an inverse," "A is nonsingular," and "A has rank n" are equivalent. Suppose that one, and hence all, are true of A and that $(b_{ij}) = [A^{-1}]_{\mathscr{B}}$. Then, since $AA^{-1} = A^{-1}A = 1$,

$$(a_{ij})(b_{ij}) = (b_{ij})(a_{ij}) = I.$$

An $n \times n$ matrix (a_{ij}) over a field F for which there exists a matrix (b_{ij}) over F such that the above equations hold is called *invertible* or *nonsingular* and (b_{ij}) is called an *inverse* of (a_{ij}). If an inverse of (a_{ij}) exists, which is the case if and only if (a_{ij}) has rank n, then clearly it is uniquely determined and we shall denote it by

$$(a_{ij})^{-1}.$$

EXERCISES

1.1. Find the rank of each of the following matrices:

(a) $\begin{bmatrix} -1 & 2 & 3 \\ 0 & 1 & 4 \\ 7 & 2 & 1 \end{bmatrix}$ over R; (b) $\begin{bmatrix} -1 & 1 & 2 \\ 3 & 1-i & 2 \\ 0 & 1 & 5i \end{bmatrix}$ over C;

(c) $\begin{bmatrix} -1 & 4 & 6 & 2 \\ 5 & -2 & 1 & 3 \\ 0 & 1 & 0 & 1 \\ -4 & 1 & -3 & 0 \end{bmatrix}$ over Z_7 (see Example 1.4.4).

1.2. What is the rank of the linear transformation A on V into W if

(a) $V = \mathsf{R}^3$, $W = \mathsf{R}^4$, and
$(x, y, z)A = (2x - y + z, x + y + z, x - 2z, 3x - y)$?

(b) $V = \mathsf{R}^3$, $W = P_3(\mathsf{R})$, and
$(a, b, c)A = (2a - b) + (a + b + c)x + (a + 2b)x^2 + (3a - b)x^3$?

(c) $V = C[a, b]$, $W = \mathsf{R}$, and

$$fA = \int_a^b f(x)\, dx?$$

1.3. Let (a_{ij}) be an $n \times n$ matrix over C. Show that the conjugate transpose $(a_{ij})^*$ of (a_{ij}) has the same rank as (a_{ij}).

1.4. (a) Show that

$$\begin{bmatrix} 1 & 0 & 1 \\ 0 & 1 & 1 \end{bmatrix} \quad \text{and} \quad \begin{bmatrix} -1 & 2 & 0 \\ 5 & 4 & 1 \end{bmatrix}$$

have the same rank.

(b) Find an invertible 2×2 matrix (p_{ij}) over R and an invertible 3×3 matrix (q_{ij}) over R such that

$$(p_{ij}) \begin{bmatrix} 1 & 0 & 1 \\ 0 & 1 & 1 \end{bmatrix} (q_{ij}) = \begin{bmatrix} -1 & 2 & 0 \\ 5 & 4 & 1 \end{bmatrix}$$

1.5. Show that there does not exist an invertible 2×2 matrix (a_{ij}) such that

$$\begin{bmatrix} 0 & 1 \\ 0 & 0 \end{bmatrix} = (a_{ij})^{-1} \begin{bmatrix} 1 & 1 \\ 0 & 0 \end{bmatrix} (a_{ij}).$$

1.6. Let $A, B \in \mathscr{L}(V, V)$, where V is a vector space over a field F. Show that if the rank of $A + B$ is equal to the sum of the ranks of A and B then

(a) $R_A \cap R_B = \{0\}$ and $R_A + R_B = R_{A+B}$;

(b) if $AB = BA$, then $AB = 0$.

1.7. Let V and W be vector spaces over a common field F of dimensions n and m, respectively. Let $A \in \mathscr{L}(V, W)$. Show that there exist ordered bases $\mathscr{B}$ and $\mathscr{B}'$ of V and W, respectively, such that, if $A \neq 0$, the matrix (a_{ij}) of A relative to $(\mathscr{B}, \mathscr{B}')$ has the following form: $a_{11} = \cdots = a_{rr} = 1$ and all other entries are zeros for some $r \leq \min(n, m)$.

REMARK. Hereafter we shall indicate a matrix of this form by

$$\begin{bmatrix} I_r & O \\ O & O \end{bmatrix}$$

1.8. Let (a_{ij}) be an $n \times m$ matrix over a field F of rank r. Show that there exists a nonsingular $n \times n$ matrix (p_{ij}) over F and a nonsingular $m \times m$ matrix (q_{ij}) over F such that

$$(p_{ij})(a_{ij})(q_{ij}) = \begin{bmatrix} I_r & O \\ O & O \end{bmatrix}$$

1.9. Let (a_{ij}) and (b_{ij}) be $n \times m$ matrices over a field F. Then (a_{ij}) is said to be *equivalent* to (b_{ij}) if and only if there exist nonsingular matrices (p_{ij}) and (q_{ij}) over F such that

$$(a_{ij}) = (p_{ij})(b_{ij})(q_{ij}).$$

Show that
(a) equivalence is a symmetric relation in the set of all $n \times m$ matrices over F;
(b) two $n \times m$ matrices over F are equivalent if and only if they have the same rank.

1.10. Show that

$$\begin{bmatrix} 0 & 1 \\ 0 & 0 \end{bmatrix} \quad \text{and} \quad \begin{bmatrix} 1 & 1 \\ 0 & 0 \end{bmatrix}$$

are equivalent matrices and find 2×2 nonsingular matrices (p_{ij}) and (q_{ij}) such that

$$\begin{bmatrix} 0 & 1 \\ 0 & 0 \end{bmatrix} = (p_{ij}) \begin{bmatrix} 1 & 1 \\ 0 & 0 \end{bmatrix} (q_{ij}).$$

2. SIMILAR LINEAR TRANSFORMATIONS AND MATRICES

Let us consider a special case of the discussion in Section 1 which culminated in the conclusion that if A, A' are linear transformations induced by the same matrix then $A' = PAQ$. Instead of treating the special case as such, it is simpler to start afresh. Suppose that (a_{ij}) is an $n \times n$ matrix over a field F, that V is an n-dimensional vector space over F, and that $\mathscr{B}_1 = (\alpha_1, \ldots, \alpha_n)$, $\mathscr{B}_2 = (\beta_1, \ldots, \beta_n)$ are ordered bases of V. The question we pose is this: If (a_{ij}) induces the linear transformation A on V relative to the $\mathscr{B}_1$-basis and the linear transformation A' on V relative to the $\mathscr{B}_2$-basis, how can their relationship be formulated explicitly? Let Q be the linear transformation on V such that

$$\alpha_i Q = \beta_i, \qquad 1 \le i \le n.$$

Then Q is nonsingular and $\beta_i Q^{-1} = \alpha_i$, $1 \le i \le n$. Since, by definition

$$\alpha_i A = \sum_{j=1}^{n} a_{ij}\,\alpha_j \quad \text{and} \quad \beta_i A' = \sum_{j=1} a_{ij}\beta_j, \qquad 1 \le i \le n,$$

$$\alpha_i(Q A') = \beta_i A' = \sum_{j=1}^{n} a_{ij}\beta_j = \sum_{j=1}^{n} a_{ij}(\alpha_j Q) = \alpha_i(AQ)$$

for $1 \le i \le n$. Hence QA' and AQ agree on a basis of V and this implies that $QA' = AQ$ or $A' = Q^{-1}AQ$.

□ **Definition.** If V is a vector space and $A, A' \in \mathscr{L}(V, V)$, then A' is called *similar* to A if and only if $A' = Q^{-1}AQ$ for some $Q \in \mathscr{L}(V, V)$.

Since similarity is a symmetric relation [if $A' = Q^{-1}AQ$, then $A = (Q^{-1})^{-1}AQ^{-1}$], we may speak simply of the similarity of two transformations. This calculation is a proof of one part of the following result.

□ **Theorem 2.1.** If V is a finite-dimensional vector space and A, A' are linear transformations on V, then A and A' are similar if and only if they can be induced by the same matrix.

PROOF. It remains to prove that if A and A' are similar then they are induced by the same matrix. Assume that $A' = P^{-1}AP$ for some $P \in \mathscr{L}(V, V)$. Let $\mathscr{B}_1 = (\alpha_1, \ldots, \alpha_n)$ be an ordered basis of V, then $\mathscr{B}_2 = (\alpha_1 P, \ldots, \alpha_n P)$ is an ordered basis of V, since P is invertible. If

$$\alpha_i A = \sum_{j=1}^{n} a_{ij}\,\alpha_j, \qquad 1 \le i \le n,$$

then

$$(\alpha_i P)A' = \alpha_i(PA') = \alpha_i(AP) = (\alpha_i A)P = \sum_{j=1}^{n} (a_{ij}\,\alpha_j)P = \sum_{j=1}^{n} a_{ij}(\alpha_j P);$$

that is, A is the linear transformation induced by (a_{ij}) relative to the $\mathscr{B}_1$-basis and A' is the linear transformation induced by the same matrix relative to the $\mathscr{B}_2$-basis. ◇

Our theorem implies that properties shared by a class of linear transformations, each of which is similar to a given linear transformation A in a finite-dimensional vector space V, are determined by the properties of the matrix of A relative to any basis of V.

EXAMPLES

2.1. Let $V = P_\infty(\mathsf{R})$ and D be the derivative operator in V. The mapping P of V into V such that $f(x)P = f(x - c)$ for a fixed real number c is an isomorphism of V onto V. Thus $P^{-1}DP$ and D are similar linear transformations.

2.2. We define the linear transformations A and A' on R^3 into R^3 as follows:

$$(x, y, z)A = (x + z, y - z, -x),$$

$$(x, y, z)A' = \tfrac{1}{2}(x - y + z, 2y, -3x + y + z).$$

Since A is induced by the matrix

$$\begin{bmatrix} 1 & 0 & -1 \\ 0 & 1 & 0 \\ 1 & -1 & 0 \end{bmatrix},$$

relative to the standard basis of R^3 and A' is induced by the same matrix relative to the basis ((1, 1, 0), (0, 1, 1), (1, 0, 1)), they are similar linear transformations.

Turning matters around, we investigate this question next: If $\mathscr{B}$ and $\mathscr{B}'$ are ordered bases of an n-dimensional vector space V and A is a linear transformation on V, how are the matrices $[A]_{\mathscr{B}}$ and $[A]_{\mathscr{B}'}$ related? Let (p_{ij}) be the matrix of $\mathscr{B}$ to $\mathscr{B}'$ (see Section 2.4). Then (p_{ij}) is invertible and for vectors α and αA of V

$$[\alpha]_{\mathscr{B}'} = [\alpha]_{\mathscr{B}}(p_{ij}), \tag{1}$$

$$[\alpha A]_{\mathscr{B}'} = [\alpha A]_{\mathscr{B}}(p_{ij}). \tag{2}$$

Further, by definition, we have

$$[\alpha A]_{\mathscr{B}} = [\alpha]_{\mathscr{B}}[A]_{\mathscr{B}}. \tag{3}$$

From (2) and (3) we deduce that

$$[\alpha A]_{\mathscr{B}'} = [\alpha]_{\mathscr{B}}[A]_{\mathscr{B}}(p_{ij})$$

and, since (1) implies that $[\alpha]_{\mathscr{B}} = [\alpha]_{\mathscr{B}'}(p_{ij})^{-1}$, we conclude that

$$[\alpha A]_{\mathscr{B}'} = [\alpha]_{\mathscr{B}'}(p_{ij})^{-1}[A]_{\mathscr{B}}(p_{ij}).$$

Thus an answer to our question is

$$[A]_{\mathscr{B}'} = (p_{ij})^{-1}[A]_{\mathscr{B}}(p_{ij}).$$

If (a_{ij}) and (b_{ij}) are two $n \times n$ matrices over a field F, then (b_{ij}) is called *similar* to (a_{ij}) if and only if there exists a nonsingular matrix (p_{ij}) over F such that

$$(b_{ij}) = (p_{ij})^{-1}(a_{ij})(p_{ij}).$$

Since similarity of matrices is a symmetric relation, we may speak simply of the similarity of two matrices. Thus our computation above may be summarized by the statement that any two matrices of a linear transformation are similar.

Turning to the converse, suppose that (a_{ij}) and (b_{ij}) are similar $n \times n$ matrices over a field F. Then, for some nonsingular matrix (p_{ij}), $(b_{ij}) = (p_{ij})^{-1}(a_{ij})(p_{ij})$. Thus $(p_{ij})(b_{ij}) = (a_{ij})(p_{ij})$, which is equivalent to the following set of relations

$$(4) \qquad \sum_{k=1}^{n} p_{ik} b_{kj} = \sum_{k=1}^{n} a_{ik} p_{kj}, \qquad 1 \leq i, j \leq n.$$

Let V be an n-dimensional vector space over F having $\mathscr{B} = (\alpha_1, \ldots, \alpha_n)$ as an ordered basis. The nonsingular matrix (p_{ij}) induces a nonsingular linear transformation P on V relative to $\mathscr{B}$ and, consequently, if $\beta_i = \alpha_i P$, $1 \leq i \leq n$, then $\mathscr{B}' = (\beta_1, \ldots, \beta_n)$, where

$$\beta_k = \sum_{j=1}^{n} p_{kj}\alpha_j, \qquad 1 \leq k \leq n$$

is an ordered basis of V. We shall show that the linear transformation A on V induced by (b_{ij}) relative to the $\mathscr{B}$-basis is equal to the linear transformation A' on V induced by (a_{ij}) relative to the $\mathscr{B}'$-basis. By definition

$$\alpha_k A = \sum_{j=1}^{n} b_{kj}\alpha_j \quad \text{and} \quad \beta_k A' = \sum_{j=1}^{n} a_{kj}\beta_j, \qquad 1 \leq k \leq n,$$

so that on the one hand

$$\beta_i A = \left(\sum_{k=1}^{n} p_{ik}\alpha_k\right) A = \sum_{k=1}^{n} p_{ik}\left(\sum_{j=1}^{n} b_{kj}\alpha_j\right) = \sum_{j=1}^{n}\left(\sum_{k=1}^{n} p_{ik} b_{kj}\right)\alpha_j$$

and on the other

$$\beta_i A' = \sum_{k=1}^{n} a_{ik}\beta_k = \sum_{k=1}^{n} a_{ik}\left(\sum_{j=1}^{n} p_{kj}\alpha_j\right) = \sum_{j=1}^{n}\left(\sum_{k=1}^{n} a_{ik}p_{kj}\right)\alpha_j .$$

In view of (4), it follows that A and A' agree on a basis of V and, hence, are equal or, in other words, that (a_{ij}) and (b_{ij}) are matrices of a single linear transformation. We summarize our findings in the next theorem.

□ **Theorem 2.2.** Two $n \times n$ matrices over a field F are similar if and only if they are matrices of some linear transformation on an n-dimensional vector space over F.

We conclude this section with a definition and a theorem for linear transformations on a unitary space and give the matrix analogue of each. The reader can extend our remarks to the case of euclidean spaces by replacing each occurrence of the word "unitary" by the word "orthogonal."

□ **Definition** Let V be a unitary vector space and A, $A' \in \mathscr{L}(V, V)$. Then A and A' are called *unitarily similar* if and only if there exists a unitary linear transformation P on V such that $A' = P^{-1}AP$. Two $n \times n$ matrices (c_{ij}) and (c'_{ij}) over C are called *unitarily similar* if and only if there exists a unitary matrix (p_{ij}) such that

$$(c'_{ij}) = (p_{ij})^{-1}(c_{ij})(p_{ij})$$

or, what is the same,

$$(c'_{ij}) = (p_{ij})^{*}(c_{ij})(p_{ij}).$$

The proofs of the following theorems are left as exercises.

□ **Theorem 2.3.** Two linear transformations on a finite-dimensional unitary space are unitarily similar if and only if they are transformations induced by a matrix relative to two orthonormal bases of the space.

□ **Theorem 2.4.** Two $n \times n$ matrices over C are unitarily similar if and only if they are the matrices of a linear transformation on an n-dimensional unitary space relative to two orthonormal bases of the space.

EXERCISES

2.1. Show that if A and B are similar linear transformations then so are A^n and B^n for each positive integer n.

2.2. Let A and B be similar linear transformations in a vector space V over a field F. Show that
(a) if there exists a nonzero vector α in V such that $\alpha A = c\alpha$ for some scalar c then there exists a nonzero vector β in V such that $\beta B = c\beta$ for the same c;
(b) if $a_k A^k + \cdots + a_1 A + a_0 = 0$, where the a_i's are elements of F, then $a_k B^k + \cdots + a_1 B + a_0 = 0$.

2.3. Let A and B be similar linear transformations on a finite-dimensional unitary space.
(a) If A is self-adjoint, is B self-adjoint?
(b) If A is unitary, is B unitary?
(c) If A is idempotent, is B idempotent?

2.4. (a) Show that if A and B are linear transformations on the same vector space and one of them is invertible then AB and BA are similar.
(b) Does the conclusion of (a) remain valid if neither A nor B is invertible?

2.5. (a) Show that similar linear transformations on a finite-dimensional vector space have the same rank.
(b) Show that the matrices

$$\begin{bmatrix} 1 & 0 \\ 0 & 0 \end{bmatrix} \quad \text{and} \quad \begin{bmatrix} 0 & 1 \\ 0 & 0 \end{bmatrix}$$

have the same rank but are not similar.

2.6. Show that the matrices

$$\begin{bmatrix} 0 & 1 \\ -1 & 0 \end{bmatrix} \quad \text{and} \quad \begin{bmatrix} i & 0 \\ 0 & -i \end{bmatrix}$$

over C are similar

2.7. For what values of a, b, and c are each of the following a pair of similar matrices?

$$\text{(a)} \quad \begin{bmatrix} 0 & a & c \\ 0 & 0 & b \\ 0 & 0 & 0 \end{bmatrix} \quad \text{and} \quad \begin{bmatrix} 0 & 1 & 0 \\ 0 & 0 & 1 \\ 0 & 0 & 0 \end{bmatrix};$$

$$\text{(b)} \quad \begin{bmatrix} a & -b \\ b & a \end{bmatrix} \quad \text{and} \quad \begin{bmatrix} c & 0 \\ 0 & -c \end{bmatrix}$$

2.8. Let (a_{ij}) be a 3×3 matrix such that $(a_{ij})^3 = O$ but $(a_{ij})^2 \neq O$. Show that (a_{ij}) is similar to

$$\begin{bmatrix} 0 & 1 & 0 \\ 0 & 0 & 1 \\ 0 & 0 & 0 \end{bmatrix}.$$

Generalize this result to $n \times n$ matrices.

2.9. Clearly the $n \times n$ identity matrix is similar only to itself. Describe all $n \times n$ matrices that have this property.

2.10. Recalling the definition of the trace of a square matrix, show that
(a) if (a_{ij}) and (b_{ij}) are $n \times n$ matrices over F then the traces of $(a_{ij})(b_{ij})$ and $(b_{ij})(a_{ij})$ are equal;
(b) similar matrices have equal traces.

2.11. If

$$(a_{ij}) = \begin{bmatrix} 1 & -2 & 0 \\ -2 & 1 & 0 \\ 0 & 0 & -1 \end{bmatrix} \quad \text{and} \quad (p_{ij}) = \begin{bmatrix} \frac{1}{\sqrt{2}} & 0 & \frac{1}{\sqrt{2}} \\ \frac{1}{\sqrt{2}} & 0 & \frac{1}{\sqrt{2}} \\ 0 & 1 & 0 \end{bmatrix},$$

show that $(p_{ij})^{-1}(a_{ij})(p_{ij})$ is a diagonal matrix (see Section 4). Using this result, devise an *easy* method to calculate $(a_{ij})^{17}$.

2.12. Show that an $n \times n$ matrix is similar to its transpose.

2.13. Show that if A and B are $n \times n$ matrices such that the $(2n) \times (2n)$ matrices

$$\begin{bmatrix} A & O \\ O & O \end{bmatrix} \quad \text{and} \quad \begin{bmatrix} B & O \\ O & O \end{bmatrix}$$

where O is the $n \times n$ zero matrix, are similar, then A and B are similar. Is the converse also true?

3. ELEMENTARY MATRICES

In Section 2.3 we introduced three types of elementary row operations for matrices. We now show that these operations can be performed via matrix multiplication. For this we prove the following preliminary result.

☐ **Lemma.** Let r be an elementary row operation for n-rowed matrices, let (a_{ik}) be a matrix of n rows, and let (b_{kj}) be a matrix such that

$(a_{ik})(b_{kj})$ is defined. Let $r(a_{ik})$ denote the matrix resulting from (a_{ik}) by the application of the operation r. Then

$$r((a_{ik})(b_{kj})) = (r(a_{ik}))(b_{kj})$$

PROOF. We compare the left- and the right-hand sides of the alleged equality for each type of row operation.

Suppose r is a type I operation, that is, r interchanges two rows. The subscript i is the row index in both (a_{ik}) and $(a_{ik})(b_{kj})$, so that an interchange of two rows in (a_{ik}) will produce the same interchange of rows in $(a_{ik})(b_{kj})$.

Suppose r is a type II operation; assume that it multiplies the ith row by a scalar $a(\neq 0)$. When the ith row of (a_{ik}) is multiplied by a, the ith row of $(r(a_{ik}))(b_{kj})$ is

$$\left(\sum_{k=1}^{m} aa_{ik}b_{k1}, \ldots, \sum_{k=1}^{m} aa_{ik}b_{kp}\right),$$

where m is the number of columns of (a_{ik}); that is, the ith row of $(a_{ik})(b_{kj})$ is multiplied by a.

Suppose r is a type III operation; say r adds c times the lth row to the ith row of a matrix. Then the ith row of $r(a_{ik})$ is

$$(a_{i1} + ca_{l1}, \ldots, a_{in} + ca_{ln}),$$

and so the ith row of $(r(a_{ik}))(b_{kj})$ is

$$\left(\sum_{k-1}^{m} (a_{ik} + ca_{lk})b_{k1}, \ldots, \sum_{k-1}^{m} (a_{ik} + ca_{lk})b_{kp}\right),$$

which is equal to

$$\left(\sum_{k=1}^{m} a_{ik}b_{k1} + c\sum_{k=1}^{m} a_{lk}b_{k1}, \ldots, \sum_{k=1}^{m} a_{ik}b_{kp} + c\sum_{k=1}^{m} a_{lk}b_{kp}\right);$$

but this is the result of applying r to $(a_{ik})(b_{kj})$. ◇

□ **Definition.** An $n \times n$ matrix obtained from the $n \times n$ identity matrix I by performing any one elementary row operation r (which is defined for n-rowed matrices) is called an *elementary matrix*. We call rI the elementary matrix associated with r; for example, each of the following is an elementary matrix:

$$\begin{bmatrix} 1 & 0 & 0 & 0 \\ 0 & 0 & 0 & 1 \\ 0 & 0 & 1 & 0 \\ 0 & 1 & 0 & 0 \end{bmatrix}, \quad \begin{bmatrix} 1 & 0 & 0 & 0 \\ 0 & 1 & 0 & 0 \\ 0 & 0 & 2 & 0 \\ 0 & 0 & 0 & 1 \end{bmatrix}, \quad \begin{bmatrix} 1 & 0 & 0 & 0 \\ 0 & 1 & 3 & 0 \\ 0 & 0 & 1 & 0 \\ 0 & 0 & 0 & 1 \end{bmatrix}.$$

The first is associated with the type I operation (for matrices of at least four rows) which interchanges the second and fourth rows. The second is associated with the type II operation which multiplies the third row by 2 and the third is associated with the type III operation which adds three times the third row to the second. The usefulness of these matrices is indicated in the next result.

□ **Theorem 3.1.** An elementary row operation r on a matrix (a_{ij}) can be effected by multiplying (a_{ij}) on the left by the associated elementary matrix.

PROOF. By virtue of the foregoing lemma, $r(a_{ij}) = r(I(a_{ij})) = (rI)(a_{ij})$. ◇

From our observations in Section 2.3 it is clear that an elementary matrix is invertible and its inverse is also an elementary matrix.

□ **Theorem 3.2.** An $n \times n$ matrix is invertible if and only if it is equal to a product of a finite number of elementery matrices.

PROOF. Assume that the $n \times n$ matrix (a_{ij}) is invertible. Then its rank is n and it follows that the echelon matrix to which it can be reduced is the $n \times n$ identity matrix I. Hence there exist elementary matrices $E_1, \ldots, E_h$ such that

$$(E_h \cdots E_1)(a_{ij}) = I,$$

and so $(a_{ij}) = E_1^{-1} \cdots E_h^{-1}$, which is a product of elementary matrices.

Conversely, assume that (a_{ij}) is a product of elementary matrices, say $(a_{ij}) = E_1 \cdots E_k$. Then $E_k^{-1} \cdots E_1^{-1}$ is the inverse of (a_{ij}). ◇

The first part of this proof supplies a practical method for computing the inverse of an invertible matrix (a_{ij}), namely, by applying to I, in the same order, those row operations that reduce (a_{ij}) to I, for if $(E_h \cdots E_1)(a_{ij}) = I$, so that $(a_{ij}) = E_1^{-1} \cdots E_h^{-1} = (E_h \cdots E_1)^{-1}$, then $(a_{ij})^{-1} = E_h \cdots E_1 = (E_h \cdots E_1)I$.

EXAMPLE

3.1. A simple way to execute this method for computing the inverse of a given invertible $n \times n$ matrix (a_{ij}) is to "augment" (a_{ij}) with the $n \times n$ identity matrix I on the left, let us say, to obtain an $n \times 2n$

matrix and apply to it those row operations that transform (a_{ij}) to I. Then, simultaneously, the augmented I will be transformed into $(a_{ij})^{-1}$. As an illustration, we compute the inverse of the invertible matrix

$$\begin{bmatrix} 1 & 1 & 1 \\ 2 & -3 & 2 \\ -1 & -3 & -2 \end{bmatrix}$$

as follows:

$$\left[\begin{array}{ccc|ccc} 1 & 0 & 0 & 1 & 1 & 1 \\ 0 & 1 & 0 & 2 & -3 & 2 \\ 0 & 0 & 1 & -1 & -3 & -2 \end{array}\right] \rightarrow \left[\begin{array}{ccc|ccc} 1 & 0 & 0 & 1 & 1 & 1 \\ -2 & 1 & 0 & 0 & -5 & 0 \\ 1 & 0 & 1 & 0 & -2 & -1 \end{array}\right]$$

$$\rightarrow \left[\begin{array}{ccc|ccc} 1 & 0 & 0 & 1 & 1 & 1 \\ -5 & 1 & -3 & 0 & 1 & 3 \\ 1 & 0 & 1 & 0 & -2 & 1 \end{array}\right]$$

$$\rightarrow \left[\begin{array}{ccc|ccc} 6 & -1 & 3 & 1 & 0 & -2 \\ -5 & 1 & -3 & 0 & 1 & 3 \\ -9 & 2 & -5 & 0 & 0 & 5 \end{array}\right]$$

$$\rightarrow \left[\begin{array}{ccc|ccc} 6 & -1 & 3 & 1 & 0 & -2 \\ -5 & 1 & -3 & 0 & 1 & 3 \\ -\frac{9}{5} & \frac{2}{5} & -1 & 0 & 0 & 1 \end{array}\right]$$

$$\rightarrow \left[\begin{array}{ccc|ccc} \frac{12}{5} & -\frac{1}{5} & 1 & 1 & 0 & 0 \\ \frac{2}{5} & -\frac{1}{5} & 0 & 0 & 1 & 0 \\ -\frac{9}{5} & \frac{2}{5} & -1 & 0 & 0 & 1 \end{array}\right]$$

Hence

$$\begin{bmatrix} 1 & 1 & 1 \\ 2 & -3 & 2 \\ -1 & -3 & -2 \end{bmatrix}^{-1} = \frac{1}{5}\begin{bmatrix} 12 & -1 & 5 \\ 2 & -1 & 0 \\ -9 & 2 & -5 \end{bmatrix}.$$

EXERCISES

3.1. Write the nonsingular matrix

$$\begin{bmatrix} 1 & 1 & 0 \\ -2 & 0 & 1 \\ 4 & 2 & 0 \end{bmatrix}$$

as a product of elementary matrices and find its inverse.

3.2. Find the inverse of each of the following matrices:

$$\text{(a)} \begin{bmatrix} 1 & 2 & 1 \\ 0 & 3 & 4 \\ -1 & 1 & 0 \end{bmatrix}, \qquad \text{(b)} \begin{bmatrix} 1 & 0 & 1 \\ 1 & 1 & 0 \\ 0 & 1 & 1 \end{bmatrix},$$

first regarding them as matrices over R and then as matrices over Z_5.

3.3. Show that each of the following linear transformations A on R^3 is nonsingular and determine $(x, y, z)A^{-1}$.
(a) $(x, y, z)A = (12x + 2y - 9z, \ -x - y + 2z, \ 5x - 5z)$,
(b) $(x, y, z)A = (x + z, \ -x + y - z, \ x + 2y)$,
(c) $(x, y, z)A = (x - z, \ y, \ x + z)$.

REMARK. For the remaining exercises we introduce the following notation for elementary matrices:
R_{ij} = an identity matrix after the interchange of row i and row j.
$R_i(c)$ = an identity matrix after multiplication of row i by $c \neq 0$.
$R_{ij}(c)$ = an identity matrix after c times the jth row is added to the ith row.

3.4. Show that $R_{ij}^{-1} = R_{ij}$, $[R_i(c)]^{-1} = R_i(c^{-1})$ and $[R_{ij}(c)]^{-1} = R_{ij}(-c)$.

3.5. Defining elementary column operations for a matrix in the obvious way, show that they may be effected by multiplication on the right with elementary matrices. Specifically, show that
$(a_{ij})R_{ij}$ = the original matrix with the i and jth columns interchanged;
$(a_{ij})R_i(c)$ = the original matrix with the ith column multiplied by c;
$(a_{ij})R_{ij}(c)$ = the original matrix with c times the ith column added to the jth column.

3.6. Let (a_{ij}) be an $n \times m$ matrix of rank r.
(a) Using the methods and results of this section, show that there exist nonsingular matrices (p_{ij}) and (q_{ij}) such that

$$(p_{ij})(a_{ii})(q_{ij}) = \begin{bmatrix} I_r & O \\ O & O \end{bmatrix}.$$

(b) Obtain the same result with a study of linear transformations induced by (a_{ij}).

4. TRIANGULAR MATRICES

In this section we discuss some special types of matrices that play important roles in later chapters. Throughout the section we consider matrices over an arbitrary field F.

□ **Definition.** The n-tuple $(a_{11}, a_{22}, \ldots, a_{nn})$ of elements of an $n \times n$ matrix (a_{ij}) over a field F is called the *(principal) diagonal* of (a_{ij}). The individual elements $a_{11}, \ldots, a_{nn}$ are called *diagonal elements* of (a_{ij}). If all nondiagonal elements of (a_{ij}) are 0, then (a_{ij}) is called a *diagonal matrix*.

The notation

$$\text{diag}\,(a_1, \ldots, a_n)$$

is used to designate the diagonal matrix having a_i as its (i, i)th entry.

A linear transformation that is induced by a diagonal matrix has a property worthy of mention. To describe it we need some background material. If V is an n-dimensional vector space over F having $\{\alpha_1, \ldots, \alpha_n\}$ as a basis, then V is the direct sum of the one-dimensional subspaces $[\alpha_1], \ldots, [\alpha_n]$ (see Section 1.3):

$$V = [\alpha_1] \oplus \cdots \oplus [\alpha_n].$$

Conversely, if $S_1 = [\beta_1], \ldots, S_n = [\beta_n]$ are n one-dimensional subspaces of V such that $V = S_1 + \cdots + S_n$, then this sum is direct and $\{\beta_1, \ldots, \beta_n\}$ is a basis of V, as the reader can easily show.

Now let V be an n-dimensional vector space over a field F and let $(\alpha_1, \ldots, \alpha_n)$ be an ordered basis of V. A matrix over F of the form $\text{diag}\,(a_1, \ldots, a_n)$ induces a linear transformation A on V such that

$$\alpha_i A = a_i \alpha_i, \qquad 1 \leq i \leq n,$$

and consequently $[\alpha_i]A \subseteq [a_i]$, $1 \leq i \leq n$. A subspace S of a vector space V is called *A-invariant* under a linear transformation A on V if and only if $SA \subseteq S$. So we can describe the observations made above as follows. If A is a linear transformation on a vector space V and is induced by a diagonal matrix, then there can be found n one-dimensional subspaces of V, each of which is A-invariant under A and whose sum (which is a direct sum) is V.

We consider the converse next. Suppose that A is a linear transformation on an n-dimensional space V and that $S_1 = [\alpha_1], \ldots, S_n = [\alpha_n]$ are one-dimensional subspaces such that V is their (direct) sum and each is invariant under A. Then the matrix of A relative to the basis $(\alpha_1, \ldots, \alpha_n)$ is obviously diagonal. We summarize our findings in the next theorem.

□ **Theorem 4.1.** Let A be a linear transformation on an n-dimensional vector space V. Then A has a diagonal matrix if and only if there exist n

one-dimensional subspaces $S_1, \ldots, S_n$ of V such that

(a) $V = S_1 + \cdots + S_n$,
(b) each S_i is invariant under A.

EXAMPLE

4.1. Consider the linear transformation A on R^3 defined by

$$(x, y, z)A = (x - 2y, \ -2x + y, \ -z).$$

It is obvious that $S_1 = [(0, 0, 1)]$ is A-invariant and that $\alpha A = -\alpha$ for $\alpha \in S_1$. Also, $(1, 1, 0)A = (-1, -1, 0) = -(1, 1, 0)$ and so $S_2 = [(1, 1, 0)]$ is A-invariant. Further, $(1, -1, 0)A = 3(1, -1, 0)$ and so $S_3 = [(1, -1, 0)]$ is A-invariant. It follows that R^3 is the direct sum of S_1, S_2, and S_3 and that the matrix of A relative to the basis $((0, 0, 1), \ (1, 1, 0), \ (1, -1, 0))$ is diag $(-1, -1, 3)$. We may also conclude that the matrix

$$\begin{bmatrix} 1 & -2 & 0 \\ -2 & 1 & 0 \\ 0 & 0 & -1 \end{bmatrix}$$

is similar to diag $(-1, -1, 3)$.

Next, let us relate a bit of the past and the future with the present. Recall that if A is a self-adjoint linear transformation on a finite-dimensional inner-product space V then the matrix (a_{ij}) of A relative to an orthonormal basis is symmetric if V is euclidean and hermitian if V is unitary. In Chapter 6 we show that for such an A there exists an orthonormal basis of V relative to which the matrix of A is diagonal and the diagonal elements are real numbers. The matrix interpretation of this is any symmetric matrix over R and any hermitian symmetric matrix is orthogonally (unitarily) similar to a diagonal matrix whose diagonal elements are real numbers.

☐ **Definition.** An $n \times n$ matrix (a_{ij}) is called *upper triangular* if and only if each entry below the principal diagonal is 0 $(a_{ij} = 0, \ i > j)$, an $n \times n$ matrix is called *lower triangular* if and only if each entry above the principal diagonal is 0.

Let (a_{ij}) be an $n \times n$ upper triangular matrix over F and let V be an n-dimensional vector space over F with $\mathscr{B} = (\alpha_1, \ldots, \alpha_n)$ as a basis. Define

$$S_{n-i+1} = [\alpha_i, \ldots, \alpha_n], \qquad 1 \leq i \leq n.$$

Then $S_1 \subset S_2 \subset \cdots \subset S_n = V$ and $\dim S_{n-i+1} = n - i + 1$, $1 \leq i \leq n$. If A is the linear transformation in V induced by (a_{ij}) relative to the $\mathscr{B}$-basis, then

$$\alpha_i A = a_{ii}\alpha_i + \cdots + a_{in}\alpha_n, \qquad 1 \leq i \leq n.$$

It follows that $S_{n-i+1}A \subseteq S_{n-i+1}$ or, in other words, that each of $S_1, \ldots, S_n$ is a subspace of V which is invariant under A.

Conversely, if A is a linear transformation on V such that for subspaces $S_1, \ldots, S_n$ of V, we have

(i) $S_1 \subset S_2 \subset \cdots \subset S_n$,

(ii) $\dim S_k = k, \qquad 1 \leq k \leq n$,

(iii) S_k is invariant under A, $1 \leq k \leq n$,

then, there exists a basis of V relative to which the matrix of A is upper triangular. The proof is left as an exercise.

Returning to the discussion following the last definition, suppose that not only (c_{ij}) is upper triangular but that, in addition, each diagonal element is 0. It will then be clear that the linear transformation A has the following properties:

$$S_k A \subseteq S_{k-1}, \qquad k = 2, \ldots, n \quad \text{and} \quad S_1 A = \{0\}.$$

It follows that $A^n = 0$ and in turn that $(a_{ij})^n = O$.

☐ **Definition.** A linear transformation A on a vector space V is called *nilpotent* if and only if there exists a positive integer k such that $A^k = 0$. The least such positive integer is called the *index* of A. An $n \times n$ matrix (a_{ij}) is called *nilpotent* if and only if there exists a positive integer k such that $(a_{ij})^k = O$. Again, the least such positive integer is called the index of (a_{ij}).

EXAMPLES

4.1. An upper or lower triangular matrix such that each diagonal element is 0 is nilpotent. A nilpotent matrix induces nilpotent transformations and, conversely, the matrices of a finite-dimensional nilpotent linear transformation are nilpotent.

4.2. Let D be the derivative operator in $P_2(\mathsf{R})$. Since $pD^3 = 0$ for each $p \in P_2(\mathsf{R})$, it follows that D is nilpotent. Since $(1 + x + x^2)D^2 = 2$, the index of D is 3. Thus the matrix of D relative to any basis of $P_2(\mathsf{R})$ is nilpotent of index 3.

4.3. In $V = P_\infty(\mathsf{R})$ the derivative operator D is not nilpotent. Let S be the subspace of V that consists of all polynomials of degree at most 2 and let S' be a complement of S. Then each element p of V has a unique representation in the form

$$p = p_1 + p_2, \qquad p_1 \in S, p_2 \in S'.$$

Define the linear transformation A in V by

$$pA = p_1.$$

Then AD is nilpotent of index 3.

4.4. The matrix

$$\begin{bmatrix} 2 & -8 & 12 & -60 \\ 2 & -5 & 9 & -48 \\ 6 & -17 & 29 & -152 \\ 1 & -3 & 5 & -26 \end{bmatrix}$$

is nilpotent of index 2. If the reader doubts this assertion, he is at liberty to investigate the matter!

EXERCISES

4.1. Show that the matrix

$$\begin{bmatrix} -1 & 2 \\ 1 & 0 \end{bmatrix}$$

is similar to a diagonal matrix.

4.2. Find a diagonal matrix for each of the following linear transformations on R^3, defined as follows:

(a) $(x, y, z)A = (z, y, x + z)$.

(b) $(x, y, z)A = (-y, x + 2y, x + y + z)$.

4.3. Show that if A is an *idempotent* linear transformation (that is, $A \neq 0$ and $A^2 = A$) on an n-dimensional vector space then some matrix of A is diagonal.

4.4. Show that a nonzero nilpotent matrix is not similar to a diagonal matrix.

4.5. Suppose that (a_{ij}) is an upper triangular matrix. Show that

$$((a_{ij}) - a_{11}I)((a_{ij}) - a_{22}I) \cdots ((a_{ij}) - a_{nn}I) = O,$$

or, equivalently, if A is a linear transformation induced by (a_{ij}), then

$$(A - a_{11})(A - a_{22}) \cdots (A - a_{nn}) = 0.$$

4.6. Suppose that A, a linear transformation on a vector space V, is nilpotent of index $k > 1$. Thus, for some vector α in V, $\alpha A^{k-1} \neq 0$. Show that $\{\alpha, \alpha A, \ldots, \alpha A^{k-1}\}$ is a linearly independent set.

4.7. Suppose that A, a linear transformation on an n-dimensional ($n > 2$) vector space, is nilpotent of index $n - 1$. Show that A has a matrix of the form

$$\begin{bmatrix} O & I_{n-1} \\ O & O \end{bmatrix}.$$

4.8. Let A be a nilpotent linear transformation. Show that if there exists a nonzero vector α_0 such that $\alpha_0 A = c\alpha_0$ then $c = 0$.

4.9. Show that

(a) the sum of two nilpotent linear transformations is not necessarily nilpotent unless they commute;

(b) the product of two nilpotent linear transformations is not necessarily nilpotent unless they commute;

(c) if A is a nilpotent linear transformation, then $1 + A$ is an isomorphism of the underlying vector space onto itself.

4.10 Show that if (a_{ij}) and (b_{ij}) are symmetric $n \times n$ matrices, then their product is symmetric if and only if (a_{ij}) and (b_{ij}) commute.

4.11 Show that if A is a self-adjoint transformation on an inner product space V and S is a subspace invariant under A, then $S^\perp$ is also invariant under A.

4.12 Suppose that A is a self-adjoint transformation on some finite-dimensional inner-product space and that B is similar to A. Is B also self-adjoint? Justify your answer.

4.13 Show that if A is a nilpotent matrix and B is similar to A then B is nilpotent of the same index as A.

4.14 Find a 2×2 orthogonal matrix (p_{ij}) such that

$$(p_{ij})^{-1} \begin{bmatrix} 2 & -2 \\ -2 & 1 \end{bmatrix} (p_{ij})$$

is diagonal.

4.15 Show that a 2×2 nilpotent matrix of index 2 is similar to

$$\begin{bmatrix} 0 & 1 \\ 0 & 0 \end{bmatrix}.$$

Extend this result to 4×4 matrices of index 4.

4.16. Let A be a linear transformation on an n-dimensional vector space V over a field F. Suppose that there exists an ordered basis $\mathscr{B} = (\beta_1, \ldots, \beta_n)$ of V relative to which the matrix of A is diag $(c_1, \ldots, c_n)$. Show that $(A - c_1) \cdots (A - c_n) = 0$.

5. DETERMINANTS

No doubt the reader is familiar with the definition of the determinant of a 2×2 matrix

$$\begin{bmatrix} a & b \\ c & d \end{bmatrix}$$

over R as the number $ad - bc$. This is simply the definition of a particular function on the set of all 2×2 matrices over R into R. In this section we present the rudiments of the theory of determinant functions of $n \times n$ matrices over an arbitrary field F and then show that our results hold for systems (commutative rings with an identity element) that need not satisfy all of the requirements for a field.

At the outset of our presentation it is convenient to interpret an $n \times n$ matrix (a_{ij}) over F as the ordered n-tuple $(\alpha_1, \ldots, \alpha_n)$, where $\alpha_i = (a_{i1}, \ldots, a_{in})$ is the ith row vector of (a_{ij}). Precisely, there is an obvious $1 - 1$ correspondence between the set of all n-tuples $(\alpha_1, \ldots, \alpha_n)$ with $\alpha_i \in F^n$, $1 \leq i \leq n$ and the set $M_{n \times n}(F)$, and we do not distinguish between corresponding members of the two sets.

□ **Definition.** Let F be a field and let $\mathscr{S}_n$ be the set of all ordered n-tuples $(\alpha_1, \ldots, \alpha_n)$ with $\alpha_i \in F^n$, $1 \leq i \leq n$. A mapping D on $\mathscr{S}_n$ into F is called a *determinant function* if and only if D has the following properties:

(a_1) For $a \in F$ and each i, $1 \leq i \leq n$,

$$(\alpha_1, \ldots, a\,\alpha_i, \ldots, \alpha_n)D = a((\alpha_1, \ldots, \alpha_i, \ldots, \alpha_n)D).$$

(a_2) For each i, $1 \leq i \leq n$, if $\alpha_i = \alpha_i' + \alpha_i''$, then

$$(\alpha_1, \ldots, \alpha_i' + \alpha_i'', \ldots, \alpha_n)D = (\alpha_1, \ldots, \alpha_i', \alpha_i', \ldots, \alpha_n)D + (\alpha_1, \ldots, \alpha_i'', \ldots, \alpha_n)D.$$

(b) If $\alpha_i = \alpha_j$, $\quad i \neq j$,

$$(\alpha_1, \ldots, \alpha_i, \ldots, \alpha_j, \ldots, \alpha_n)D = 0.$$

(c) If $\{\epsilon_1, \ldots, \epsilon_n\}$ is the standard basis of F^n, then

$$(\epsilon_1, \ldots, \epsilon_n)D = 1.$$

As defined, a determinant function D on $\mathscr{S}_n$ is a function of n arguments (variables). Collectively, properties (a_1) and (a_2) mean that D is linear in each of its n arguments. It follows that

$$\left(\alpha_1, \ldots, \sum_{j=1}^{k} a_j\beta_j, \ldots, \alpha_n\right)D = \sum_{j=1}^{k} a_j(\alpha_1, \ldots, \beta_j, \ldots, \alpha_n)D,$$

where the indicated linear combination is the ith argument of D.

If (a_{ij}) is the $n \times n$ matrix over F that corresponds to $(\alpha_1, \ldots, \alpha_n)$ in $\mathscr{S}_n$ [thus $\alpha_i = (a_{i1}, \ldots, a_{in})$, $1 \leq i \leq n$] and there is no confusion about the determinant function D under consideration, then we may write

$$\det(a_{ij}) \quad \text{or} \quad |a_{ij}| \quad \text{for} \quad (\alpha_1, \ldots, \alpha_n)D.$$

Of course, at this point the *existence* of determinant functions for a given n has not been established. This is disposed of in due time; our first objective is to show that if such functions exist then, for each n, there exists only one.

□ **Theorem 5.1.** If D is a determinant function on the set $\mathscr{S}_n$ of all ordered n-tuples $(\alpha_1, \ldots, \alpha_n)$ with $\alpha_i \in F^n$, $1 \leq i \leq n$, then

(d) $(\alpha_1, \ldots, \alpha_i + \sum_{k \neq i} a_k \alpha_k, \ldots, \alpha_n)D = (\alpha_1, \ldots, \alpha_i, \ldots, \alpha_n)D$ for scalars $a_1, \ldots, a_{i-1}, a_{i+1}, \ldots, a_n$;

(e) $(\alpha_1, \ldots, \alpha_n)D = 0$ if $\{\alpha_1, \ldots, \alpha_n\}$ is a linearly dependent set;

(f) If $(\alpha_1, \ldots, \alpha_j, \ldots, \alpha_i, \ldots, \alpha_n)$ is the n-tuple obtained from $(\alpha_1, \ldots, \alpha_i, \ldots, \alpha_j, \ldots, \alpha_n)$ on interchanging α_i and $\alpha_j (i \neq j)$, then

$$(\alpha_1, \ldots, \alpha_j, \ldots, \alpha_i, \ldots, \alpha_n)D = -((\alpha_1, \ldots, \alpha_i, \ldots, \alpha_j, \ldots, \alpha_n)D).$$

PROOF. The linearity of D in its ith argument implies that

$$\left(\alpha_1, \ldots, \alpha_i + \sum_{k \neq i} a_k\alpha_k, \ldots, \alpha_n\right)D = (\alpha_1, \ldots, \alpha_i, \ldots, \alpha_n)D$$
$$+ \sum_{k \neq i} a_k(\alpha_1, \ldots, \alpha_k, \ldots, \alpha_n)D,$$

where, in the latter sum, α_k is the ith argument. According to (b), each term of this sum, hence the sum itself, is 0. Thus (d) is true.

For (e) assume that $\{\alpha_1, \ldots, \alpha_n\}$ is linearly dependent. Then for some i, $1 \leq i \leq n$, $\alpha_i = \sum_{k \neq i} a_k \alpha_k$. Hence

$$(\alpha_1, \ldots, \alpha_i, \ldots, \alpha_n)D = \left(\alpha_1, \ldots, \sum_{k \neq i} a_k \alpha_k, \ldots, \alpha_n\right)D$$

$$= \sum_{k \neq i} a_k(\alpha_1, \ldots, \alpha_k, \ldots, \alpha_n)D = 0,$$

using (b) again.

The following computation disposes of (f).

$$\begin{aligned}(\alpha_1, \ldots, a_j, \ldots, \alpha_i, \ldots, \alpha_n)D &= -\,[(\alpha_1, \ldots, \alpha_j, \ldots, -\alpha_i, \ldots, \alpha_n)D] \\ &= -[(\alpha_1, \ldots, \alpha_j, \ldots, \alpha_j - \alpha_i, \ldots, \alpha_n)D] \\ &= -\,[(\alpha_1, \ldots, \alpha_j - (\alpha_j - \alpha_i), \ldots, \\ &\qquad\qquad \alpha_j - \alpha_i, \ldots, \alpha_n)D] \\ &= -\,[(\alpha_1, \ldots, \alpha_i, \ldots, \alpha_j - \alpha_i, \ldots, \alpha_n)D] \\ &= -\,[(\alpha_1, \ldots, \alpha_i, \ldots, \alpha_j, \ldots, \alpha_n)D]. \ \diamond\end{aligned}$$

□ **Theorem 5.2.** For each positive integer n there is at most one determinant function D on the set $\mathscr{S}_n$.

PROOF. Suppose that D_1 and D_2 are determinant functions on $\mathscr{S}_n$. Let $E = (\epsilon_1, \ldots, \epsilon_n)$, that is, the n-tuple in which the elements of the standard basis of F^n appear in their natural order. Then $ED_1 = 1 = ED_2$. Further, if E' results from E by the interchange of ϵ_i and ϵ_j ($i \neq j$) then $E'D_1 = -1 = E'D_2$. Since any arrangement E^* of $\{\epsilon_1, \ldots, \epsilon_n\}$ can be obtained from the natural arrangement E by a finite succession of interchanges of pairs $\{\epsilon_i, \epsilon_j\}$, we infer that $E^*D_1 = E^*D_2$.

Now consider any element $(\alpha_2, \ldots, \alpha_n)$ of $\mathscr{S}_n$; suppose that

$$\alpha_i = a_{i1}\epsilon_1 + \cdots + a_{in}\epsilon_n, \qquad 1 \leq i \leq n.$$

Then

$$(\alpha_1, \ldots, \alpha_n)D_1 = \left(\sum_{j_1=1}^{n} a_{1j_1}\epsilon_{j_1}, \ldots, \sum_{j_n=1}^{n} a_{nj_n}\epsilon_{j_n}\right)D_1$$

$$= \sum_{j_1=1}^{n} \cdots \sum_{j_n=1}^{n} a_{1j_1} \cdots a_{nj_n}(\epsilon_{j_1}, \ldots, \epsilon_{j_n})D_1,$$

which is a sum of n^n terms of which $n^n - n!$ are zero by virtue of part (b) of the definition of a determinant function. In view of the result stated in the preceding paragraph $(\epsilon_{j_1}, \ldots, \epsilon_{j_n})D_1 = (\epsilon_{j_1}, \ldots, \epsilon_{j_n})D_2$. Replacing the former by the latter in the right-hand member of the last equation, we obtain the expanded version of $(\alpha_1, \ldots, \alpha_n)D_2$. Thus $(\alpha_1, \ldots, \alpha_n)D_1 = (\alpha_1, \ldots, \alpha_n)D_2$ for all $(\alpha_1, \ldots, \alpha_n) \in \mathscr{S}_n$ and the proof is complete. ◇

EXAMPLES

5.1. The set $\mathscr{S}_1$ consists of all ordered 1-tuples (a), where $a \in F$. Define the function D on $\mathscr{S}_1$ into F by $(a)D = a$. Then it is easily seen that D qualifies as a determinant function, hence is the only one.

5.2. Let (α_1, α_2), where

$$\alpha_1 = a_{11}\epsilon_1 + a_{12}\epsilon_2, \qquad \alpha_2 = a_{21}\epsilon_1 + a_{22}\epsilon_2,$$

be an element of $\mathscr{S}_2$. If D is a determinant function on $\mathscr{S}_2$, then

$$\begin{aligned}(\alpha_1, \alpha_2)D &= (a_{11}\epsilon_1 + a_{12}\epsilon_2, a_{21}\epsilon_2 + a_{22}\epsilon_2)D \\ &= a_{11}(\epsilon_1, a_{21}\epsilon_1 + a_{22}\epsilon_2)D + a_{12}(\epsilon_2, a_{21}\epsilon_1 + a_{22}\epsilon_2)D \\ &= a_{11}[a_{22}(\epsilon_1, \epsilon_2)D + a_{21}(\epsilon_1, \epsilon_1)D] + a_{12}[a_{21}(\epsilon_2, \epsilon_1)D \\ &\qquad + a_{22}(\epsilon_2, \epsilon_2)D] \\ &= a_{11}a_{22}(\epsilon_1, \epsilon_2)D + a_{12}a_{21}(\epsilon_2, \epsilon_1)D \\ &= a_{11}a_{22} - a_{12}a_{21}.\end{aligned}$$

Thus we have an explicit form for the only possible determinant function on $\mathscr{S}_2$. Again it may be shown that this function *is a* (hence *the*) determinant function on $\mathscr{S}_2$. So we may write

$$\det \begin{bmatrix} a_{11} & a_{12} \\ a_{21} & a_{22} \end{bmatrix} = \begin{vmatrix} a_{11} & a_{12} \\ a_{21} & a_{22} \end{vmatrix} = a_{11}a_{22} - a_{12}a_{21}.$$

□ **Theorem 5.3.** For each positive integer n there exists a determinant function on $\mathscr{S}_n$ (that is, $M_{n\times n}(F)$). Thus for each n there exists exactly one determinant function.

PROOF. A preliminary definition is required. If (a_{ij}) is an $n \times n$ matrix $(n \geq 2)$, let A_{ij} be the $(n-1) \times (n-1)$ matrix obtained from (a_{ij}) by deleting the ith row and the jth column of (a_{ij}).

The proof of the theorem is by induction on n. Example 5.1 disposes of the case $n = 1$. For $n > 1$ assume that there exists a (hence exactly one) determinant function on $\mathscr{S}_{n-1}$. We now define a function D on $\mathscr{S}_n$ into F. Let $(\alpha_1, \ldots, \alpha_n) \in \mathscr{S}_n$, where $\alpha_i = (a_{i1}, \ldots, a_{in})$, $1 \le i \le n$, and let (a_{ij}) be the corresponding $n \times n$ matrix. By our induction hypothesis *the* determinant D_{ij} of the $(n-1) \times (n-1)$ matrix A_{ij} is defined for $1 \le i, j \le n$. Let

$$(\alpha_1, \ldots, \alpha_n)D = (-1)^{1+j} a_{1j} D_{1j} + \cdots + (-1)^{n+j} a_{nj} D_{nj},$$

where $1 \le j \le n$. Then D has property (c) of a determinant function, since

$$(\epsilon_1, \ldots, \epsilon_n)D = (-1)^{j+j} \cdot 1 \cdot D_{jj} = 1.$$

To prove that D has property (b) suppose that $\alpha_i = \alpha_k$, where $i < k$. Then, if $r \ne i$ and $r \ne k$, A_{rj} has two equal rows, hence $D_{rj} = 0$. Thus

$$(\alpha_1, \ldots, \alpha_i, \ldots, \alpha_k, \ldots, \alpha_n)D = (-1)^{i+j} a_{ij} D_{ij} + (-1)^{k+j} a_{ij} D_{kj}.$$

We continue by showing that the indicated sum on the right-hand side of this equation has the value 0. Between the $(i-1)$st row of A_{ij} and its $(k-1)$st row (which is α_i with a_{ij} deleted) there are $k - i - 1$ rows. Interchanging the $(k-1)$st row successively with each of $k - i - 1$ rows immediately above it yields A_{kj}. According to property (f) of Theorem 5.1 it follows that

$$D_{kj} = (-1)^{k-i-1} D_{ij} \quad \text{or} \quad D_{ij} = (-1)^{k-i-1} D_{kj}.$$

Substitution of this value for D_{ij} in the equation under consideration results in the value 0. Thus D has property (b).

We turn next to property (a_1). By the definition of D

$$(\alpha_1 \ldots, a\,\alpha_i, \ldots, \alpha_n)D = (-1)^{1+j} a_{1j} D'_{1j} + \cdots + (-1)^{i+j} a a_{ij} D'_{ij} + \cdots + (-1)^{n+j} a_{nj} D'_{nj},$$

where D'_{kj} is the determinant of the $(n-1) \times (n-1)$ matrix obtained from $(\alpha_1, \ldots, a\alpha_i, \ldots, \alpha_n)$ by deleting the kth row and the jth column of the matrix $(\alpha_1, \ldots, a\alpha_i, \ldots, \alpha_n)$. Since, by the induction hypothesis,

$$D'_{kj} = aD_{kj}, \qquad k \ne i \quad \text{and} \quad D'_{ij} = D_{ij},$$

we have

$$\begin{aligned}(\alpha_1, \ldots, a\alpha_i, \ldots, \alpha_m)D &= a[(-1)^{1+j} a_{1j} D_{1j} + \cdots + (-1)^{i+j} a_{ij} D_{ij} \\ &\quad + \cdots + (-1)^{n+j} a_{nj} D_{nj}] \\ &= a[(\alpha_1, \ldots, \alpha_i, \ldots, \alpha_n)D].\end{aligned}$$

Thus D satisfies (a_1).

Finally, we show that D satisfies (a_2). If

$$\alpha_i' = (a_{i1}', \ldots, a_{in}') \quad \text{and} \quad \alpha_i'' = (a_{i1}'', \ldots, a_{in}''),$$

let D_{kj}' and D_{kj}'' denote the determinants of the $(n-1) \times (n-1)$ matrices obtained from $(\alpha_1, \ldots, \alpha_i', \ldots, \alpha_n)$ and $(\alpha_1, \ldots, \alpha_i'', \ldots, \alpha_n)$, respectively, by deleting the kth row and the jth column of each. Then

$$D_{kj} = D_{kj}' + D_{kj}'', \qquad i \neq k, \quad \text{and} \quad D_{ij} = D_{ij}' = D_{ij}''.$$

Hence

$$\begin{aligned}(\alpha_1, \ldots, \alpha_i' + \alpha_i'', \ldots, \alpha_n)D &= (-1)^{1+j}a_{1j}(D_{1j}' + D_{1j}'') \\ &\quad + \cdots + (-1)^{i+j}(a_{ij}' + a_{ij}'')D_j + \cdots \\ &\quad + (-1)^{n+j}a_{nj}(D_{nj}' + D_{nj}'') \\ &= [(-1)^{1+j}a_{1j}D_{1j}' + \cdots + (-1)^{i+j}a_{ij}'D_{ij}' + \cdots \\ &\quad + (-1)^{n+j}a_{nj}D_{nj}'] + [(-1)^{1+j}a_{ij}D_{1j}'' + \cdots \\ &\quad + (-1)^{i+j}a_{ij}''D_{ij}'' + \cdots + (-1)^{n+j}a_{nj}D_{nj}''] \\ &= (\alpha_1, \ldots, \alpha_i', \ldots, \alpha_n)D + (\alpha_1, \ldots, \alpha_i'', \ldots, \alpha_n)D.\end{aligned}$$

It follows that (a_2) holds for D and so the proof is complete. ◇

Observe that the definition of D in this proof provides an algorithm for determining its value at a given matrix or, as one says, computing the determinant $\det(a_{ij})$ of (a_{ij}). Since the expression

$$(-1)^{1+j}a_{1j}D_{1j} + \cdots + (-1)^{n+j}a_{nj}D_{nj}$$

for $\det(a_{ij})$ is independent of j, we have established, in view of Theorem 5.2, the equality of n different determinant functions. The one that employs the jth column of (a_{ij}) is referred to as "the development of a determinant (of a matrix) by the elements of the jth column." Also, $(-1)^{i+j}D_{ij}$ is called the *cofactor* of the element a_{ij}.

EXAMPLES

5.3. Properties (f), (a_1), and (d) of the determinant function on $M_{n \times n}(F)$ indicate the relationship of $\det(a_{ij})$ to the determinant of the matrix obtained from (a_{ij}) by the application of an elementary row operation of types I, II, and III, respectively. We can take advantage of the simplicity of this relationship for each type of row operation when forced to evaluate a determinant. Following is a modest example.

$$\det\begin{bmatrix} 2 & 0 & 1 \\ 1 & 1 & 2 \\ -1 & 3 & 4 \end{bmatrix} = \det\begin{bmatrix} 2 & 0 & 1 \\ 1 & 1 & 2 \\ 0 & 4 & 6 \end{bmatrix} = \det\begin{bmatrix} 0 & -2 & -3 \\ 1 & 1 & 2 \\ 0 & 4 & 6 \end{bmatrix}$$

$$= -\det\begin{bmatrix} -2 & -3 \\ 4 & 6 \end{bmatrix} = -\det\begin{bmatrix} -2 & -3 \\ 0 & 0 \end{bmatrix} = 0.$$

5.4. By definition

$$\begin{vmatrix} a_{11} & a_{12} & a_{13} \\ a_{21} & a_{22} & a_{23} \\ a_{31} & a_{32} & a_{33} \end{vmatrix} = a_{11}\begin{vmatrix} a_{22} & a_{23} \\ a_{32} & a_{33} \end{vmatrix} - a_{21}\begin{vmatrix} a_{12} & a_{13} \\ a_{32} & a_{33} \end{vmatrix} + a_{31}\begin{vmatrix} a_{12} & a_{13} \\ a_{22} & a_{23} \end{vmatrix}$$

$$= a_{11}a_{22}a_{33} + a_{12}a_{23}a_{31} + a_{13}a_{21}a_{32}$$

$$- a_{11}a_{23}a_{32} - a_{12}a_{21}a_{33} - a_{13}a_{22}a_{31}.$$

Observe that the expression on the right-hand side is the sum of all terms of the form

$$\pm a_{1i}a_{2j}a_{3k},$$

where (i, j, k) is an arrangement of $\{1, 2, 3\}$; equivalently, it is the sum of all signed products of three elements of the matrix (a_{ij}) such that every row and every column is "represented." An explicit definition of the determinant of an $n \times n$ matrix can be given in this way—the only troublesome feature is the formulation of a rule for determining the sign of each product.

5.5. We shall prove by induction that if two columns of an $n \times n$ matrix are identical then its determinant is 0. Clearly this is true for 2×2 matrices. Assuming that it is true for $(n-1) \times (n-1)$ matrices, consider an $n \times n$ matrix with identical jth and kth columns. Developing $\det(a_{ij})$ by elements of its ith column, where $i \neq j, k$ gives

$$\det(a_{ij}) = (-1)^{1+i}a_{ij}D_{ij} + \cdots + (-1)^{n+i}a_{ni}D_{ni},$$

which is equal to 0, since $D_{ij} = \cdots = D_{ni} = 0$.

☐ **Theorem 5.4.** The determinant of the transpose of an $n \times n$ matrix is equal to the determinant of the matrix.

PROOF. For an $n \times n$ matrix (a_{ij}) over a field F define

$$\alpha_i = (a_{i1}, \ldots, a_{in}) \quad \text{and} \quad \gamma_i = (a_{1i}, \ldots, a_{ni}), \qquad 1 \leq i \leq n.$$

Let D denote the determinant function on $n \times n$ matrices over F and define the function D^* on the same set by

$$(\gamma_1, \ldots, \gamma_n)D^* = (\alpha_1, \ldots, \alpha_n)D.$$

If we can show that D^* has the defining properties of a determinant function, the desired result will follow, since then we will have

$$(\gamma_1, \ldots, \gamma_n)D = (\gamma_1, \ldots, \gamma_n)D^* = (\alpha_1, \ldots, \alpha_n)D$$

for all (a_{ij}) in $M_{n \times n}(F)$.

Since

$$\begin{aligned}(\gamma_1, \ldots, a\gamma_i, \ldots, \gamma_n)D^* &= \begin{vmatrix} a_{11} & \cdots & aa_{1i} & \cdots & a_{1n} \\ \vdots & & & & \\ a_{n1} & \cdots & aa_{ni} & \cdots & a_{nn} \end{vmatrix} \\ &= a[(-1)^{1+i}a_{1i}D_{1i} + \cdots + (-1)^{n+i}a_{ni}D_{ni}] \\ &= a(\alpha_1, \ldots, \alpha_n)D \\ &= a(\gamma_1, \ldots, \gamma_n)D^*,\end{aligned}$$

D^* has property (a_1) of a determinant function.

Also it has property (a_2) since,

$$\begin{aligned}&(\gamma_1, \ldots, \gamma_i' + \gamma_i'', \ldots, \gamma_n)D^* \\ &= \begin{vmatrix} a_{11} & \cdots & a_{1i}' + a_{1i}'' & \cdots & a_{in} \\ \vdots & & & & \\ a_{ni} & \cdots & a_{ni}' + a_{ni}'' & \cdots & a_{nn} \end{vmatrix} \\ &= (a_{1i}' + a_{1i}'')(-1)^{1+i}D_{1i} + \cdots + (a_{ni}' + a_{ni}'')(-1)^{n+i}D_{ni} \\ &= (\alpha_1, \ldots, \alpha_i', \ldots, \alpha_n)D + (\alpha_i, \ldots, \alpha_i'', \ldots, \alpha_n)D \\ &= (\gamma_1, \ldots, \gamma_i', \ldots, \gamma_n)D^* + (\gamma_i, \ldots, \gamma_i'', \ldots, \gamma_n)D^*.\end{aligned}$$

Next, if $\gamma_i = \gamma_k$ for $i \neq k$, then

$$(\gamma_1, \ldots, \gamma_i, \ldots, \gamma_k, \ldots \gamma_n)D^* = (\alpha_1, \ldots, \alpha_i, \ldots, \alpha_k, \ldots, \alpha_n)D = 0,$$

for the assumption implies that (a_{ij}) has two identical columns, and, hence, that $(\alpha_1, \ldots, \alpha_n)D = 0$ in view of Example 5.5. Thus D^* has property (b). Finally, it is immediate that it satisfies the remaining property of a determinant function, for if $\gamma_i = \epsilon_i$, $1 \le i \le n$, then $\alpha_1 = \epsilon_1$, $1 \le i \le n$, and so $(\epsilon_1, \ldots, \epsilon_n)D^* = (\epsilon_1, \ldots, \epsilon_n)D = 1$. ◇

☐ **Corollary.** All properties of the determinant function of $n \times n$ matrices, when regarded as a function of the n-tuples, which are the rows of a matrix, hold when it is regarded as a function of the n-tuples, which are the columns of a matrix. In particular, for $1 \le i \le n$, $\det(a_{ij})$ may be developed by elements of its ith row:

$$\det(a_{ij}) = (-1)^{i+1}a_{i1}D_{i1} + \cdots + (-1)^{i+n}a_{in}D_{in}.$$

As a "working" version of the corollary we might take this: Rules that pertain to evaluating a determinant by manipulating its rows also hold for columns. For the proof of still another property of determinants we need an auxiliary result.

☐ **Lemma.** Let F be a field and let $\mathscr{S}_n$ be the set of all ordered n-tuples $(\alpha_1, \ldots, \alpha_n)$ with $\alpha_i \in F^n$, $1 \le i \le n$. If f is a function on $\mathscr{S}_n$ into F, which satisfies conditions (a_1), (a_2), and (b) for the determinant function D, then

$$(\alpha_1, \ldots, \alpha_n)f = (\alpha_1, \ldots, \alpha_n)D\,(\epsilon_1, \ldots, \epsilon_n)f.$$

PROOF. If $(\epsilon_1, \ldots, \epsilon_n)f = 1$, then $f = D$ by the uniqueness of D (for a fixed n) and the relation in question clearly holds. Suppose that $(\epsilon_1, \ldots, \epsilon_n)f \neq 1$ and consider the function D' on $\mathscr{S}_n$ into F such that

$$(\alpha_1, \ldots, \alpha_n)D' = \frac{(\alpha_1, \ldots, \alpha_n)D - (\alpha_1, \ldots, \alpha_n)f}{1 - (\epsilon_1, \ldots, \epsilon_n)f}.$$

Then D' satisfies all the defining properties of a determinant function (the details are left as an exercise); hence $D' = D$. Setting $D' = D$ in the displayed equation gives the desired result. ◇

☐ **Theorem 5.5.** If (a_{ij}) and (b_{ij}) are $n \times n$ matrices, then

$$\det[(a_{ij})(b_{ij})] = \det(a_{ij}) \cdot \det(b_{ij}).$$

PROOF. Let (a_{ij}) and (b_{ij}) be $n \times n$ matrices and let B be the linear transformation in F^n induced by (b_{ij}) relative to the standard basis. Then, if $\alpha_i = (a_{i1}, \ldots, a_{in})$, $1 \leq i \leq n$,

$$\alpha_i B = \left(\sum_{k=1}^{n} a_{ik} b_{k1}, \ldots, \sum_{j=1}^{n} a_{ik} b_{kn} \right),$$

which is the ith row of $(a_{ij})(b_{ij})$.

Now define $f: \mathscr{S}_n \to F$ by

$$(\alpha_1, \ldots, \alpha_n) f = (\alpha_1 B, \ldots, \alpha_n B) D.$$

In view of the linearity of B, it is clear that f satisfies the conditions of the lemma. Hence

$$(\alpha_1, \ldots, \alpha_n) f = (\alpha_1, \ldots, \alpha_n) D \cdot (\epsilon_1, \ldots, \epsilon_n) f$$

and so

$$(\alpha_1, \ldots, \alpha_n) D \cdot (\epsilon_1, \ldots, \epsilon_n) f = (\alpha_1 B, \ldots, \alpha_n B) D.$$

This equation is the desired conclusion because (i)

$$(\alpha_1, \ldots, \alpha_n) D = \det(a_{ij}),$$

(ii) by definition of f,

$$(\epsilon_1, \ldots, \epsilon_n) f = (\epsilon_1 B, \ldots, \epsilon_n B) D = \det(b_{ij}),$$

and (iii),

$$(\alpha_1 B, \ldots, \alpha_n B) D = \det[(a_{ij})(b_{ij})]. \ \diamond$$

□ **Theorem 5.6.** An $n \times n$ matrix (a_{ij}) over a field is nonsingular (=invertible) if and only if $\det(a_{ij}) \neq 0$.

PROOF. Assume that (a_{ij}) is nonsingular. Then $(a_{ij})^{-1}(a_{ij}) = I$ and, hence,

$$\det(a_{ij})^{-1} \cdot \det(a_{ij}) = \det I = 1$$

which implies that $\det(a_{ij}) \neq 0$.

Conversely, suppose that $\det(a_{ij}) \neq 0$ and let $\alpha_i = (a_{i1}, \ldots, a_{in})$, $1 \leq i \leq n$. Then, according to property (e) of a determinant function (see

Theorem (5.1), $\{\alpha_1, \ldots, \alpha_n\}$ is a linearly independent set of elements of F^n. It follows that the rank of (a_{ij}) is n and, hence, that (a_{ij}) is nonsingular. ◇

In this proof appears the identity

$$\det(a_{ij})^{-1} = [\det(a_{ij})]^{-1}$$

for a nonsingular matrix (a_{ij}). From this and Theorem 5.6 it follows that if (a_{ij}) is any $n \times n$ matrix and (p_{ij}) is an invertible $n \times n$ matrix then

$$\begin{aligned}\det[(p_{ij})^{-1}(a_{ij})(p_{ij})] &= \det(p_{ij})^{-1} \cdot \det(a_{ij}) \cdot \det(p_{ij}) \\ &= \det(a_{ij}),\end{aligned}$$

so that *similar matrices have equal determinants.* Since any two matrices of a linear transformation A in an n-dimensional vector space over a field are similar, this result makes possible the definition of the determinant of A, in symbols,

$$\det A,$$

as the determinant of any matrix of A. In view of Theorem 5.6, a linear transformation in an n-dimensional space is nonsingular if and only if its determinant is nonzero.

Almost all of the foregoing introduction to the theory of determinants is valid for matrices whose entries are elements of a commutative ring with identity element. This paragraph is devoted to an elaboration of the foregoing statement. A *commutative ring with identity element* is an algebraic system that satisfies all of the defining properties of a field (see Section 1.4) with the possible exception of $M4$ (the existence of a multiplicative inverse for nonzero elements). The system of integers, together with the familiar operations of addition and multiplication, qualifies as a commutative ring with identity element. A class of commutative rings with identity elements (rings of polynomials over a field), which is important in the study of linear transformations, is discussed in Chapter 6. Computations may be carried out in such a system, just as in a field, with one exception: since the presence of the inverse of a nonzero element is not assured, the operation of division is, in general, not possible. The reader who is sufficiently conscientious to review this section in detail will find that all results are valid for matrices with entries in a commutative ring with identity element with the following two exceptions: part (e) of Theorem 5.1 and that part of Theorem 5.6 which asserts that if $\det(a_{ij}) \neq 0$ then (a_{ij}) is nonsingular [its proof relies on property (e)].

EXERCISES

5.1. Find the determinant of each of the following matrices:

$$\text{(a)} \begin{bmatrix} 2 & 0 & 1 \\ 3 & 1 & 0 \\ 0 & 2 & 3 \end{bmatrix} \quad \text{(b)} \begin{bmatrix} 1+i & i & 2 \\ 3-i & 0 & -i \\ 1 & -2 & 2i \end{bmatrix} \quad \text{(c)} \begin{bmatrix} 3 & 5 & 2 & 12 \\ 2 & 4 & 1 & 6 \\ 2 & -3 & 0 & 2 \\ 2 & 4 & 2 & 7 \end{bmatrix}.$$

5.2. Show that the determinant of a triangular matrix is equal to the product of its diagonal elements.

5.3. Suppose that an $n \times n$ matrix has the form

$$\begin{bmatrix} A & O \\ C & B \end{bmatrix},$$

where A is an $r \times r$ matrix, B is an $(n-r) \times (n-r)$ matrix, C is an $(n-r) \times r$ matrix, and O denotes the $r \times (n-r)$ zero matrix. Show that

$$\det \begin{bmatrix} A & O \\ C & B \end{bmatrix} = \det A \det B.$$

5.4. Show that the absolute value of the determinant of an orthogonal matrix is equal to 1.

5.5. Show that the Vandermonde determinant

$$V(x_1, \ldots, x_n) = \det \begin{bmatrix} 1 & 1 & \cdots & 1 \\ x_1 & x_2 & \cdots & x_n \\ x_1^2 & x_2^2 & \cdots & x_n^2 \\ \vdots & & & \\ x_1^{n-1} & x_2^{n-1} & \cdots & x_n^{n-1} \end{bmatrix}$$

is equal to the product of all differences $x_i - x_j$, with $i > j$; that is

$$V = \prod_{1 \le j < i \le n} (x_i - x_j).$$

5.6. Let (a_{ij}) be a 2×2 matrix over R. Prove that $\det[I + (a_{ij})] = 1 + \det(a_{ij})$ if and only if $a_{11} + a_{22} = 0$.

5.7. Show that an equation of the line through the distinct points $(x_1 y_1)$ and (x_2, y_2) in E_2 is given by

$$\det \begin{bmatrix} x & y & 1 \\ x_1 & y_1 & 1 \\ x_2 & y_2 & 1 \end{bmatrix} = 0.$$

5.8. Show that the area of the triangle in E_2 with vertices (a_1, a_2), (b_1, b_2), (c_1, c_2) is given by the absolute value of

$$(\tfrac{1}{2}) \det \begin{bmatrix} a_1 & a_2 & 1 \\ b_1 & b_2 & 1 \\ c_1 & c_2 & 1 \end{bmatrix}.$$

5.9. Let $\alpha_1, \ldots, \alpha_n$ be vectors in R^n; say $\alpha_i = (a_{i1}, \ldots, a_{in})$, $1 \leq i \leq n$. Prove that $\alpha_1, \ldots, \alpha_n$ lie on a linear manifold of dimension $< n-1$ if and only if

$$\det \begin{bmatrix} x_1 & x_2 & \cdots & x_n & 1 \\ a_{11} & a_{12} & \cdots & a_{1n} & 1 \\ \vdots & & & & \\ a_{n1} & a_{n2} & \cdots & a_{nn} & 1 \end{bmatrix} = 0$$

for all $x_1, \ldots, x_n$ in R.

5.10. Let A be a linear transformation on an n-dimensional vector space V over and $x \in F$. Show that $\det(xI - A) = 0$ if and only if there exists a nonzero vector α in V such that $\alpha A = c\alpha$ for a scalar c.

5.11. Referring to the result displayed in the corollary to Theorem 5.4, show that if $k \neq i$ then

$$(-1)^{i+1} a_{i1} D_{k1} + \cdots + (-1)^{i+n} a_{in} D_{kn} = 0.$$

5.12. Let (a_{ij}) be an $n \times n$ matrix and let (b_{ij}) be the matrix such that b_{ij} is the cofactor of a_{ji} for $1 \leq i, j \leq n$. Show that

$$(a_{ij})(b_{ij}) = (b_{ij})(a_{ij}) = (\det(a_{ij}))I,$$

the scalar matrix with $\det(a_{ij})$ appearing in its principal diagonal. Deduce that if $\det(a_{ij}) \neq 0$, then

$$(a_{ij})^{-1} = \frac{1}{\det(a_{ij})} (b_{ij})$$

5.13. Let $B \in V$, the vector space of all 2×2 matrices over a field F. Let L_B be the map of V into V such that $AL_B = BA$ for all $A \in V$.

(a) Show that $L_B \in \mathscr{L}(V, V)$.

(b) Show that $\det L_B = (\det B)^2$.

(c) Generalize the results in (a) and (b) to the $n \times n$ case.

5.14. Let V be an n-dimensional vector space, $W = \mathscr{L}(V, V)$, and $B \in W$. Define the mapping $f_B : W \to W$ by

$$Af_B = BA - AB.$$

(a) Show that $f_B \in \mathscr{L}(W, W)$.

(b) What is $\det f_B$? Under what circumstances does $\det f_B = 0$?

6 Algebraic Properties of Linear Transformations

This is a long and important chapter. Its primary purpose is to introduce the reader to tools (cyclic subspaces, minimal polynomial, characteristic polynomial) that are useful in analyzing a given linear transformation on a finite-dimensional vector space. The definitive results that we obtain with these tools include a necessary and sufficient condition that some matrix of a linear transformation be diagonal and a corresponding condition that some matrix of a linear transformation on an inner-product space be triangular. Also, self-adjoint transformations succumb to our treatment. Investigations along these lines are continued in Chapter 8, in which more significant results appear. Section 1 is devoted to a presentation of basic properties of polynomials, regarded as elements of a commutative ring with identity element. An understanding of this section is essential to what follows.

1. POLYNOMIAL RINGS

In this section we sketch properties of a class of commutative rings with identity elements (see Section 5.5), which have applications in this and the following chapters.

Let F be a field and let P be the set of all infinite sequences

$$(a_0, a_1, a_2, \ldots)$$

of elements of F such that only a finite number of the a_i's are different from 0. We define addition and multiplication of elements of P as follows:†

† Since a sequence is a function, the definition of equality of two sequences is at hand: $(a_0, a_1, a_2, \ldots) = (b_0, b_1, b_2, \ldots)$ if and only if $a_i = b_i$, $i = 0, 1, 2, \ldots$.

$$(a_0, a_1, a_2, \ldots) + (b_0, b_1, b_2, \ldots) = (a_0 + b_0, a_1 + b_1, a_2 + b_2, \ldots)$$
$$(a_0, a_1, a_2, \ldots) \cdot (b_0, b_1, b_2, \ldots) = (c_0, c_1, c_2, \ldots),$$

where

$$c_i = \sum_{j+k=i} a_j b_k, \qquad i = 0, 1, 2, \ldots.$$

Clearly addition and multiplication are binary operations in P. It is an easy exercise to verify that addition satisfies properties A1 to A4 of Section 1.4. Computations that are somewhat more tedious lead to the conclusion that multiplication satisfies properties M1 to M3 and the distributive law D given in Section 1.4. It follows that P, together with the operations of addition and multiplication, is a commutative ring with identity element $(1, 0, 0, \ldots)$.

We now direct our attention to deriving a more familiar representation of the elements of P. The set F' of all elements of P with the form

$$a' = (a, 0, 0, \ldots), \qquad a \in F,$$

together with the operations of addition and multiplication in P, is a field. The mapping $a \to a' = (a, 0, 0, \ldots)$ of F into P is 1-1, onto F', and preserves addition and multiplication: that is,

$$(a + b)' = a' + b' \quad \text{and} \quad (ab)' = a'b'.$$

This situation is summarized by saying that F' is isomorphic to F. Let us identify the element a of F with the sequence $(a, 0, 0, \ldots)$ of F' which corresponds to a under the above mapping; that is, we write

$$a = (a, 0, 0, \ldots).$$

Now let us set

$$x = (0, 1, 0, \ldots).$$

Then $x^2 = (0, 0, 1, 0, \ldots)$ and, in general,

$$x^k = (0, 0, \ldots, 0, 1, 0, \ldots),$$

where the only nonzero entry of x^k is in the $(k+1)$st position. Further,

$$ax^k = (a, 0, 0, \ldots)(0, \ldots, 0, 1, 0, \ldots) = (0, \ldots, 0, a, 0, \ldots) = x^k a,$$

and so an arbitrary element $(a_0, a_1, \ldots, a_n, 0, \ldots)$ of P may be written in the form

$$a_0 + a_1 x + \cdots + a_n x^n.$$

At this point we adopt the name "the polynomial $a_0 + a_1 x + \cdots + a_n x^n$, with coefficients $a_0, a_1, \ldots, a_n$ in F" for the sequence $(a_0, a_1, \ldots, a_n, 0, \ldots)$, the notation $F[x]$ in place of P, and the name "the ring of polynomials in x over F" for this ring.

We call 0 $[=(0, 0, 0, \ldots)]$ the *zero polynomial.* If the polynomial $f \neq 0$ and

$$f = a_0 + a_1x + \cdots + a_n x^n, \qquad a_n \neq 0,$$

we define the degree of f to be n (in symbols, $\deg f$). If $a_n = 1$, we refer to f as a *monic polynomial.* A proof of the following theorem and its corollary is left as an exercise.

□ **Theorem 1.1.** Let f and g be nonzero polynomials over F. Then
(a) fg is a nonzero polynomial and

$$\deg(fg) = \deg f + \deg g;$$

(b) if $f + g \neq 0$, then

$$\deg(f+g) \leq \max(\deg f, \deg g).$$

□ **Corollary.** Let $f, g, h \in F[x]$. Then $fg = 0$ implies that $f = 0$ or $g = 0$. If $fg = hg$ and $g \neq 0$, then $f = h$.

Having devised a representation for elements of $F[x]$, which has the form of the value of a polynomial function, an explanation of the difference between the two possible interpretations of the expression

$$a_0 + a_1x + \cdots + a_n x^n \tag{1}$$

is in order. As a polynomial, (1) is simply a denotation of the sequence $(a_0, a_1, \ldots, a_n, 0, \ldots)$. In this notation x denotes the sequence $(0, 1, 0, \ldots)$, a_i denotes $(a_i, 0, 0, \ldots)$, the symbol + denotes addition of sequences, and juxtaposition and exponents denote multiplication of sequences. The merit of this notation for a polynomial is the fact that it makes possible the calculation of sums and products of polynomials in accordance with the rules of elementary algebra. Now the polynomial (1) defines a polynomial function on F into F in a natural way, namely, the function whose value at c (in F) is

$$a_0 + a_1c + \cdots + a_n c^n. \tag{2}$$

If the polynomial (1) is denoted by f, then the polynomial function defined by f is denoted by $f(x)$† and its value (2) at c, by $f(c)$.

† This is a departure from our policy of designating functions simply by an alphabetic character; we do it solely for the sake of clarity.

The set $P_\infty(F)$ of all polynomial functions on F into F, together with the operations of addition and multiplication of functions, is a commutative ring with identity. Properties of the mapping of $F[x]$ onto $P_\infty(F)$ already defined are given below.

□ **Theorem 1.2.** The mapping φ of $F[x]$ onto $P_\infty(F)$ such that $f\varphi = f(x)$ preserves sums and products; that is, if $f, g \in F[x]$ and $c \in F$, then

$$(f+g)(c) = f(c) + g(c) \quad \text{and} \quad (fg)(c) = f(c)\,g(c).$$

PROOF. Let

$$f = a_0 + a_1 x + \cdots + a_n x^n \quad \text{and} \quad g = b_0 + b_1 x + \cdots + b_m x^m.$$

Then

$$\begin{aligned}(f+g)(c) &= (a_0 + b_0) + (a_1 + b_1)c + \cdots \\ &= (a_0 + a_1 c + \cdots) + (b_0 + b_1 c + \cdots) \\ &= f(c) + g(c).\end{aligned}$$

Further,

$$\begin{aligned}(fg)(c) &= a_0 b_0 + (a_0 b_1 + a_1 b_0)c + (a_0 b_2 + a_1 b_1 + a_2 b_0)c^2 + \cdots \\ &= a_0(b_0 + b_1 c + b_2 c^2 + \cdots) + a_1(b_0 c + b_1 c^2 + \cdots) \\ &\qquad\qquad + a_2(b_0 c^2 + \cdots) + \cdots \\ &= a_0(b_0 + b_1 c + b_2 c^2 + \cdots) + a_1 c(b_0 + b_1 c + b_2 c^2 + \cdots) \\ &\qquad\qquad + a_2 c^2(b_0 + \cdots) + \cdots \\ &= (a_0 + a_1 c + a_2 c^2 + \cdots)(b_0 + b_1 x + b_2{}^2 c^2 + \cdots) = f(c)\,g(c). \quad \diamond\end{aligned}$$

The properties of φ are summarized by calling it a *homomorphism* of $F[x]$ onto $P_\infty(F)$. In general, φ is not 1-1 (when it is, we call it an *isomorphism*); for example, the unequal polynomials $x^2 + x^5$ and $x + x^6$ of $Z_5[x]$ determine the same polynomial function over Z_5, as may be verified directly. If, however, F is a field of infinitely many elements, then φ is an isomorphism. Since we have no need for this result a proof is not given.

Our reason for giving the proof of Theorem 1.2 in such detail is that it has a wider range of applicability than to the replacement of the symbol x by a field element. To show this a definition is required.

☐ **Definition.** Let F be a field, V be a vector space over F, A be a linear transformation on V, and $f = a_0 + a_1 x + \cdots + a_n x^n \in F[x]$. Then, by $f(A)$ we shall understand the linear transformation

$$a_0 + a_1 A + \cdots + a_n A^n$$

on V. (Observe that the scalars $a_0, a_1, \ldots, a_n$ now play the role of scalar transformations.) Further, if (a_{ij}) is an $m \times m$ matrix over F, then by $f(a_{ij})$ we shall understand the matrix

$$a_0 I + a_1(a_{ij}) + \cdots + a_n(a_{ij})^n,$$

where I is the $m \times m$ identity matrix.

☐ **Theorem 1.3.** Let F be a field, V a vector space over F, A a linear transformation in V, and $f, g \in F[x]$. Then

$$(f + g)(A) = f(A) + g(A) \quad \text{and} \quad (fg)(A) = f(A)\,g(A).$$

Similar results hold for matrices.

PROOF. The proof of Theorem 1.2 may be repeated. In the proof of the second assertion (concerning fg) we now rely on the fact that scalar transformations commute with all linear transformations to justify some of the algebraic manipulations. ◇

It is this result that is the basis of our interest in polynomials in connection with the study of a linear transformation A, for we shall find that properties of polynomials yield properties of polynomials in a linear transformation A (via Theorem 1.3) that play an important role in the analysis of its structure. We proceed now to statements of those properties of polynomials that are pertinent.

☐ **Theorem 1.4** (Division algorithm). Let $f, g \in F[x]$, with $g \neq 0$. Then there exist uniquely determined polynomials q, r in $F[x]$ (called the quotient and remainder, respectively, on division of f by g) such that

$$f = gq + r,$$

where either $r = 0$ or $\deg r < \deg g$.

The proof of this theorem, as well as most of those that follow, is omitted; the proofs that the reader should have seen in his elementary algebra text carry over to the case of polynomials as now defined.

On setting $g = x - a$, $a \in F$, it is an easy matter to establish the following two corollaries of Theorem 1.4.

☐ **Corollary 1** (Remainder theorem). Let $f \in F[x]$ and $a \in F$. Then

$$f = (x - a)q + f(a).$$

☐ **Corollary 2** (Factor theorem). Let $f \in F[x]$ and $a \in F$. Then $f = (x - a)q$ for some $q \in F[x]$ if and only if $f(a) = 0$.

If $f \in F[x]$, an element a of F such that $f(a) = 0$ is called a *zero* of f; a zero of f is also called a *root* of the polynomial equation $f(x) = 0$.

In order to present the remaining theorems that we shall need about polynomials, some definitions are required. If f and g are polynomials in $F[x]$, such that for some $h \in F[x]$, $g = fh$, then we say that "*f divides g*" or "f is a *factor* of g" or "g is a *multiple* of f" and write

$$f \mid g.$$

An element u of $F[x]$ is called a *unit* if and only if $u \mid 1$; clearly the units of $F[x]$ are the polynomials of degree 0. A polynomial p is called *irreducible* (or *prime* over F) if it is neither 0 nor a unit and $p = fg$ for polynomials f and g in $F[x]$ implies that either f or g is a unit. Two distinct polynomials are *relatively prime* if and only if their only common divisors are units. Finally, a polynomial d is called a *greatest common divisor* (gcd) of polynomials $f_1, \ldots, f_k$ if and only if $d \mid f_i$, $1 \le i \le k$, and, if d' is a polynomial such that $d' \mid f_i$, $1 \le i \le k$, then $d' \mid d$.

☐ **Theorem 1.5.** Let $f_1, \ldots, f_k$ be nonzero polynomials in $F[x]$. Then $f_1, \ldots, f_k$ possess *a* gcd d that is uniquely determined within a unit factor and can be expressed in the form

$$d = h_1 f_1 + \cdots + h_k f_k$$

for suitable polynomials $h_1, \ldots, h_k$ in $F[x]$.

Proof. Let S be the set of all polynomials of the form $\sum_1^k g_i f_i$ with $g_i \in F[x]$, $1 \le i \le k$. Then three basic properties of S are: each f_i is in S, if $p \in S$ and $q \in F[x]$, then $pq \in S$ and, if $p, q \in F[x]$, then $p + q \in F[x]$.

Consider now the (nonempty) set of all (nonnegative) integers that occur as the degree of a nonzero element of S. This set has a least number and there is a polynomial d in S of this degree. Thus

$$d = \sum_{i=1}^{k} h_i f_i$$

and $\deg d \le \deg d'$ for a nonzero element d' of S.

We prove now that $d \mid f_i$, $1 \le i \le k$. By the division algorithm $f_i = dq_i + r_i$, where $r_i = 0$ or deg $r_i <$ deg d. The latter possibility can be ruled out by using the three basic properties of S. Hence $r_i = 0$ and $d \mid f_i$.

Next, let d' be another common divisor of $f_1, \ldots, f_k$. Thus $f_i = g_i d'$ for a polynomial g_i, $1 \le i \le k$, and so

$$d = \sum_{i=1}^{k} h_i f_i = \sum_{i=1}^{k} h_i g_i d' = \left(\sum_{i=1}^{k} h_i g_i\right) d'.$$

Thus $d' \mid d$ and we have proved that d is a gcd.

Finally, let e be another gcd of $f_1, \ldots, f_\alpha$. Then $d \mid e$ and $e \mid d$. Henee there exists polynomials r and s such that $e = rd$ and $d = se$. It follows, in turn, that $e = rse$, $e(1 - rs) = 0$, and $1 - rs = 0$ (by the corollary to Theorem 1.1); but $1 - rs = 0$ implies that r and s are units, using part (a) of Theorem 1.1. ◇

☐ **Corollary 1.** Suppose that $f_1, \ldots, f_k$ are polynomials in $F[x]$ whose only common factors are units. Then there exist polynomials $h_1, \ldots, h_k$ in $F[x]$ such that

$$\sum_{i=1}^{k} h_i f_i = 1.$$

☐ **Corollary 2.** Let p be an irreducible polynomial in $F[x]$ and suppose that $p \mid fg$, where $f, g \in F[x]$. Then either $p \mid f$ or $p \mid g$.

Complementary to the notion of a gcd is that of a *least common multiple* (lcm) of polynomials $f_1, \ldots, f_k$ as a polynomial m, which is a multiple of each f_i, and such that if m' is a common multiple of each f_i then $m \mid m'$. It is left as an exercise to show that if d is a gcd of $f_1, \ldots, f_k$ then there is an lcm m of $f_1, \ldots, f_k$ such that

$$md = f_1 \cdots f_k.$$

☐ **Theorem 1.6** (Unique factorization theorem). A polynomial f of $F[x]$ of positive degree can be expressed uniquely, apart from the order of the factors, in the form

$$f = ap_1^{k_1} \cdots p_r^{k_r},$$

where a is the leading coefficient of f, the p_i's are irreducible monic polynomials, and the k_i's are positive integers.

PROOF. The existence of some factorization of f into a product of irreducible polynomials is easily proved by induction on $\deg f$.

To prove the uniqueness property we show first that if

$$f = p_1 \cdots p_s = q_1 \cdots q_t, \tag{3}$$

where the p's and q's are irreducible, then $s = t$ and, for a suitable indexing of the p's and q's,

$$p_1 = u_1 q_1, \ldots, p_s = u_s q_s,$$

where the u's are units. The proof is by induction on s. The result is trivial if $s = 1$. Starting with (3) for $s > 1$, we conclude from Corollary 2 of Theorem 1.5 that p_1 divides some q_j and we may assume that $j = 1$. Then $q_1 = u_1 p_1$ for a unit u_1. Thus

$$p_1 \cdots p_s = u_1 p_1 q_2 \cdots q_t,$$

and by the cancellation law

$$p_2 \cdots p_s = u_1 q_2 \cdots q_t = q_2' \cdots q_t.$$

The result then follows by the principle of induction.

The necessary "tidying up" to obtain the statement in the theorem is left as an exercise. ◇

In the event that F is the field C of complex numbers, the only irreducible monic polynomials have the form $x - a$†, and so the preceding theorem reads as follows:

□ **Theorem 1.7.** An element f of C$[x]$ of positive degree can be expressed uniquely, apart from the order of the factors, in the form

$$f = a(x - c_1)^{k_1} \cdots (x - c_r)^{k_r},$$

where a is the leading coefficient of f, the c_i are distinct complex numbers, and the k_i are positive integers.

EXERCISES

1.1. Prove that the system P satisfies (M1) to (M3) and (D) of Section 1.4.

1.2. Prove Theorem 1.1 and its corollary.

† This fact is a consequence of the "fundamental theorem of algebra" which is the assertion that every polynomial $f \in$ C$[x]$ of positive degree has at least one zero in C.

1.3. Give an example (other than the one in the text!) of two distinct polynomials that determine the same polynomial function.
1.4. Using Theorem 1.3, show that two polynomials in a linear transformation commute.
1.5. Prove Theorem 1.4.
1.6. Give proofs for the corollaries to Theorem 1.5.
1.7. Show that if d is a gcd of polynomials $f_1, \ldots, f_k$, then there exists a lcm m of $f_1, \ldots, f_k$ such that $md = f_1 \cdots f_k$.
1.8. Complete the proof of Theorem 1.7.

2. MINIMAL POLYNOMIALS

Let V be an n-dimensional vector space over a field F. Then, since $\mathscr{L}(V, V)$ is of dimension n^2 (see Section 4.3), if $A \in \mathscr{L}(V, V)$, $\{1, A, \ldots, A^{n^2}\}$ is a linearly dependent set. Consequently, there exist scalars $c_0, \ldots, c_{n^2}$ not all zero such that

$$c_0 + c_1 A + \cdots + c_{n^2} A^{n^2} = 0.$$

In other words, there exists a polynomial f of positive degree in $F[x]$ such that $f(A) = 0$. These remarks launch us into the proof of the theorem that follows.

□ **Theorem 2.1.** Assume that V is an n-dimensional vector space over a field F and that $A \in \mathscr{L}(V, V)$. Let r be the least positive integer such that the set $\{1, A, \ldots, A^r\}$ is linearly dependent. Then there exist scalars $b_0, \ldots, b_{r-1}$ such that

$$A^r = b_0 + b_1 A + \cdots + b_{r-1} A^{r-1}$$

and the polynomial

$$m = x^r + a_{r-1} x^{r-1} + \cdots + a_1 x + a_0,$$

where $a_i = -b_i$, $1 \leq i \leq r-1$, has the following properties:

(a) The degree of m is positive and $m(A) = 0$.
(b) If $f \in F[x]$ and $f(A) = 0$, then $m \mid f$.
(c) m is a monic polynomial.

PROOF. The existence of a monic polynomial m with the property stated in (a) follows from our preliminary remarks. To prove (b) suppose that $f \in F[x]$ and $f(A) = 0$. We apply the division algorithm to obtain

$$f = mq + u,$$

where q, $u \in F[x]$ and $u = 0$ or $\deg u < \deg m$. In view of Theorem 1.3, we may infer that

$$f(A) = m(A)\,q(A) + u(A)$$

and in turn that $u(A) = 0$. It follows that we must have $u = 0$ and consequently that $m \mid f$. ◇

☐ **Definition.** Let V be a vector space over a field F and $A \in \mathscr{L}(V, V)$. A polynomial $m \in F[x]$ is called a *minimal polynomial* of A if and only if m has properties (a), (b), (c) given in Theorem 2.1.

It is left as an exercise to show that (i) if a linear transformation has a minimal polynomial then it is unique and (ii) if A is a linear transformation for which there exists a polynomial f such that $f(A) = 0$ then A has a minimal polynomial. In terms of our definition we may restate the primary part of Theorem 2.1 as follows:

☐ **Theorem 2.2.** A linear transformation on a finite-dimensional vector space has a (unique) minimal polynomial.

EXAMPLES

2.1. The derivative operator D on the space $P_\infty(R)$ has no minimal polynomial, since, for each positive integer s, $x^{s+1}D^s \neq 0$. On the other hand, when regarded as a linear transformation on $P_n(R)$, D has a minimal polynomial.

2.2. Let S be a subspace of a vector space V and suppose S' is a complement of S. Then, if $\alpha \in V$, $\alpha = \alpha_S + \alpha_{S'}$. Define $A: V \to V$ by

$$\alpha A = \alpha_S \quad \text{if} \quad \alpha = \alpha_S + \alpha_{S'}.$$

Then $A \in \mathscr{L}(V, V)$ and $A^2 = A$. Thus $f(A) = 0$ if $f = x^2 - x$. It follows that A has a minimal polynomial m and that $m \mid f$. It is left as an exercise to show that $m = f$ if and only if S is a proper subspace of V. Observe that $m = x$ if $S = \{0\}$ and $m = x - 1$ if $S = V$.

2.3. Let A be the linear transformation on R^3 such that

$$(x, y, z)A = (x + y, y, 2z).$$

Then $m = (x - 1)^2(x - 2)$ is the minimal polynomial of A. This may be proved by showing that $m(A) = 0$ and that for each factor f of m, $f(A) \neq 0$.

2.4. Let A be the linear transformation on R^3 induced by the matrix

$$(a_{ij}) = \begin{bmatrix} 1 & -1 & 0 \\ -1 & 1 & 0 \\ 0 & 0 & 2 \end{bmatrix}$$

relative to the standard basis. Thus

$$(x, y, z)A = (x - y, -x + y, 2z).$$

Then $m = x(x - 2)$ is the minimal polynomial of A. Moreover, $(a_{ij})^2 - 2(a_{ij})$ is the zero matrix.

□ **Theorem 2.3.** If A and B are similar linear transformations on a vector space V and if A has a minimal polynomial m, then m is also the minimal polynomial of B.

PROOF. Assume that A and B are similar and that A has a minimal polynomial

$$m = x^r + a_{r-1}x^{r-1} + \cdots + a_1 x + a_0 .$$

Suppose that the linear transformation P demonstrates the similarity of A and B; thus $B = P^{-1}AP$. Then

$$\begin{aligned} m(B) = m(P^{-1}AP) &= (P^{-1}AP)^r + a_{r-1}(P^{-1}AP)^{r-1} + \cdots \\ &\qquad + a_1(P^{-1}AP) + a_0 \\ &= P^{-1}A^rP + a_{r-1}P^{-1}A^{r-1}P + \cdots + a_1(P^{-1}AP) + P^{-1}a_0 P \\ &= P^{-1}(m(A))P = 0. \end{aligned}$$

It follows that B has a minimal polynomial m' and that $m' \mid m$. If we reverse the roles of A and B in the displayed computations, it follows that $m \mid m'$. Thus, since m and m' are both monic, $m = m'$. ◇

Theorem 2.1 is true for an $n \times n$ matrix (a_{ij}) over a field F. Thus we can define the minimal polynomial of a square matrix as the polynomial m with the properties (a), (b), and (c) stated in Theorem 2.1. Now suppose that V is an n-dimensional vector space, $A \in \mathscr{L}(V, V)$, and m is the minimal polynomial of A. If (a_{ij}) is a matrix of A, then $m((a_{ij})) = 0$ by virtue of Theorem 4.3.3. Conversely, if m' is the minimal polynomial of (a_{ij}) and A is a linear transformation on V induced by (a_{ij}), the same theorem implies that $m'(A) = 0$. It then follows from Theorem 2.1 that a linear transformation on a finite-dimensional vector space and each of its matrices has the same minimal polynomial. Observe that in view of Theorem 5.2.1 we now have an alternative proof of Theorem 2.3 for the case of linear transformations on finite-dimensional spaces.

Since we have discussed an effective procedure for extracting a maximal linearly independent set from a set of vectors, there is an effective procedure for determining the minimal polynomial of a linear transformation A on a finite-dimensional vector space: determine a matrix (a_{ij}) of A and then the first member of the sequence

$$\{I, (a_{ij})\}, \{I, (a_{ij}), (a_{ij})^2\}, \ldots,$$

which is linearly dependent.

EXAMPLE

2.5. Define the linear transformation A on R^3 as follows:

$$(x, y, z)A = (y, x - z, -y).$$

The matrix of A relative to the standard basis of R^3 is

$$(a_{ij}) = \begin{bmatrix} 0 & 1 & 0 \\ 1 & 0 & -1 \\ 0 & -1 & 0 \end{bmatrix}.$$

A calculation shows that $\{I, (a_{ij}), (a_{ij})^2\}$ is linearly independent and that $(a_{ij})^3 = 2(a_{ij})$. Thus, since

$$(a_{ij})^3 = 0I + 2(a_{ij}) + 0(a_{ij})^2,$$

$x^3 - 2x$ is the minimal polynomial of both (a_{ij}) and A.

With minimal polynomials still in mind, we recall the earlier definition of an A-invariant subspace of a vector space V (here A is a linear transformation on V) as a subspace such that $\sigma A \in S$ for all $\sigma \in S$; if S is A-invariant then A induces a linear transformation on S. Consider now, for a vector α of a vector space V and a given linear transformation A on V, the collection $\mathscr{C}$ of all subspaces of V that contain α and are A-invariant. Since $V \in \mathscr{C}$, $\mathscr{C}$ is nonempty and the intersection $C_{\alpha,A}$ over $\mathscr{C}$ exists. It is immediate that $C_{\alpha,A} \in \mathscr{C}$, hence that $C_{\alpha,A}$ is the smallest subspace of V that contains α and is A-invariant. The space $C_{\alpha,A}$ is called the *cyclic subspace of* α *relative to* A. Clearly, $\{\alpha, \alpha A, \alpha A^2, \ldots\}$ span $C_{\alpha,A}$. In case V in finite-dimensional, we can establish the following more definitive results:

□ **Theorem 2.4.** Let A be a linear transformation on the finite-dimensional vector space V, and for the nonzero α in V let s be the least positive integer such that $\{\alpha, \alpha A, \ldots, \alpha A^s\}$ is linearly dependent. Then there exist scalars $b_0, \ldots, b_{s-1}$ not all zero such that

$$\alpha A^s = b_0 \alpha + b_1 \alpha A + \cdots + b_{s-1} \alpha A^{s-1}. \tag{1}$$

The polynomial

$$m_{\alpha,A} = x^s - b_{s-1}x^{s-1} - \cdots - b_1 x - b_0$$

is the minimal polynomial of the linear transformation induced on the space $C_{\alpha,A}$ by A. Also, dim $C_{\alpha,A} = s$.

PROOF. There should be no questions in regard to the existence of the integer s and the relation (1). Let $N = [\alpha, \alpha A, \ldots, \alpha A^{s-1}]$. Since $\{\alpha, \alpha A, \ldots, \alpha A^{s-1}\}$ is linearly independent, dim $N = s$. Also, $N \subseteq C_{\alpha,A}$ and, on the other hand, since $\alpha \in N$ and N is A-invariant [in view of (1)], $C_{\alpha,A} \subseteq N$. Hence $C_{\alpha,A} = N$ and so dim $C_{\alpha,A} = s$.

Since $\alpha[m_{\alpha,A}(A)] = 0$, it follows that $\beta[m_{\alpha,A}(A)] = 0$ for all $\beta \in C_{\alpha,A}$. Hence, if p is the minimal polynomial of the linear transformation A_α induced on $C_{\alpha,A}$ by A, then $p \mid m_{\alpha,A}$ or, in other words, $m_{\alpha,A} = pq$ for some q in $F[x]$. Now let $t = \deg p$ and assume that $t < s$. Then there exist scalars $c_0, \ldots, c_{t-1}$ such that $\alpha A^t + c_{t-1}\alpha A^{t-1} + \cdots + c_1 \alpha A + c_0 \alpha = 0$, contrary to our choice of s as the least integer for which such a linear combination is 0. Hence $\deg p = s$ and it follows that $m_{\alpha,A} = p$. ◇

Next we single out an extreme, but important, instance of Theorem 2.4.

□ **Theorem 2.5.** Let V be an n-dimensional vector space, $A \in \mathscr{L}(V, V)$, and α a vector in V such that $C_{\alpha,A} = V$. Then $m_{\alpha,A}$ is the minimal polynomial of A and $(\alpha, \alpha A, \ldots, \alpha A^{n-1})$ is an ordered basis of V. If $m_{\alpha,A} = x^n - b_{n-1}x^{n-1} - \cdots - b_1 x - b_0$, then the matrix of A relative to this basis is

$$\begin{bmatrix} 0 & 1 & 0 & \cdots & 0 \\ 0 & 0 & 1 & \cdots & 0 \\ \vdots & & & & \\ 0 & 0 & 0 & \cdots & 1 \\ b_0 & b_1 & b_2 & \cdots & b_{n-1} \end{bmatrix}.$$

The next theorem offers an alternative to the method suggested after Theorem 2.3 for computing the minimal polynomial of a linear transformation on a finite-dimensional vector space.

☐ **Theorem 2.6.** Let V be an n-dimensional vector space and $\{\alpha_1, \ldots, \alpha_n\}$ be a basis of V. If $A \in \mathscr{L}(V, V)$, then the minimal polynomial m of A is the monic lcm of $m_{\alpha_1,A}, \ldots, m_{\alpha_n,A}$ (in the notation of Theorem 2.4).

PROOF. Let f be the monic lcm of $m_{\alpha_1,A}, \ldots, m_{\alpha_n,A}$. Then $f = m_{\alpha_1,A} g_i$ for a suitable polynomial g_i, $1 \le i \le n$. Also,

$$\alpha_i f(A) = \alpha_i m_{\alpha_1,A}(A)\, g_i(A) = 0, \qquad 1 \le i \le n.$$

Hence $f(A) = 0$ and f is a multiple of m. On the other hand,

$$\alpha_i m(A) = 0, \qquad 1 \le i \le n,$$

and so m is a multiple of each $m_{\alpha_i,A}$. In turn, m is a multiple of f; hence $m = f$, since both are monic polynomials. ◇

EXAMPLES

2.6. Let A be the linear transformation on R^3 induced by the matrix

$$\begin{bmatrix} 0 & -1 & 1 \\ 1 & 1 & 0 \\ -1 & 0 & 1 \end{bmatrix}$$

relative to the standard basis of R^3. Then

$$\begin{aligned} \epsilon_1 A &= (0, -1, 1) = -\epsilon_2 + \epsilon_3, \\ \epsilon_2 A &= (1, 1, 0) \quad = \epsilon_1 + \epsilon_2, \\ \epsilon_3 A &= (-1, 0, 1) = -\epsilon_1 + \epsilon_3. \end{aligned}$$

Hence $\epsilon_1 A^2 = (-2, -1, 1)$, and since

$$(-2, -1, 1) = (0, -1, 1) - 2(1, 0, 0)$$

we conclude that

$$m_{\epsilon_1,A} = x^2 - x + 2.$$

Further,

$$\epsilon_2 A = (1, 1, 0), \qquad \epsilon_2 A^2 = (1, 0, 1), \qquad \epsilon_2 A^3 = (-1, -1, 2).$$

Since $\{\epsilon_2, \epsilon_2 A, \epsilon_2 A^2\}$ is linearly independent and

$$\epsilon_2 A^3 = (-1, -1, 2) = 2(1, 0, 1) - 3(1, 1, 0) + 2(0, 1, 0),$$

we conclude that

$$m_{\epsilon_2,A} = x^3 - 2x^2 + 3x - 2.$$

Since $C_{\epsilon_2,A} = \mathsf{R}^3$, it follows that $m_{\epsilon_2,A}$ is the minimum polynomial of A. According to Theorem 2.6, each $m_{\epsilon_i,A}$ divides the minimum polynomial of A. In the case at hand it follows that $m_{\epsilon_1,A} | m_{\epsilon_2,A}$; indeed, $m_{\epsilon_2,A} = (x-1)m_{\epsilon_1,A}$.

2.7. Let us find the minimal polynomial of

$$(a_{ij}) = \begin{bmatrix} 1 & -2 & 0 \\ -2 & 1 & 0 \\ 0 & 0 & -1 \end{bmatrix}.$$

If A is the linear transformation induced by (a_{ij}) on R^3 relative to the standard basis, then

$$(x, y, z)A = (x - 2y, -2x + y, -z).$$

Computations like those in the preceding example yield the following results:

$$m_{\epsilon_1,A} = m_{\epsilon_2,A} = x^2 - 2x - 3, \qquad m_{\epsilon_3,A} = x + 1.$$

Hence $x^2 - 2x - 3$ is the minimum polynomial of A and of (a_{ij}).

Thus, in turn,

$$(a_{ij})^2 - 2(a_{ij}) - 3I = O,$$

$$3I = (a_{ij})[(a_{ij}) - 2I],$$

$$I = (a_{ij})[\tfrac{1}{3}(a_{ij}) - \tfrac{2}{3}I].$$

We infer that (a_{ij}) is invertible and

$$(a_{ij})^{-1} = \tfrac{1}{3}(a_{ij}) - \tfrac{2}{3}I = \tfrac{1}{3}\begin{bmatrix} -1 & -2 & 0 \\ -2 & -1 & 0 \\ 0 & 0 & -3 \end{bmatrix}.$$

We conclude this section with two remarks about linear transformations that have minimal polynomials. If

$$m = x^r + a_{r-1}x^{r-1} + \cdots + a_1 x + a_0$$

is the minimal polynomial of A, then A has an inverse if and only if $a_0 \neq 0$ and, in that event,

$$A^{-1} = -\frac{1}{a_0}(A^{r-1} + a_{r-1}A^{r-2} + \cdots + a_1).$$

Also, since

$$A^r = -(a_{r-1}A^{r-1} + \cdots + a_1 A + a_0),$$

integral powers A^s, with $s > r$, can be expressed as linear combinations of $\{1, A, \ldots, A^{r-1}\}$.

EXERCISES

2.1. Show that if the minimal polynomial of a linear transformation exists it is unique.

2.2. Show that if A is a linear transformation and p is a polynomial such that $p(A) = 0$ the minimal polynomial of A exists.

2.3. Show that a nilpotent linear transformation has a minimal polynomial and find it.

2.4. For each of the matrices

$$\begin{bmatrix} 1 & -1 & 0 \\ -1 & 2 & 3 \\ 0 & 3 & 2 \end{bmatrix}, \quad \begin{bmatrix} 1+i & 2 \\ 2 & 2-i \end{bmatrix}, \quad \begin{bmatrix} -1 & 0 & 1 & 0 \\ 1 & 2 & 0 & 1 \\ -1 & 0 & 0 & -1 \\ 0 & 1 & 0 & 0 \end{bmatrix},$$

find
(a) the minimal polynomial,
(b) the inverse, if it exists,
(c) the fifth power.

2.5. Find the minimal polynomial of the linear transformation A such that
(a) $A \in \mathscr{L}(\mathsf{R}^3, \mathsf{R}^3)$ and $(x, y, z)A = (3x - y, x + z, x - y + z)$.
(b) $A \in \mathscr{L}(\mathsf{C}^3, \mathsf{C}^3)$ and
$(x, y, z)A = (x - iy - z, ix + 2y + (1 + i)z, -x + (1 - i)y + z)$.

2.6. Let $V = P_n(\mathsf{R})$ and D be the differentiation operator. What is the minimal polynomial of D? What is the minimal polynomial of the linear transformation A on V defined by $p(x)A = p(x + 1)$?

2.7. Prove that the minimal polynomial of a 2×2 matrix over a field F has degree at most 2 by showing that if

$$(a_{ij}) = \begin{bmatrix} a & b \\ c & d \end{bmatrix}$$

and $p = x^2 - (a + d)x + (ad - bc)$, then $p((a_{ij})) = O$.

2.8. Prove or disprove that if two 2×2 matrices over R have the same minimal polynomial they are similar.

2.9. Find two 3×3 matrices over R that have the same minimal polynomial but are not similar.

2.10. Show that two idempotent linear transformations on a finite dimensional vector space are similar if and only if they have the same rank. Find two nonsimilar idempotent linear transformations on R^3 that have the same minimal polynomial.

2.11. Let V be a finite-dimensional vector space, let A be a linear transformation on V, and $\alpha, \beta \in V$.
(a) Show that if $m_{\alpha,A}$ and $m_{\beta,A}$ are relatively prime, then $C_{\alpha,A} \cap C_{\beta,A} = \{0\}$ and $m_{\alpha+\beta,A} = m_{\alpha,A} m_{\beta,A}$.
(b) Show that if h is a monic divisor of $m_{\alpha,A}$ then there exists $\gamma \in V$ such that $h = m_{\gamma,A}$.
(c) Show that if q is an irreducible factor of the minimal polynomial m of A and $m = q^m k$, where q and k are relatively prime, then there exists $\gamma \in V$ such that $m_{\gamma,A} = q^m$.
(d) If m is the minimal polynomial of A, does there always exist a vector α in V such that $m = m_{\alpha,A}$?

2.12. Suppose that $m = x^r + a_{r-1}x^{r-1} + \cdots + a_1 x + a_0$ is the minimal polynomial of a linear transformation A in a vector space V. Show that A has an inverse if and only if $a_0 \neq 0$ and, if $a_0 \neq 0$, then

$$A^{-1} = -\frac{1}{a_0}(A^{r-1} + a_{r-1}A^{r-2} + \cdots + a_1).$$

2.13. Suppose that m is the minimal polynomial of a linear transformation A on a vector space V and that c is a root of the polynomial equation $m(x) = 0$. Show that there exists a nonzero α in V such that $\alpha A = c\alpha$.

2.14. Show that two nilpotent linear transformations in a three-dimensional vector space are similar if and only if they have the same minimal polynomial.

2.15. Let A be a linear transformation in a finite-dimensional inner-product space. Show that A and A^* have the same minimal polynomial.

2.16. Show by an example that if A and B are linear transformations on a vector space V then AB and BA need not have the same minimal polynomial.

2.17. Let

$$(a_{ij}) = \begin{bmatrix} 1 & 1 & 0 \\ 0 & -1 & 0 \\ 0 & 0 & 1 \end{bmatrix}.$$

(a) Prove that if (b_{ij}) is a 3×3 matrix over C such that $(b_{ij})^2 = (a_{ij})$, then the minimal polynomial of (b_{ij}) is a factor of $x^4 - 1$.
(b) Determine all matrices (b_{ij}) such that $(b_{ij})^2 = (a_{ij})$.

3. CHARACTERISTIC VALUES AND VECTORS

We begin with definitions of the concepts named in the heading of this section.

☐ **Definition.** Let V be a vector space, $A \in \mathscr{L}(V, V)$, and c a scalar such that

$$\alpha A = c\alpha$$

for a nonzero vector α in V. Then c is called a *characteristic value* of A and α is called a *characteristic vector* of A belonging to the characteristic value c.

Clearly, a scalar c is a characteristic value of the linear transformation A on a vector space V if and only if the linear transformation $c - A$ is singular. Further, we can prove that A has a characteristic value if and only if V has a one-dimensional A-invariant subspace. Indeed, if c is a characteristic value of A and α is a characteristic vector belonging to c, then the one-dimensional space $[\alpha]$ is A-invariant. Conversely, let S be a one-dimensional A-invariant subspace of V. Then there exists a nonzero vector α in S and $S = [\alpha]$. Since $\alpha A \in S$, there exists a scalar c such that $\alpha A = c\alpha$. Thus c is a characteristic value of A. In passing, we note that an easy calculation shows that this characteristic value of A is independent of α in S.

A scalar c of a field F is called a characteristic value of an $n \times n$ matrix (a_{ij}) over F if and only if the matrix $cI - (a_{ij})$ is singular. It is left as an exercise to show that a scalar c is a characteristic value of (a_{ij}) if and only if c is a characteristic value of any linear transformation induced by (a_{ij}).

☐ **Theorem 3.1.** Let V be a vector space and A be a linear transformation on V. If A has a minimal polynomial m, then a scalar c is a characteristic value of A if and only if c is a root of the polynomial equation $m(x) = 0$.

Proof. Let $m = x^r + a_{r-1}x^{r-1} + \cdots + a_1 x + a_0$ and assume that c is a characteristic value of A. Then there exists a nonzero vector α of V such that $\alpha A = c\alpha$. It follows that for each scalar a and each positive integer

i, $\alpha(aA^i) = (ac^i)\alpha$. Hence

$$0 = \alpha m(A) = \alpha(A^r + a_{r-1}A^{r-1} + \cdots + a_1 A + a_0)$$
$$= (c^r + a_{r-1}c^{r-1} + \cdots + a_1 c + a_0)\alpha.$$

Since $\alpha \neq 0$, we may infer that

$$c^r + a_{r-1}c^{r-1} + \cdots + a_1 c + a_0 = 0$$

or, in other words, $m(c) = 0$.

Conversely, assume that $m(c) = 0$. Then m has $x - c$ as a factor and, hence, has the form $m = q(x - c)$ where q is a polynomial of degree $r - 1$. Hence $q(A) \neq 0$, and so there exists a vector β such that $\beta q(A) \neq 0$. Then

$$0 = \beta m(A) = (\beta q(A))(A - c)$$

and, in turn,

$$(\beta q(A))A = c(\beta q(A)),$$

which means that c is a characteristic value of A. ◇

EXAMPLES

3.1. In Example 2.3 we remarked that $m = (x - 1)^2(x - 2)$ is the minimal polynomial of the linear transformation A on R^3 such that

$$(x, y, z)A = (x + y, y, 2z).$$

This result may be established by using the method employed in Example 2.5. We find that $m_{\epsilon_1,A} = x - 1$, $m_{\epsilon_2,A} = (x - 1)^2$, and $m_{\epsilon_2,A} = x - 2$. It follows that the characteristic values of A are 1 and 2.

3.2. Let V be the vector space of all functions on R into R that have derivatives of all orders. Then every scalar c is a characteristic value of the derivative operator D on V, since $e^{cx}D = ce^{cx}$.

3.3. Let A be the linear transformation on R^2 induced by

$$\begin{bmatrix} \cos\theta & -\sin\theta \\ \sin\theta & \cos\theta \end{bmatrix}, \qquad 0 < \theta < \pi,$$

relative to the standard basis. Then $m = x^2 - (2\cos\theta)x + 1$ and A has no characteristic values in R. This conclusion also follows from the interpretation of elements of R^2 as points of E_2. Then A defines a rotation through θ radians, and, clearly, E_2 has no one-dimensional A-invariant subspaces.

3.4. Suppose that c is a characteristic value of a linear transformation A in a vector space V. Let

$$S_c = \{\alpha \in V \mid \alpha A = c\alpha\}$$

It is then clear that S_c is an A-invariant subspace of V and that the nonzero elements of S_c are the characteristic vectors of A belonging to c. If S_c is finite-dimensional, the matrix of the linear transformation induced in S_c by A is the scalar matrix cI relative to any basis of S_c.

☐ **Lemma 3.1.** Let V be a vector space, $A \in \mathscr{L}(V, V)$ and let $\{c_i \mid i \in I\}$ be a collection of distinct characteristic values of A. (Here I is a set that serves to index the characteristic values under consideration.) If α_i is a characteristic vector belonging to c_i, $i \in I$, then $\{\alpha_i \mid i \in I\}$ is a linearly independent set of vectors in V.

PROOF. Assume that $\{\alpha_i \mid i \in I\}$ is not linearly independent. Then there exists a finite subset S that is linearly dependent. With no loss of generality, we may assume that $S = \{\alpha_1, \ldots, \alpha_k\}$, where no proper subset of S is linearly dependent. Therefore there exist scalars $a_1, \ldots, a_{k-1}$, not all 0, such that

$$\alpha_k = a_1\alpha_1 + \cdots + a_{k-1}\alpha_{k-1}.$$

Hence

$$\alpha_k A = a_1(\alpha_1 A) + \cdots + a_{k-1}(\alpha_{k-1}A)$$

or

$$c_k \alpha_k = a_1 c_1 \alpha_1 + \cdots + a_{k-1} c_{k-1} \alpha_{k-1}.$$

Substitution in this equation of the above representation of α_k yields

$$a_1(c_1 - c_k)\alpha_1 + \cdots + a_{k-1}(c_{k-1} - c_k)\alpha_{k-1} = 0,$$

where not all the scalar coefficients are 0. Hence $\{\alpha_1, \ldots, \alpha_{k-1}\}$ is linearly dependent. This conclusion, together with the assumption that no proper subset of S is linearly dependent, is a contradiction. It follows that $\{\alpha_i \mid i \in I\}$ is linearly independent. ◇

□ **Corollary 1.** If V is an n-dimensional vector space, then a linear transformation on V has at most n distinct characteristic values.

□ **Corollary 2.** If V is an n-dimensional vector space, $A \in \mathscr{L}(V, V)$, and A has n distinct characteristic values, then A has a diagonal matrix.

□ **Corollary 3.** If an $n \times n$ matrix has n distinct characteristic values, it is similar to a diagonal matrix.

□ **Definition.** A linear transformation is said to be *diagonable* if and only if one of its matrices is diagonal. A (square) matrix is said to be *diagonable* if and only if it is similar to a diagonal matrix.

□ **Lemma 3.2.** Let $c_1, \ldots, c_k$ be distinct characteristic values of the linear transformation A on a vector space V and let

$$S_{c_i} = \{\alpha \in V \mid \alpha A = c_i \alpha\}, \qquad 1 \le i \le k.$$

Further, set

$$p = (x - c_1) \cdots (x - c_k).$$

Then, if $\beta \in V$, $\beta p(A) = 0$ if and only if $\beta \in S_{c_1} + \cdots + S_{c_k}$.

PROOF. Assume that $\beta \in S_{c_1} + \cdots + S_{c_k}$. Then $\beta = \delta_1 + \cdots + \delta_k$, where $\delta_i \in S_{c_i}$, $1 \le i \le k$. Hence

$$\begin{aligned} \beta p(A) &= (\delta_1 + \cdots + \delta_k)(A - c_1) \cdots (A - c_k) \\ &= \delta_1(A - c_1) \cdots (A - c_k) + \delta_2(A - c_2)(A - c_1) \cdots (A - c_k) \\ &\quad + \cdots + \delta_k(A - c_k)(A - c_1) \cdots (A - c_{k-1}) \\ &= 0 + 0 + \cdots + 0 = 0. \end{aligned}$$

The proof of the converse, that if $\beta p(A) = 0$, then $\beta \in S_{c_1} + \cdots + S_{c_k}$, is by induction on k. Let $k = 1$ and suppose that $\beta p(A) = 0$. With $p = x - c_1$, this means that $\beta(A - c_1) = 0$, hence $\beta \in S_{c_1}$. Assume that the statement is true for the case of $k - 1$ factors and consider the case of k factors. So let $p = (x - c_1) \cdots (x - c_k)$ and suppose that $\beta p(A) = 0$. Set

$$\delta = \beta(A - c_1) \cdots (A - c_{k-1}). \tag{1}$$

Then, by assumption, $\delta(A - c_k) = 0$, hence $\delta \in S_{c_k}$. Further

$$\begin{aligned}\delta(A - c_1)\cdots(A - c_{k-1}) &= (\delta A - c_1\delta)(A - c_2)\cdots(A - c_{k-1})\\ &= (c_k - c_1)\,\delta(A - c_2)\cdots(A - c_{k-1})\\ &= \cdots\\ &= (c_k - c_1)(c_k - c_2)\cdots(c_k - c_{k-1})\delta.\end{aligned}$$

Now define c by

$$c = (c_k - c_1)\cdots(c_k - c_{k-1}).$$

Then $c \neq 0$ and

$$\delta(A - c_1)\cdots(A - c_{k-1}) = c\delta. \tag{2}$$

From (1) and (2) we deduce that

$$(\delta - c\beta)(A - c_1)\cdots(A - c_{k-1}) = 0.$$

Hence, by our induction hypothesis, $\delta - c\beta \in S_{c_1} + \cdots + S_{c_{k-1}}$ and this implies that $\beta \in S_{c_1} + \cdots + S_{c_k}$, since $c \neq 0$ and $\delta \in S_{c_k}$. ◇

□ **Theorem 3.2.** Let A be a linear transformation on a finite-dimensional vector space V over a field F. Then A is diagonable if and only if the minimal polynomial n of A is a product of distinct linear factors in $F[x]$:

$$m = (x - c_1)\cdots(x - c_k), \qquad c_i \in F \quad \text{and} \quad c_i \neq c_j \quad \text{if} \quad i \neq j.$$

Proof. Assume that m has the form stated in the theorem. Then for all β in V, $\beta m(A) = 0$ and Lemma 3.2 may be applied to conclude that $\beta \in S_{c_1} + \cdots + S_{c_k}$. Let $\mathscr{B}_1, \ldots, \mathscr{B}_k$ be bases of $S_{c_1}, \ldots, S_{c_k}$, respectively. Then $\mathscr{B} = \mathscr{B}_1 \cup \cdots \cup \mathscr{B}_k$ is a basis of V (why?) and the matrix of A relative to B is diagonal.

Conversely, assume that the matrix of A relative to the basis $(\beta_1, \ldots, \beta_n)$ of V is diagonal. Hence

$$\beta_i A = c_i \beta_i, \qquad 1 \leq i \leq n.$$

Without loss of generality, we may assume that $c_1, \ldots, c_k$ are the distinct scalars that occur in $\{c_1, \ldots, c_n\}$. Then $\beta_i(A - c_1)\cdots(A - c_k) = 0, 1 \leq i \leq n$, and so $\beta(A - c_1)\cdots(A - c_k) = 0$ for all $\beta \in V$. Hence the minimal polynomial m of A divides $(x - c_1)\cdots(x - c_k)$; but c_i, $1 \leq i \leq k$, is a root of $m(x) = 0$. Thus $m = (x - c_1)\cdots(x - c_k)$. ◇

□ **Corollary.** An $n \times n$ matrix is diagonal if and only if its minimal polynomial can be expressed as a product of distinct linear factors with scalars in the field at hand.

EXAMPLES

3.5. Let (a_{ij}) be an $n \times n$ matrix over a field F. Let A be the linear transformation on F^n induced by (a_{ij}) relative to the standard basis of F^n, that is,

$$\epsilon_i A = \sum_{j=1}^{n} a_{ij}\epsilon_j, \qquad 1 \leq i \leq n.$$

Suppose that

$$\beta_i = (p_{i1}, \ldots, p_{in}), \qquad 1 \leq i \leq n,$$

are characteristic vectors of A belonging to the (not necessarily distinct) characteristic values $c_n, \ldots, c_n$ of A. Then the linear transformation P on F^n such that

$$\epsilon_i P = \beta_i = \sum_{j=1}^{n} p_{ij}\epsilon_j, \qquad 1 \leq i \leq n,$$

has the matrix

$$\begin{bmatrix} p_{11} & \cdots & p_{1n} \\ \vdots & & \\ p_{n1} & \cdots & p_{nn} \end{bmatrix} = (p_{ij}),$$

relative to the standard basis. Further, by the definition of P

$$\epsilon_i(PA) = \beta_i A = c_i \beta_i, \qquad 1 \leq i \leq n. \tag{3}$$

If $(\beta_1, \ldots, \beta_n)$ is a basis of F^n, then P is nonsingular and the equations (3) imply that

$$\epsilon_i(PAP^{-1}) = c_i \beta_i P^{-1} = c_i \epsilon_i, \qquad 1 \leq i \leq n.$$

This means that

$$(p_{ij})(a_{ij})(p_{ij})^{-1} = \text{diag}(c_1, \ldots, c_n).$$

3.6. Since the minimal polynomial of

$$(a_{ij}) = \begin{bmatrix} 1 & -2 & 0 \\ -2 & 1 & 0 \\ 0 & 0 & -1 \end{bmatrix}$$

is $(x-3)(x+1)$, this matrix is diagonable. The linear transformation A it induces on R^3 relative to the standard basis is given by

$$(x, y, z)A = (x-2y, \ -2x+y, \ -z).$$

Using the notation of Lemma 3.2, we have

$$\begin{aligned} S_{-1} &= \{(x, y, z) \in \mathsf{R}^3 \mid (x, y, z)A = -(x, y, z)\} \\ &= \{(x, y, z) \in \mathsf{R}^3 \mid x = y\}. \end{aligned}$$

Since R^3 happens to be a euclidean space, it is possible to construct an orthonormal basis of S_{-1} (relative to the standard inner product). We find that

$$S_{-1} = \left[\left(\frac{1}{\sqrt{2}}, \frac{1}{\sqrt{2}}, 0\right), (0, 0, 1)\right],$$

where the basis exhibited is orthonormal. Similarly,

$$\begin{aligned} S_3 &= \{(x, y, z) \in \mathsf{R}^3 \mid (x, y, z)A = 3(x, y, z)\} \\ &= \{(x, y, z) \in \mathsf{R}^3 \mid x + y = 0, z = 0\} \\ &= \left[\left(\frac{1}{\sqrt{2}}, \frac{-1}{\sqrt{2}}, 0\right)\right]. \end{aligned}$$

Since the orthonormal basis $\{(1/\sqrt{2}, -1/\sqrt{2}, 0)\}$ of S_3 is orthogonal to the basis of S_{-1}, the union of the two bases,

$$\left(\left(\frac{1}{\sqrt{2}}, \frac{1}{\sqrt{2}}, 0\right), (0, 0, 1), \left(\frac{1}{\sqrt{2}}, \frac{-1}{\sqrt{2}}, 0\right)\right)$$

forms an orthonormal basis of R^3. The matrix of A relative to this basis is diag $(-1, -1, 3)$.

The similarity of the given matrix to diag $(-1, -1, 3)$ may be shown directly via the matrix

$$(p_{ij}) = \begin{bmatrix} \frac{1}{\sqrt{2}} & \frac{1}{\sqrt{2}} & 0 \\ 0 & 0 & 1 \\ \frac{1}{\sqrt{2}} & \frac{-1}{\sqrt{2}} & 0 \end{bmatrix}.$$

Then $(p_{ij})^{-1} = (p_{ij})^t$ (why?) and

$$(p_{ij})(a_{ij})(p_{ij})^{-1} = \text{diag}\,(-1, -1, 3).$$

3.7. The minimal polynomial of the matrix

$$\begin{bmatrix} 0 & -1 & 1 \\ 1 & 1 & 0 \\ -1 & 0 & 1 \end{bmatrix}$$

is $m = x^3 - 2x^2 + 3x - 2$. A decomposition of m into polynomials that are irreducible over R is

$$m = (x - 1)(x^2 - x + 2).$$

Hence this matrix is not diagonable over R. However,

$$m = (x - 1)\left(x - \frac{1}{2} - \frac{\sqrt{7}}{2}i\right)\left(x - \frac{1}{2} + \frac{\sqrt{7}}{2}i\right),$$

when regarded as an element of $\mathsf{C}[x]$. Thus, if the field C of complex numbers is the underlying field of scalars, m decomposes into a product of distinct linear factors and the matrix is similar (over C) to diag $(1, \frac{1}{2} + (\sqrt{7}/2)i, \frac{1}{2} - (\sqrt{7}/2)i)$.

We turn now to descriptions that can be given at this point of characteristic values of various types of linear transformations which we have defined.

□ **Singular transformations.** If a linear transformation in a vector space is singular, then 0 is one of its characteristic values. This statement (as well as its converse) follows directly from the definitions of the concepts involved.

□ **Unitary or orthogonal transformations.** Each characteristic value of a unitary or orthogonal transformation has absolute value 1. Let us prove this for the case of a characteristic value of a unitary transformation A in a unitary space. Let α be a characteristic vector belonging to c. Then

$$(\alpha \mid \alpha) = (\alpha A \mid \alpha A) = (c\alpha \mid c\alpha) = c\bar{c}(\alpha \mid \alpha).$$

Hence $c\bar{c} = |c|^2 = 1$ or $|c| = 1$. In the orthogonal case, this means $c = \pm 1$.

□ **Nilpotent transformations.** The only characteristic value of a nilpotent linear transformation is 0, for if A is a nilpotent linear transformation of index k then the minimal polynomial of A is x^k.

Since we can give deeper and more exhaustive results in the case of self-adjoint linear transformations, we devote the next section to this topic.

EXERCISES

3.1. Let V be a two-dimensional vector space over R with $\{\alpha_1, \alpha_2\}$ as a basis. Find all characteristic values and corresponding characteristic vectors for the linear transformation on V defined as follows.

(a) $\alpha_1 A = \alpha_1 + \alpha_2$, $\alpha_2 A = \alpha_1 - \alpha_2$.

(b) $\alpha_1 A = 5\alpha_1 + 6\alpha_2$, $\alpha_2 A = -7\alpha_2$.

(c) $\alpha_1 A = \alpha_1 + 2\alpha_2$, $\alpha_2 A = 3\alpha_1 + b\alpha_2$.

3.2. With V defined as the vector space in Exercise 3.1, suppose that the linear transformation A on V has the following form:

$$\alpha_1 A = a\alpha_1 + b\alpha_2, \quad \alpha_2 A = c\alpha_1 + d\alpha_2.$$

Find necessary and sufficient conditions in terms of a, b, c, and d that

(a) 0 is a characteristic value of A,

(b) A has two distinct real characteristic values.

3.3. Show that a scalar c of a field F is a characteristic value of an $n \times n$ matrix (a_{ij}) over F if and only if c is a characteristic value of any linear transformation A induced by (a_{ij}) in any n-dimensional vector space over F.

3.4. Let V be the vector space of all continuous functions f on R into R. Let A be the linear transformation in V defined by

$$f(x)A = \int_0^x f(t)\,dt.$$

Show that A has no characteristic values.

3.5. Is the matrix

$$\begin{bmatrix} 3 & 1 & -1 \\ 2 & 2 & -1 \\ 2 & 2 & 0 \end{bmatrix}$$

similar to a diagonal matrix over R? Is it similar to a diagonal matrix over C?

3.6. Find a 3×3 nonsingular matrix (p_{ij}) such that $(p_{ij})^{-1}(a_{ij})(p_{ij})$ is a diagonal matrix if

$$\text{(a) } (a_{ij}) = \begin{bmatrix} 2 & 0 & 0 \\ 0 & 2 & 1 \\ 0 & 0 & 1 \end{bmatrix}, \qquad \text{(b) } (a_{ij}) = \begin{bmatrix} 1 & 1 & -1 \\ -1 & 3 & -1 \\ -1 & 1 & 1 \end{bmatrix}.$$

3.7. Show that if $a^3 = 1$ and $a \neq 1$ the matrices

$$\begin{bmatrix} 0 & 1 & 0 \\ 0 & 0 & 1 \\ 1 & 0 & 0 \end{bmatrix} \quad \text{and} \quad \begin{bmatrix} 1 & 0 & 0 \\ 0 & a & 0 \\ 0 & 0 & a^2 \end{bmatrix}$$

are similar.

3.8. Find the minimal polynomial, the characteristic values c, and the associated subspaces $S_{c,A}$, for the linear transformation A on $\mathbf{R}^3$ induced by the matrix

$$\begin{bmatrix} 1 & 1 & 1 \\ -1 & 2 & 1 \\ -1 & 1 & 2 \end{bmatrix}$$

relative to the standard basis of $\mathbf{R}^3$.

3.9. Let (a_{ij}) be an $n \times n$ matrix such that $\sum_{j=1}^{n} a_{ij} = c$ for $1 \leq i \leq n$. Show that c is a characteristic value of (a_{ij}).

3.10. Prove or disprove each of the following statements.

(a) If V is a vector space over a field F, $A \in \mathscr{L}(V, V)$, $p \in F[x]$, and $p(A) = 0$, then every zero of p is a characteristic value of A.

(b) If the minimal polynomial of a linear transformation A on an n-dimensional space has degree n, then A is diagonable.

3.11. Show that the converse of Corollary 2 of Lemma 3.1 is false.

3.12. Let $\{c_i \mid i \in I\}$ be a set of distinct real numbers. Show that $\{e^{c_i x} \mid i \in I\}$ is a linearly independent set of functions in the vector space $C[a, b]$.

4. DIAGONALIZATION OF SELF-ADJOINT TRANSFORMATIONS

Let A be a self-adjoint linear transformation on an inner-product space V, so that $(\alpha A \mid \beta) = (\alpha \mid \beta A)$ for all α, β. Then all characteristic values (if any exist) of A are real numbers, and, if c_1, c_2, are distinct characteristic values of A, then $S_{c_1} = \{\alpha \in V \mid \alpha A = c_1 \alpha\}$ is *orthogonal* to $S_{c_2} = \{\alpha \in V \mid \alpha A = c_2 \alpha\}$; that is, each vector of one subspace is orthogonal to every vector of the other subspace. To establish the first assertion let c be a characteristic value of A and α be a characteristic vector belonging to c. Then

$$c(\alpha \mid \alpha) = (c\alpha \mid \alpha) = (\alpha A \mid \alpha) = (\alpha \mid \alpha A) = (\alpha \mid c\alpha) = \bar{c}(\alpha \mid \alpha),$$

which implies that $c = \bar{c}$. To prove the second assertion let $\alpha_1 \in S_{c_1}$ and $\alpha_2 \in S_{c_2}$. Then

$$\begin{aligned} c_1(\alpha_1 \mid \alpha_2) &= (c_1\alpha_1 \mid \alpha_2) = (\alpha_1 A \mid \alpha_2) = (\alpha_1 \mid \alpha_2 A) \\ &= (\alpha_1 \mid c_2 \alpha_2) = c_2(\alpha_1 \mid \alpha_2), \end{aligned}$$

which implies that $(\alpha_1 \mid \alpha_2) = 0$, since $c_1 \neq c_2$.

We continue with further observations in connection with a self-adjoint linear transformation A. From the foregoing calculations we may infer that if c is a characteristic value of A then all characteristic vectors belonging to other characteristic values are in $S_c^{\perp}$. Now $S_c^{\perp}$ is A-invariant, since if $\beta \in S_c^{\perp}$ and $\alpha \in S_c$, then

$$(\alpha \mid \beta A) = (\alpha A \mid \beta) = c(\alpha \mid \beta) = 0,$$

which implies that $\beta A \in S_c^{\perp}$. The restriction of A to $S_c^{\perp}$ is self-adjoint and its characteristic values and vectors are also characteristic values and vectors of A in V. If V is finite-dimensional, then $V = S_c \oplus S_c^{\perp}$ (see Theorem 3.4.1).

To continue with the analysis of a self-adjoint linear transformation A on a finite-dimensional inner-product space V we must distinguish between the case where the field of scalars is C (and therefore V is unitary) and where it is R (and therefore V is euclidean). If the field of scalars is C, then the minimal polynomial of A is a product of linear factors and (see Theorem 3.1) characteristic values of A exist. Recalling our first observation about self-adjoint transformations, we may conclude that characteristic values of a self-adjoint transformation exist and that they are real numbers.

The same is true in the case of finite-dimensional euclidean spaces. To show this we must rely on the fact that every polynomial in $\mathsf{R}[x]$ can be expressed as a product of linear and/or irreducible quadratic factors in $\mathsf{R}[x]$. A simple computation shows that a monic irreducible quadratic polynomial in $\mathsf{R}[x]$ can be written in the form $q = (x - a)^2 + b^2$ with $a, b \in \mathsf{R}$ and $b \neq 0$. Assume that $q = (x - a)^2 + b^2$ is a factor of the minimal polynomial of a self-adjoint linear transformation A on a finite-dimensional euclidean space V. Then $m = qh$, where $\deg h < \deg m$. Hence there exists a vector γ such that $\gamma\, h(A) \neq 0$. Then $\beta = \gamma\, h(A)$ is a nonzero vector such that $\beta q(A) = 0$. Since the linear transformation A is self-adjoint, it follows that

$$\begin{aligned} 0 = (\beta q(A) \mid \beta) &= (\beta(A - a)^2 \mid \beta) + b^2(\beta \mid \beta) \\ &= (\beta(A - a) \mid \beta(A - a)) + b^2(\beta \mid \beta). \end{aligned}$$

Since $(\beta(A - a) \mid \beta(A - a)) \geq 0$ and $(\beta \mid \beta) > 0$, b must equal 0 and, hence, $q = (x - a)^2$. Thus m is a product of linear factors in $\mathsf{R}[x]$. It follows that characteristic values of A exist and all characteristic values are real numbers.

□ **Theorem 4.1.** Let A be a self-adjoint linear transformation on a finite-dimensional inner-product space V.

(a) Characteristic values of A exist and all characteristic values of A are real numbers.

(b) If $\{c_1, \ldots, c_k\}$ is the set of all distinct characteristic values of A, then $\{S_{c_1}, \ldots, S_{c_k}\}$ is a set of mutually orthogonal subspaces of V.

(c) If $\mathscr{B}_1, \ldots, \mathscr{B}_k$ are orthonormal bases of $S_{c_1}, \ldots, S_{c_k}$, respectively, then $\mathscr{B} = \mathscr{B}_1 \cup \cdots \cup \mathscr{B}_k$ is an orthonormal basis of V.

(d) The matrix of A relative to $\mathscr{B}$ has the form

$$\operatorname{diag}(c_1, \ldots, c_1, \ldots, c_k, \ldots, c_k),$$

where the number of occurrences of c_i is equal to the dimension of S_{c_i}, $1 \leq i \leq k$.

PROOF. We have proved (a) and (b). Statement (d) follows directly from (c) and the definition of S_{ci}. For (c) let

$$S = S_{c_1} + \cdots + S_{c_k}.$$

Then $V = S + S^{\perp}$, since V is finite-dimensional. We contend that $S^{\perp}$ is A-invariant. For proof let $\beta \in S^{\perp}$ and consider $(\alpha \mid \beta A)$, where $\alpha \in S$. Since $\alpha = \alpha_1 + \cdots + \alpha_k$, with $\alpha_i \in S_{c_i}$ for $1 \leq i \leq k$,

$$(\alpha \mid \beta A) = (\alpha A \mid \beta) = \left(\sum_{i=1}^{k} \alpha_i A \mid \beta \right) = \sum_{i=1}^{k} c_i(\alpha_i \mid \beta) = 0.$$

Thus $\beta A \in S^{\perp}$ and $S^{\perp}$ is A-invariant. It follows that if $S^{\perp} \neq \{0\}$ then the restriction of A to $S^{\perp}$ is a self-adjoint linear transformation on $S^{\perp}$ for which there exist characteristic values. But each such number is a characteristic value of A; that is, if $S^{\perp} \neq \{0\}$, then A has characteristic values other than $c_1, \ldots, c_k$. Since, by assumption, $c_1, \ldots, c_k$ exhaust the set of distinct characteristic values of A, it follows that $S^{\perp} = \{0\}$ and, hence, that $V = S_{c_1} + \cdots + S_{c_k}$. The assertion (c) then follows in view of (b). ◇

□ **Corollary 1.** If (c_{ij}) is an $n \times n$ hermitian matrix, then there exists a unitary matrix (u_{ij}) such that

$$(u_{ij})(c_{ij})(u_{ij})^* = \operatorname{diag}(r_1, \ldots, r_n),$$

where each r_i is a real number.

□ **Corollary 2.** If (a_{ij}) is a real symmetric $n \times n$ matrix, then there exists an orthogonal matrix (p_{ij}) (of real numbers) such that

$$(p_{ij})(a_{ij})(p_{ij})^t = \text{diag}\,(r_1, \ldots, r_n),$$

where each r_i is a real number.

□ **Corollary 3.** The minimal polynomial m of a self-adjoint linear transformation on a finite-dimensional inner-product space has the form

$$m = (x - c_1) \cdots (x - c_k)$$

for distinct real numbers $c_1, \ldots, c_k$. The same is true of a hermitian matrix as well as a real symmetric matrix.

EXAMPLES

4.1. Let A be the linear transformation in R^3 given by

$$(x, y, z)A = (x + y + z,\ x + y + z,\ x + y + z).$$

The matrix of A relative to the standard basis of R^3 is

$$\begin{bmatrix} 1 & 1 & 1 \\ 1 & 1 & 1 \\ 1 & 1 & 1 \end{bmatrix}.$$

It is obvious (why?) that 0 and 3 are characteristic values of A. Since

$$S_0 = \{(x, y, z) \in \mathsf{R}^3 \mid x + y + z = 0\}$$

$$= \left[\left(\frac{1}{\sqrt{2}}, 0, \frac{-1}{\sqrt{2}}\right), \left(\frac{1}{\sqrt{6}}, \frac{-2}{\sqrt{6}}, \frac{1}{\sqrt{6}}\right)\right],$$

$\dim S_0 = 2$ and therefore the minimal polynomial of A is $x(x - 3)$ (why?). Further,

$$S_3 = \left[\left(\frac{1}{\sqrt{3}}, \frac{1}{\sqrt{3}}, \frac{1}{\sqrt{3}}\right)\right],$$

so that

$$\mathscr{B} = \left(\left(\frac{1}{\sqrt{2}}, 0, \frac{-1}{\sqrt{2}}\right), \left(\frac{1}{\sqrt{6}}, \frac{-2}{\sqrt{6}}, \frac{1}{\sqrt{6}}\right), \left(\frac{1}{\sqrt{3}}, \frac{1}{\sqrt{3}}, \frac{1}{\sqrt{3}}\right)\right)$$

is an orthonormal basis of R^3, regarded as an inner-product space with standard inner product. The matrix of A relative to $\mathscr{B}$ is diag (0, 0, 3). Indeed,

$$\begin{bmatrix} \frac{1}{\sqrt{2}} & 0 & \frac{-1}{\sqrt{2}} \\ \frac{1}{\sqrt{6}} & \frac{-2}{\sqrt{6}} & \frac{1}{\sqrt{6}} \\ \frac{1}{\sqrt{3}} & \frac{1}{\sqrt{3}} & \frac{1}{\sqrt{3}} \end{bmatrix} \begin{bmatrix} 1 & 1 & 1 \\ 1 & 1 & 1 \\ 1 & 1 & 1 \end{bmatrix} \begin{bmatrix} \frac{1}{\sqrt{2}} & 0 & \frac{-1}{\sqrt{2}} \\ \frac{1}{\sqrt{6}} & \frac{-2}{\sqrt{6}} & \frac{1}{\sqrt{6}} \\ \frac{1}{\sqrt{3}} & \frac{1}{\sqrt{3}} & \frac{1}{\sqrt{3}} \end{bmatrix}^t = \mathrm{diag}\,(0, 0, 3).$$

4.2. The minimal polynomial of the hermitian matrix

$$\begin{bmatrix} 1 & i & 0 \\ -i & 0 & 1 \\ 0 & 1 & 1 \end{bmatrix}$$

is $m = (x + 1)(x - 1)(x - 2)$. Further,

$$S_{-1} = \left[\left(\frac{1}{\sqrt{6}}, \frac{-2i}{\sqrt{6}}, \frac{i}{\sqrt{6}}\right)\right],$$

$$S_1 = \left[\left(\frac{1}{\sqrt{2}}, 0, \frac{-i}{\sqrt{2}}\right)\right],$$

$$S_2 = \left[\left(\frac{1}{\sqrt{3}}, \frac{i}{\sqrt{3}}, \frac{i}{\sqrt{3}}\right)\right].$$

Hence

$$\begin{bmatrix} \frac{1}{\sqrt{6}} & \frac{-2i}{\sqrt{6}} & \frac{i}{\sqrt{6}} \\ \frac{1}{\sqrt{2}} & 0 & \frac{-i}{\sqrt{2}} \\ \frac{1}{\sqrt{3}} & \frac{i}{\sqrt{3}} & \frac{i}{\sqrt{3}} \end{bmatrix} \begin{bmatrix} 1 & i & 0 \\ -i & 0 & 1 \\ 0 & 1 & 1 \end{bmatrix} \begin{bmatrix} \frac{1}{\sqrt{6}} & \frac{1}{\sqrt{2}} & \frac{1}{\sqrt{3}} \\ \frac{2i}{\sqrt{6}} & 0 & \frac{-i}{\sqrt{3}} \\ \frac{-i}{\sqrt{6}} & \frac{i}{\sqrt{2}} & \frac{-i}{\sqrt{3}} \end{bmatrix} = \begin{bmatrix} -1 & 0 & 0 \\ 0 & 1 & 0 \\ 0 & 0 & 2 \end{bmatrix}$$

EXERCISES

4.1. Suppose that A is the linear transformation on R^3 induced by the matrix

$$(a_{ij}) = \begin{bmatrix} 2 & 2 & 0 \\ 2 & -1 & 0 \\ 0 & 0 & 2 \end{bmatrix}$$

relative to the standard basis of R^3.

(a) Find all characteristic values of A.

(b) Find an orthonormal basis of R^3 relative to which the matrix of A is diagonal.

(c) Find a 3×3 orthogonal matrix (p_{ij}) such that $(p_{ij})(a_{ij})(p_{ij})^{-1}$ is the diagonal matrix obtained in (b).

4.2. Find all characteristic values of

$$(a_{ij}) = \begin{bmatrix} 4 & -1 & 2 \\ -1 & -2 & -\frac{5}{2} \\ 2 & -\frac{5}{2} & 1 \end{bmatrix}$$

and a 3×3 orthogonal matrix (p_{ij}) such that $(p_{ij})(a_{ij})(p_{ij})^t$ is a diagonal matrix.

4.3. Let V be an n-dimensional vector space over R and $A \in \mathscr{L}(V, V)$. Show that an inner product can be introduced in V such that A is self-adjoint if and only if A has n linearly independent characteristic vectors.

4.4. Suppose that A is a self-adjoint linear transformation on a finite-dimensional euclidean vector space V. Let c_1 and c_2 be characteristic values of A having the largest and the smallest absolute values, respectively. Show that $|c_1| = \max \|\alpha A\|$ and $|c_2| = \min \|\alpha A\|$, where α ranges over the set of unit vectors of V.

4.5. For a linear transformation of the sort described in the preceding exercise show further that

$$\max_{\|\alpha\| = \|\beta\| = 1} |(\alpha A \mid \beta)| = \max_{\|\alpha\| = 1} |(\alpha A \mid \alpha)|.$$

4.6. Let (a_{ij}) be a real symmetric matrix having c as a characteristic value of smallest absolute value. Show that (a_{ij}) has an inverse if and only if $c \neq 0$ and, in that event, that $1/|c|$ is the absolute value of the largest characteristic value of $(a_{ij})^{-1}$.

4.7. A self-adjoint linear transformation A on an inner-product space V is said to be *positive definite* if and only if each of its characteristic values is positive.

Show that
(a) A is positive definite if and only if $(\alpha A \mid \alpha) > 0$ for all nonzero vectors α in V and that
(b) if A is positive definite, then so is its inverse.

4.8. Let (a_{ij}) be a real symmetric matrix, all of whose characteristic values are nonnegative. Prove that there exists a real symmetric matrix (b_{ij}) such that $(b_{ij})^2 = (a_{ij})$ and $(b_{ij})(a_{ij}) = (a_{ij})(b_{ij})$.

4.9. Let (a_{ij}) be a nonsingular hermitian matrix whose characteristic values are $c_1, \ldots, c_k$. What are the characteristic values of $(a_{ij})^{-1}$.

5. CHARACTERISTIC POLYNOMIALS

In view of Theorem 3.1, a scalar c is a characteristic value of a linear transformation on a finite-dimensional vector space (or a square matrix) if and only if it is a root of the minimal polynomial equation $m(x) = 0$, where m is the minimal polynomial involved. In this section we define for each such linear transformation (and square matrix) a polynomial that may be different from the minimal polynomial but has the property of a minimal polynomial mentioned above.

□ **Definition.** If (a_{ij}) is an $n \times n$ matrix over a field F, then the $n \times n$ matrix

$$xI - (a_{ij}) = \begin{bmatrix} x - a_{11} & -a_{12} & \cdots & -a_{1n} \\ -a_{21} & x - a_{22} & \cdots & -a_{2n} \\ \vdots & & & \\ -a_{n1} & -a_{n2} & \cdots & x - a_{nn} \end{bmatrix}$$

over $F[x]$ is called the *characteristic matrix* of (a_{ij}). The determinant of this matrix is called the *characteristic polynomial* of (a_{ij}). If A is a linear transformation on a finite-dimensional vector space over F, the characteristic polynomial of any matrix of A is called the *characteristic polynomial* of A.

Some comments on these definitions are in order. First, recalling that $F[x]$ is a commutative ring with identity and that our theory of determinants applies to such systems (see the last paragraph of Section 5.5), the characteristic polynomial of a square matrix is defined. Second, similar matrices have the same characteristic polynomial, for if

$$(b_{ij}) = (p_{ij})^{-1}(a_{ij})(p_{ij}),$$

then

$$xI - (b_{ij}) = (p_{ij})^{-1}(xI - (a_{ij}))(p_{ij}),$$

hence

$$\begin{aligned}\det(xI - (b_{ij})) &= \det(p_{ij})^{-1}\det(xI - (a_{ij}))\det(p_{ij})\\ &= \det[(p_{ij})^{-1}(p_{ij})]\det(xI - (a_{ij}))\\ &= \det(xI - (a_{ij})).\end{aligned}$$

Since any two matrices of a linear transformation (on a finite-dimensional space) are similar, all matrices of a given linear transformation have the same characteristic polynomial, and so our definition of the characteristic polynomial of a linear transformation is unambiguous.

The characteristic polynomial h of an $n \times n$ matrix (a_{ij}) is a monic polynomial of degree n; this is easily seen from the development of $\det(xI - (a_{ij}))$ by elements of the first column. Further, the roots of $h(x) = 0$ coincide with the roots of $m(x) = 0$, where m is the minimal polynomial of (a_{ij}). To prove this let c be a characteristic value of (a_{ij}). Then, in turn, $cI - (a_{ij})$ is singular, $\det(cI - (a_{ij})) = 0$, and c is a root of $h(x) = 0$. Reversing these steps gives a proof of the converse. The fact that the minimal and characteristic polynomials of a linear transformation in a finite-dimensional vector space are equal to, respectively, the minimal and characteristic polynomials of a matrix of the linear transformation establishes the same result for linear transformations.

EXAMPLES

5.1. In Example 3.7 we stated that the minimal polynomial of the matrix

$$\begin{bmatrix} 0 & -1 & 1 \\ 1 & 1 & 0 \\ -1 & 0 & 1 \end{bmatrix}$$

is $m = x^3 - 2x^2 + 3x - 2$. This is also the characteristic polynomial

$$\det\begin{bmatrix} x & 1 & -1 \\ -1 & x-1 & 0 \\ 1 & 0 & x-1 \end{bmatrix}$$

of the matrix, as the reader can show.

5.2. The characteristic polynomial of the matrix

$$\begin{bmatrix} 1 & -2 & 0 \\ -2 & 1 & 0 \\ 0 & 0 & -1 \end{bmatrix}$$

is

$$\det\begin{bmatrix} x-1 & 2 & 0 \\ 2 & x-1 & 0 \\ 0 & 0 & x-1 \end{bmatrix} = (x-3)(x+1)^2.$$

In Example 3.6 we found the minimal polynomial of the same matrix to be $(x-3)(x+1)$.

5.3. Let A be the linear transformation on R^4, whose matrix relative to the standard basis is

$$(a_{ij}) = \begin{bmatrix} 1 & -2 & 1 & 0 \\ 0 & 2 & 1 & -1 \\ 0 & 1 & 0 & 6 \\ 2 & -1 & -1 & 6 \end{bmatrix}$$

A computation produces $x^4 - 9x^3 + 24x^2 - 33x + 51$ as the characteristic polynomial of A.

5.4. In Theorem 2.5 we noted that the minimal polynomial of the matrix

$$\begin{bmatrix} 0 & 1 & 0 & \cdots & 0 \\ 0 & 0 & 1 & \cdots & 0 \\ \vdots & & & & \\ 0 & 0 & 0 & \cdots & 1 \\ b_0 & b_1 & b_2 & \cdots & b_{n-1} \end{bmatrix}$$

is $m = x^n - b_{n-1}x^{n-1} - \cdots - b_1x - b_0$. Let us now prove that the characteristic polynomial of this matrix is also m; that is,

$$\det\begin{bmatrix} x & -1 & 0 & \cdots & 0 \\ 0 & x & -1 & \cdots & 0 \\ \vdots & & & & \\ 0 & 0 & 0 & \cdots & -1 \\ -b_0 & -b_1 & -b_2 & \cdots & x-b_{n-1} \end{bmatrix} = m.$$

The proof is by induction on n. If $n = 1$, the matrix is (b_0), hence $\det(x - b_0) = x - b_0$ as desired. Assume that the assertion holds for

$(n-1) \times (n-1)$ matrices and consider the case of $n \times n$ matrices. Developing

$$\det \begin{bmatrix} x & -1 & 0 & \cdots & 0 \\ 0 & x & -1 & \cdots & 0 \\ \vdots & & & & \\ -b_0 & -b_1 & -b_2 & \cdots & x - b_{n-1} \end{bmatrix}$$

by the elements of the first column, we obtain for its value

$$x \det \begin{bmatrix} x & -1 & \cdots & 0 \\ 0 & x & \cdots & 0 \\ \vdots & & & \\ -b_1 & -b_2 & \cdots & x - b_n \end{bmatrix}$$

$$+ (-1)^{n+1}(-b_0) \det \begin{bmatrix} -1 & 0 & \cdots & 0 \\ x & -1 & \cdots & 0 \\ \vdots & & & \\ 0 & 0 & \cdots & -1 \end{bmatrix}$$

$$= x(x^{n-1} - b_{n-1}x^{n-2} - \cdots - b_1) + (-1)^{n+1}(-b_0)(-1)^{n-1}$$

$$= x^n - b_{n-1}x^{n+1} - \cdots - b_0 = m.$$

This completes the proof and the stage is set to prove the celebrated (but antique) Cayley-Hamilton theorem. It is Part (c) of the result which follows.

□ **Theorem 5.1.** Let A be a linear transformation on an n-dimensional vector space V. Let m be the minimal polynomial of A and let h be its characteristic polynomial. Then

(a) $m \mid h$,
(b) every zero of h is a zero of m,
(c) $h(A) = 0$.

The same results hold for square matrices.

PROOF. We direct our attention to a proof of (c), since (a) is an immediate consequence of (c) and (b) was proved earlier in this section.

Let α be a nonzero vector in V and let s be the least positive integer such that $\{\alpha, \alpha A, \ldots, \alpha A^s\}$ is linearly dependent. Then, according to Theorem 2.4, there exist scalars $b_0, \ldots, b_{s-1}$, such that

$$\alpha A^s = b_0 \alpha + b_1 \alpha A + \cdots + b_{s-1} \alpha A^{s-1},$$

and the polynomial

$$m_{\alpha,A} = x^s - b_{s-1}x^{s-1} - \cdots - b_1 x - b_0$$

has the following property: $\alpha m_{\alpha,A}(A) = 0$.

Now let us extend the linearly independent set $(\alpha, \alpha A, \ldots, \alpha A^{s-1})$ to a basis $\mathscr{B}$ of V. The matrix (b_{ij}) of A relative to $\mathscr{B}$ may be regarded as having the form

$$(b_{ij}) = \begin{bmatrix} C & D \\ E & F \end{bmatrix},$$

where C, D, E, and F are matrices having, among others, the following properties: C is the $s \times s$ matrix

$$\begin{bmatrix} 0 & 1 & 0 & \cdots & 0 \\ 0 & 0 & 1 & \cdots & 0 \\ \vdots & & & & \\ 0 & 0 & 0 & \cdots & 1 \\ b_0 & b_1 & b_2 & \cdots & b_{s-1} \end{bmatrix},$$

D is a zero matrix, and F is an $(n - s) \times (n - s)$ matrix. Thus the characteristic matrix of (b_{ij}) has the form

$$xI - (b_{ij}) = \left[\begin{array}{c|ccc} & 0 & \cdots & 0 \\ xI - C & & \cdots & \\ & 0 & \cdots & 0 \\ \hline -E & & xI - F & \end{array}\right]$$

and, in turn (see Exercise 5.5.3),

$$h = \det(xI - (b_{ij})) = \det(xI - C)\det(xI - F).$$

Using the result obtained in Example 5.4, it follows that $h = m_{\alpha,A}q$, where $q = \det(xI - F)$. (Of course, if $s = n$, then $q = 1$.) We then have that

$$\alpha h(A) = [\alpha m_{\alpha,A}(A)]\, q(A) = 0.$$

Since α is any nonzero vector in V, it follows that $h(A)$ is the zero transformation in V. This completes the proof. ◇

EXERCISES

5.1. Find the characteristic polynomial of each of the following matrices:

(a) $\begin{bmatrix} 1 & 2 & 3 \\ 4 & 1 & 0 \\ 1 & 6 & 4 \end{bmatrix}$ over $\mathbf{Z}_7$,

(b) $\begin{bmatrix} 1 & -1 & 2 \\ 0 & 2 & 1 \\ 4 & 1 & 3 \end{bmatrix}$ over $\mathbf{R}$,

(c) $\begin{bmatrix} 1 & -2 & 1 & 0 \\ -2 & 1 & 1 & 0 \\ 1 & 1 & -2 & 4 \\ 0 & 0 & 4 & -1 \end{bmatrix}$ over $\mathbf{R}$.

5.2. Find the characteristic polynomial of a triangular matrix with $d_1, d_2, \ldots, d_n$ as its diagonal entries.

5.3. Find all values of c such that the following system of linear equations has a nontrivial solution:

$$-x + 2y + 2z = cx,$$

$$2x + 2y + 2z = cy,$$

$$-3x - 6y - 6z = cz.$$

5.4. Let (a_{ij}) and (b_{ij}) be $n \times n$ matrices with (a_{ij}) nonsingular. Show that $(a_{ij})(b_{ij})$ and $(b_{ij})(a_{ij})$ have the same characteristic polynomial.

5.5. Find two 3×3 matrices over $\mathbf{R}$ that have the same characteristic polynomial but are not similar.

5.6. Compute the characteristic polynomial of a nilpotent linear transformation in an n-dimensional vector space.

5.7. Compute the characteristic polynomial of an idempotent linear transformation A [thus $A^2 = A$] of rank r in an n-dimensional vector space.

5.8. Show that the trace of an $n \times n$ matrix is the negative of the coefficient of x^{n-1} in the characteristic polynomial of the matrix. Show that similar matrices have equal traces.

5.9. Let V be a vector space over $\mathbf{R}$ whose dimension is an odd integer. Show that for every linear transformation A on V there exists a one-dimensional A-invariant subspace of V.

5.10. (a) Find a 3×3 orthogonal matrix (p_{ij}) such that

$$(p_{ij})\begin{bmatrix} -1 & 0 & 2 \\ 0 & -1 & 1 \\ 2 & 1 & 3 \end{bmatrix}(p_{ij})^t$$

is diagonal.

(b) Find a 2×2 unitary matrix (u_{ij}) such that

$$(u_{ij})\begin{bmatrix} 1 & 1+i \\ 1-i & 3 \end{bmatrix}(u_{ij})^*$$

is diagonal.

5.11. Find $(a_{ij})^{19}$, where

$$(a_{ij}) = \begin{bmatrix} \frac{1}{3} & \frac{2}{\sqrt{6}} & \frac{-2}{\sqrt{18}} \\ \frac{2}{\sqrt{6}} & 0 & \frac{1}{\sqrt{3}} \\ \frac{-2}{\sqrt{18}} & \frac{1}{\sqrt{3}} & \frac{2}{3} \end{bmatrix}$$

What is $(a_{ij})^k$ when k is a positive odd integer? What is $(a_{ij})^k$ when k is a positive even integer? Find the minimal polynomial of (a_{ij}) without going through any computations.

5.12. Let h be the characteristic polynomial and let m be the minimal polynomial of a linear transformation A on a finite-dimensional vector space. Show that if q is an irreducible factor of h then q is also a factor of m.

5.13. Find two linear transformations on a finite-dimensional vector space which have the same minimal and characteristic polynomials, yet are not similar. *Hint.* Consider 4×4 nilpotent matrices.

5.14. If A is a self-adjoint linear transformation in an n-dimensional inner-product space, show that its characteristic polynomial is

$$(x - c_1)^{h_1} \cdots (x - c_k)^{h_k},$$

where $c_1, \ldots, c_k$ are distinct scalars, $h_1 + \cdots + h_k = n$, and $h_i = \dim S_{c_i}$, $1 \leq i \leq k$.

5.15. Prove that two self-adjoint linear transformations on a finite-dimensional inner-product space are similar if and only if they have the same characteristic polynomial.

6. TRIANGULABLE LINEAR TRANSFORMATIONS

We have established the fact that some linear transformations are not diagonable; that is, there exists a linear transformation A such that no matrix of A is diagonal. For an important class of linear transformations, however, there exist triangular matrices. If some matrix of a linear transformation A is triangular, we call A *triangulable* (and as a companion definition call a square matrix *triangulable* if and only if it is similar to a triangular matrix). For many purposes these matrices are as satisfactory as diagonal matrices. In this connection there is no need to distinguish between upper and lower triangular matrices; for if the matrix of a linear transformation A is upper (lower) triangular relative to a basis $(\alpha_1, \ldots, \alpha_n)$, then the matrix of A relative to the basis $(\alpha_n, \ldots, \alpha_1)$ is lower (upper) triangular.

If the linear transformation A has a triangular matrix with $c_1, \ldots, c_k$ as the distinct elements appearing in the principal diagonal, the characteristic polynomial h of A has the form

$$h = (x - c_1)^{r_1} \cdots (x - c_k)^{r_k}$$

and $\{c_1, \ldots, c_k\}$ is the set of distinct characteristic values of A. In this section we present a theorem due to I. Schur which is concerned with the converse of the foregoing observation.

□ **Theorem 6.1.** Let A be a linear transformation on a finite-dimensional inner-product space V over a field F. If the characteristic polynomial is expressible in the form $h = (x - c_1)^{r_1} \cdots (x - c_k)^{r_k}$, where $c_1, \ldots, c_k$ are (distinct) scalars in F, then there exists an orthonormal basis of V relative to which the matrix of A is triangular.

Before giving a proof some comments are in order. First, observe that the premise of the theorem is satisfied if V is a unitary space; it is only in the case of a euclidean space that a restriction is involved. Thus every linear transformation in a finite-dimensional unitary space is triangulable. Second, observe that the theorem may appear to be a generalization of Theorem 4.1. This is a fact; in an exercise the reader is asked to show that when the constructive proof of the theorem is applied to a self-adjoint linear transformation, the resulting triangular matrix is diagonal. Finally, to reinforce the significance of the theorem we list as corollaries two of its immediate consequences.

☐ **Corollary 1.** If (c_{ij}) is a square matrix over C, then there exists a unitary matrix (u_{ij}) such that $(u_{ij})(c_{ij})(u_{ij})^*$ is triangular; in other words, (c_{ij}) is unitarily similar to a triangular matrix.

☐ **Corollary 2.** If (a_{ij}) is a square matrix over R and if its characteristic polynomial decomposes into a product of linear factors over R, then there exists an orthogonal matrix (p_{ij}) such that $(p_{ij})(a_{ij})(p_{ij})^t$ is triangular; in other words, (a_{ij}) is orthogonally similar to a triangular matrix.

PROOF (of the theorem). The proof is by induction on the dimension of V. If $\dim V = 1$, there is nothing to prove. Assume that the theorem is true in spaces of dimension $n - 1$ for linear transformations that satisfy the hypothesis of the theorem. Let V be an n-dimensional inner-product space and let A be a linear transformation on V with $h = (x - c_1)^{r1} \cdots (x - c_k)^{r_k}$ as its characteristic polynomial. Let δ_1 be a characteristic vector of unit length belonging to the characteristic value c_1. Let $\mathscr{B}_1 = (\delta_1, \gamma_2, \dots \gamma_n)$ be an extension of $\{\delta_1\}$ to an orthonormal basis of V. Then the matrix of A relative to $\mathscr{B}_1$ has the form

$$\begin{bmatrix} c_1 & 0 & \cdots & 0 \\ * & * & \cdots & * \\ \cdot & & & \\ \cdot & & & \\ \cdot & & & \\ * & * & \cdots & * \end{bmatrix}$$

and that of A^*, the adjoint of A, has the form (see Section 4.6)

$$\begin{bmatrix} \bar{c}_1 & * & \cdots & * \\ 0 & * & \cdots & * \\ \cdot & & & \\ \cdot & & & \\ \cdot & & & \\ 0 & * & \cdots & * \end{bmatrix}.$$

Hence the subspace $S = [\gamma_2, \dots, \gamma_n] = [\delta_1]^\perp$ is A^*-invariant. As a vector space, S is an inner-product space of dimension $n - 1$, and the restriction A_S^* of A^* to S is a linear transformation on S.

If V and, hence, S is a euclidean space, then h is also the characteristic polynomial of A^* (a matrix and its transpose have the same determinant) and it is clear that the characteristic polynomial of A_S^* is $h/(x - c_1)$. Hence A_S^* satisfies the induction hypothesis. If V, hence S, is a unitary space, then A_S^* automatically satisfies the induction hypothesis. Thus in both cases there exists an orthonormal basis $(\delta_2, \dots, \delta_n)$ of S relative to which the matrix of A_S^* is upper triangular. It follows, in turn, that $\mathscr{B} = (\delta_1, \dots, \delta_n)$

is an orthonormal basis of V, the matrix of A^* relative to $\mathscr{B}$ is upper triangular, and the matrix of A relative to $\mathscr{B}$ is lower triangular. ◇

The method of proof of this theorem may be made the basis of an effective procedure for constructing an orthonormal basis relative to which the matrix of a linear transformation satisfying the hypothesis is triangular. Example 6.1, which follows, illustrates this procedure.

EXAMPLES

6.1. In the euclidean space R^4 with standard inner product consider the linear transformation A whose matrix relative to the standard basis of R^4 is

$$(a_{ij}) = \begin{bmatrix} 8 & 9 & -9 & 0 \\ 0 & 2 & 0 & 2 \\ 4 & 6 & -4 & 0 \\ 0 & 0 & 0 & 3 \end{bmatrix}.$$

Since the characteristic polynomial of A is $(x-2)^3(x-3)$, it has a triangular matrix. To find one we begin with the characteristic value $c_1 = 3$. Our choice is dictated by the observation that ϵ_4 is an accompanying characteristic vector. In the notation of the proof of Theorem 6.1 we let $\mathscr{B}_1 = (\epsilon_4, \epsilon_1, \epsilon_2, \epsilon_3)$. The matrix of A relative to $\mathscr{B}_1$ is

$$\begin{bmatrix} 3 & 0 & 0 & 0 \\ 0 & 8 & 9 & -9 \\ 2 & 0 & 2 & 0 \\ 0 & 4 & 6 & -4 \end{bmatrix}$$

and that of A^* is the transpose of this matrix. Hence

$$\begin{aligned} \epsilon_4 A^* &= 3\epsilon_4 + 2\epsilon_2 \\ \epsilon_1 A^* &= 8\epsilon_1 \qquad\quad + 4\epsilon_3 \\ \epsilon_2 A^* &= 9\epsilon_1 + 2\epsilon_2 + 6\epsilon_3 \\ \epsilon_3 A^* &= -9\epsilon_1 \qquad\quad - 4\epsilon_3, \end{aligned}$$

and so in turn $S = [\epsilon_1, \epsilon_2, \epsilon_3]$ is A^*-invariant and the matrix of A_S^* relative to the basis $(\epsilon_1, \epsilon_2, \epsilon_3)$ of S is

$$\begin{bmatrix} 8 & 0 & 4 \\ 9 & 2 & 6 \\ -9 & 0 & -4 \end{bmatrix}.$$

This completes the first iteration in the reduction procedure.

We commence the second by observing that $(x-2)^3$ is the characteristic polynomial of A_S^* and, since

$$(\epsilon_2+\epsilon_3)A_S^* = 2(\epsilon_2+\epsilon_3),$$

the vector

$$\delta_1 = \frac{1}{\sqrt{2}}(\epsilon_2+\epsilon_3)$$

is a unit characteristic vector belonging to 2. One extension of $\{\delta_1\}$ to an orthonormal basis of S is

$$\mathscr{B}_2 = (\delta_1, \delta_2, \delta_3),$$

where

$$\delta_2 = \frac{1}{\sqrt{2}}(\epsilon_2-\epsilon_3), \qquad \delta_3 = \epsilon_1.$$

Then

$$\epsilon_1 = \delta_3,\ \epsilon_2 = \frac{1}{\sqrt{2}}(\delta_1+\delta_2), \qquad \epsilon_3 = \frac{1}{\sqrt{2}}(\delta_1-\delta_2).$$

Since

$$\begin{aligned}
\delta_1 A_S^* &= 2\delta_1,\\
\delta_2 A_S^* &= \frac{1}{\sqrt{2}}(\epsilon_2-\epsilon_3)A_S^* = \cdots = 6\delta_1 - 4\delta_2 + 9\sqrt{2}\delta_3,\\
\delta_3 A_S^* &= \epsilon_1 A^* = 8\epsilon_1 + 4\epsilon_3 = 2\sqrt{2}\delta_1 - 2\sqrt{2}\delta_2 + 8\delta_3,
\end{aligned}$$

the matrix of A_S^* relative to $\mathscr{B}_2$ is

$$\begin{bmatrix} 2 & 0 & 0\\ 6 & -4 & 9\sqrt{2}\\ 2\sqrt{2} & -2\sqrt{2} & 8 \end{bmatrix}.$$

The matrix $(A_S^*)^* = B$, let us say, relative to $\mathscr{B}_2$ is

$$\begin{bmatrix} 2 & 6 & 2\sqrt{2}\\ 0 & -4 & -2\sqrt{2}\\ 0 & 9\sqrt{2} & 8 \end{bmatrix},$$

and, hence,

$$\begin{aligned}
\delta_1 B &= 2\delta_1 + 6\delta_2 + 2\sqrt{2}\delta_3\\
\delta_2 B &= \qquad - 4\delta_2 - 2\sqrt{2}\delta_3\\
\delta_3 B &= \qquad 9\sqrt{2}\delta_2 + \quad 8\delta_3.
\end{aligned}$$

Thus $T = [\delta_2, \delta_3]$ is B-invariant and the matrix of B_T relative to (δ_2, δ_3) is

$$\begin{bmatrix} -4 & -2\sqrt{2} \\ 9\sqrt{2} & 8 \end{bmatrix}.$$

This brings us to the third and final iteration (since a 2×2 matrix is involved). The characteristic polynomial of B_T is $(x - 2)^2$ and

$$\gamma_1 = \frac{1}{\sqrt{11}}(3\delta_2 + \sqrt{2}\delta_3)$$

is found to be a unit characteristic vector belonging to the characteristic value 2. We extend $\{\gamma_1\}$ to an orthonormal basis of T by adjoining

$$\gamma_2 = \frac{1}{\sqrt{11}}(\sqrt{2}\delta_2 - 3\delta_3).$$

To assist with the computation of $\gamma_2 B_T$ we use the following relations:

$$\delta_2 = \frac{1}{\sqrt{11}}(3\gamma_1 + \sqrt{2}\gamma_2),$$

$$\delta_3 = \frac{1}{\sqrt{11}}(\sqrt{2}\gamma_1 - 3\gamma_2).$$

Then we find that

$$\gamma_1 B_T = 2\gamma_1$$

$$\gamma_2 B_T = \frac{1}{\sqrt{11}}(\sqrt{2}\delta_2 - 3\delta_3)B_T = \cdots = -11\sqrt{2}\gamma_1 + 2\gamma_2,$$

and so the matrix of B_T relative to (γ_1, γ_2) is

$$\begin{bmatrix} 2 & 0 \\ -11\sqrt{2} & 2 \end{bmatrix}.$$

The earlier technique of turning next to the adjoint of the linear transformation under consideration may be replaced in the case of a 2×2 matrix by interchanging the order of the basis elements. In the case at hand the matrix of B_T relative to (γ_2, γ_1) is

$$\begin{bmatrix} 2 & -11\sqrt{2} \\ 0 & 2 \end{bmatrix}.$$

Having reached an upper triangular matrix, it remains to extend the basis (γ_2, γ_1) of T to one of R^4 in such a way that we preserve triangularity. The first step is the determination of the matrix of B relative to the basis $\mathscr{B}_2' = (\delta_1, \gamma_2, \gamma_1)$. Since

$$\delta_1 B = 2\delta_1 + 6\delta_2 + 2\sqrt{2}\delta_3 = \cdots = 2\delta_1 + 0\gamma_2 + 2\sqrt{11}\gamma_1,$$

the matrix of B relative to $\mathscr{B}_2'$ is

$$\begin{bmatrix} 2 & 0 & 2\sqrt{11} \\ 0 & 2 & -11\sqrt{2} \\ 0 & 0 & 2 \end{bmatrix},$$

and so that of A_S^* relative to $\mathscr{B}_2'$ is

$$\begin{bmatrix} 2 & 0 & 0 \\ 0 & 2 & 0 \\ 2\sqrt{11} & -11\sqrt{2} & 2 \end{bmatrix}.$$

Hence the matrix of A_S^* relative to $(\gamma_1, \gamma_2, \delta_1)$ is

$$\begin{bmatrix} 2 & -11\sqrt{2} & 2\sqrt{11} \\ 0 & 2 & 0 \\ 0 & 0 & 2 \end{bmatrix}.$$

Further, the matrix of A^* relative to the orthonormal basis $(\epsilon_4, \gamma_1, \gamma_2, \delta_1)$ of R^4 is

$$\begin{bmatrix} 3 & \dfrac{6}{\sqrt{22}} & \dfrac{2}{\sqrt{11}} & \sqrt{2} \\ 0 & 2 & -11\sqrt{2} & 2\sqrt{11} \\ 0 & 0 & 22 & 0 \\ 0 & 0 & 0 & 2 \end{bmatrix},$$

since

$$\epsilon_4 A^* = 3\epsilon_4 + 2\epsilon_2 = \cdots = 3\epsilon_4 + \frac{6}{\sqrt{22}}\gamma_1 + \frac{2}{\sqrt{11}}\gamma_2 + 2\delta_1.$$

Thus the matrix of A relative to the same basis is the transpose of the foregoing matrix.

To conclude matters we find an orthogonal matrix (p_{ij}) that reduces (a_{ij}) to the triangular matrix we have found. We have

$$\epsilon_4 = (0, 0, 0, 1),$$

$$\gamma_1 = \frac{1}{\sqrt{11}}(3\delta_2 + 2\delta_3) = \frac{1}{\sqrt{11}}\left(\frac{3}{\sqrt{2}}\epsilon_2 - \frac{3}{\sqrt{2}}\epsilon_3 + \sqrt{2}\,\epsilon_1\right)$$

$$= \left(\frac{2}{\sqrt{22}}, \frac{3}{\sqrt{22}}, \frac{-3}{\sqrt{22}}, 0\right),$$

$$\gamma_2 = \frac{1}{\sqrt{11}}(2\delta_2 - 3\delta_3) = \frac{1}{\sqrt{11}}(\epsilon_2 - \epsilon_3 - 3\epsilon_1)$$

$$= \left(\frac{-3}{\sqrt{11}}, \frac{1}{\sqrt{11}}, \frac{-1}{\sqrt{11}}, 0\right).$$

$$\delta_1 = \frac{1}{\sqrt{2}}(\epsilon_2 + \epsilon_3) = \left(0, \frac{1}{\sqrt{2}}, \frac{1}{\sqrt{2}}, 0\right)$$

Thus our choice of (p_{ij}) is

$$(p_{ij}) = \begin{bmatrix} 0 & 0 & 0 & 1 \\ \frac{2}{\sqrt{22}} & \frac{3}{\sqrt{22}} & \frac{-3}{\sqrt{22}} & 0 \\ \frac{-3}{\sqrt{11}} & \frac{1}{\sqrt{11}} & \frac{-1}{\sqrt{11}} & 0 \\ 0 & \frac{1}{\sqrt{2}} & \frac{1}{\sqrt{2}} & 0 \end{bmatrix}$$

and

$$(p_{ij})(a_{ij})(p_{ij})^t = \begin{bmatrix} 3 & 0 & 0 & 0 \\ \frac{6}{\sqrt{22}} & 2 & 0 & 0 \\ \frac{2}{\sqrt{11}} & -11\sqrt{2} & 2 & 0 \\ 2 & 2\sqrt{11} & 0 & 2 \end{bmatrix}.$$

6.2. Using the matrix (a_{ij}) of the preceding example, we illustrate a second iterative method for triangularizing a matrix. It is left as an exercise to check some of the details and to show that the method can be generalized to provide an alternative proof of Theorem 6.1. We begin, as before, with the characteristic value 3 and extend the accompanying characteristic vector ϵ_4 to the orthonormal basis $\mathscr{B}_1 = (\epsilon_4, \epsilon_1, \epsilon_2, \epsilon_3)$ of R^4. Then

$$(p_{ij}) = \begin{bmatrix} 0 & 0 & 0 & 1 \\ 1 & 0 & 0 & 0 \\ 0 & 1 & 0 & 0 \\ 0 & 0 & 1 & 0 \end{bmatrix}$$

is an orthogonal matrix and

$$(p_{ij})(a_{ij})(p_{ij})^t = \begin{bmatrix} 3 & 0 & 0 & 0 \\ 0 & 8 & 9 & -9 \\ 2 & 0 & 2 & 0 \\ 0 & 4 & 6 & -4 \end{bmatrix} = (b_{ij}),$$

let us say. Now delete the first row and first column of (b_{ij}) to obtain the 3×3 matrix

$$\begin{bmatrix} 8 & 9 & -9 \\ 0 & 2 & 0 \\ 4 & 6 & -4 \end{bmatrix}$$

with characteristic polynomial $(x - 2)^3$. One unit characteristic vector belonging to the characteristic value 2 is $\dfrac{1}{\sqrt{13}}(2, 0, -3)$. This may be extended to the orthonormal basis

$$\mathscr{B}_2 = \left(\frac{1}{\sqrt{13}}(2, 0, -3), (0, 1, 0), \frac{1}{\sqrt{13}}(3, 0, 2)\right)$$

of R^3. Then the matrix

$$(q_{ij}) = \begin{bmatrix} \dfrac{2}{\sqrt{13}} & 0 & \dfrac{-3}{\sqrt{13}} \\ 0 & 1 & 0 \\ \dfrac{3}{\sqrt{13}} & 0 & \dfrac{2}{\sqrt{13}} \end{bmatrix}$$

is orthogonal and

$$(q_{ij})\begin{bmatrix} 8 & 9 & -9 \\ 0 & 2 & 0 \\ 4 & 6 & -4 \end{bmatrix}(q_{ij})^t = \begin{bmatrix} 2 & 0 & 0 \\ 0 & 2 & 0 \\ 13 & 3\sqrt{13} & 2 \end{bmatrix}.$$

Since the resulting matrix is triangular, we do not repeat the first step but continue by concocting a basis $\mathscr{B}_3$ of R^4 from the basis $\mathscr{B}_2$ of R^3 in the following way: as the first element of $\mathscr{B}_3$ choose (1, 0, 0, 0). Then extend each 3-tuple of $\mathscr{B}_2$ to a 4-tuple by supplying a 0 in front of the initial coordinate of the 3-tuple. Thus

$$\mathscr{B}_3 = \left((1, 0, 0, 0), \left(0, \frac{2}{\sqrt{13}}, 0, \frac{-3}{\sqrt{31}}\right), (0, 0, 1, 0), \left(0, \frac{3}{\sqrt{13}}, 0, \frac{2}{\sqrt{13}}\right)\right)$$

By its method of construction, clearly $\mathscr{B}_3$ is an orthonormal basis of R^4. Hence

$$(r_{ij}) = \begin{bmatrix} 1 & 0 & 0 & 0 \\ 0 & \frac{2}{\sqrt{13}} & 0 & \frac{-3}{\sqrt{13}} \\ 0 & 0 & 1 & 0 \\ 0 & \frac{3}{\sqrt{13}} & 0 & \frac{2}{\sqrt{13}} \end{bmatrix}$$

is an orthogonal matrix and

$$(r_{ij})(b_{ij})(r_{ij})^t = \begin{bmatrix} 3 & 0 & 0 & 0 \\ 0 & 2 & 0 & 0 \\ 2 & 0 & 2 & 0 \\ 0 & 13 & 3\sqrt{13} & 2 \end{bmatrix} = (t_{ij}).$$

Since $(b_{ij}) = (p_{ij})(a_{ij})(p_{ij})^t$, it follows that

$$[(r_{ij})(p_{ij})](a_{ij})[(r_{ij})(p_{ij})]^t = (t_{ij}).$$

Further, since the product of orthogonal matrices is an orthogonal matrix,

$$(r_{ij})(p_{ij}) = \begin{bmatrix} 0 & 0 & 0 & 1 \\ \frac{2}{\sqrt{13}} & 0 & \frac{-3}{\sqrt{13}} & 0 \\ 0 & 1 & 0 & 0 \\ \frac{3}{\sqrt{13}} & 0 & \frac{2}{\sqrt{13}} & 0 \end{bmatrix}$$

is orthogonal and, hence

$$\left((0, 0, 0, 1), \left(\frac{2}{\sqrt{13}}, 0, \frac{-3}{\sqrt{13}}, 0\right), (0, 1, 0, 0), \left(\frac{3}{\sqrt{13}}, 0, \frac{2}{\sqrt{13}}, 0\right)\right)$$

is an orthonormal basis of R^4 relative to which A has the triangular matrix (t_{ij}).

We conclude this section with a theorem concerning the simultaneous diagonalization and triangularization of sets of linear transformations and sets of matrices in accordance with the following definition. A collection $\{A_i \mid i \in I\}$ of linear transformations on a finite-dimensional vector space V is *simultaneously triangulable* (*diagonable*) if and only if there exists an ordered basis of V relative to which the matrix of each A_i is triangular (diagonal); a collection of $n \times n$ matrices is simultaneously triangulable (diagonable) if and only if there exists an $n \times n$ invertible matrix (p_{ij}) such that for each (a_{ij}) in the collection $(p_{ij})(a_{ij})(p_{ij})^{-1}$ is triangular (diagonal).

□ **Lemma.** Let A and B be commuting, triangulable, linear transformations on a finite-dimensional vector space V. Then A and B have a common characteristic vector.

PROOF. Let c be a characteristic value of A and $S_c = \{\alpha \in V \mid \alpha A = cA\}$. If $\alpha \in S_c$, then

$$(\alpha B)A = \alpha(BA) = \alpha(AB) = (\alpha A)B = (c\alpha)B = c(\alpha B).$$

Thus $\alpha B \in S_c$ for all $\alpha \in S_c$, which means that S_c is B-invariant. Consider now the restriction B', let us call it, of B on S_c. Since the minimal polynomial of B' is a factor of that of B and since B is triangulable, it follows that B' has a characteristic value c' which is a characteristic value of B. Thus there exists in S_c a vector β that is a characteristic vector of B' and, hence, of B. Such a vector is a characteristic vector of both A and B. ◇

□ **Theorem 6.2.** If A and B are commuting, diagonable, linear transformations on a finite-dimensional vector space V, then they are simultaneously diagonable. If A and B are commuting, triangulable, linear transformations on a finite-dimensional inner-product space, they are simultaneously unitarily (or orthogonally) triangulable.

PROOF. Assume that the commuting linear transformations A and B are diagonable. Let $c_1, \ldots, c_k$ be the distinct characteristic values of A. Then $S_{c_i} = \{\alpha \in V \mid \alpha A = c_i \alpha\}$, $1 \le i \le k$, is B-invariant (see the proof of the lemma given above). Further, we contend that the restriction of B to S_{c_i}, $1 \le i \le k$, is diagonable. This is a consequence of (i) the fact that if S is a B-invariant subspace of V then the minimal polynomial of the restriction of B to S is a factor of the minimal polynomial of B and (ii) the assumption that B is diagonable. Next, choose a basis $\mathscr{B}_{ci}$ of S_{c_i} such that the matrix of the restriction of B to S_{c_i} is diagonal, $1 \le i \le k$. Then the union $\mathscr{B}$ of the bases $\mathscr{B}_{c_i}$, $1 \le i \le k$, is a basis of V relative to which A and B have diagonal matrices.

Next, assume that A and B are commuting, triangulable, linear transformations on V, which we now assume is a finite-dimensional inner-product space. Then A^* and B^* are triangulable and $A^*B^* = B^*A^*$. If $\dim V = 1$, the result in question is trivially true. Assuming that it is true for commuting linear transformations on inner-product spaces of dimension $n - 1$, we consider the case where $\dim V = n$.

By the lemma given above, A and B have a common unit characteristic vector δ_1. Let $(\delta_1, \gamma_2, \ldots, \gamma_n)$ be an extension of $\{\delta_1\}$ to an orthonormal basis of V. Then $S = [\gamma_2, \ldots, \gamma_n]$ is both A^*- and B^*-invariant. Since A_S^* and B_S^* commute and both are triangulable (a consequence of the observation that A^* and B^* commute), it follows from the induction hypothesis that there exists an orthonormal basis $(\delta_2, \ldots, \delta_n)$ of S relative to which the matrices of A_S^* and B_S^* are triangular. Hence the matrices of A and B relative to the orthogonal basis $(\delta_1, \ldots, \delta_n)$ of V are triangular. ◇

☐ **Corollary 1.** Let $A_1, \ldots, A_k$ be diagonable (triangulable) linear transformations in a finite-dimensional inner-product space. If $A_i A_j = A_j A_i$ for $1 \le i, j \le k$, then $A_1, \ldots, A_k$ are simultaneously diagonable (triangulable).

☐ **Corollary 2.** Let $(a_{ij})_1, \ldots, (a_{ij})_k$ be a set of $n \times n$ diagonable matrices over a field F. If each pair of this set commutes, then there exists a nonsingular matrix (p_{ij}) over F such that $(p_{ij})(a_{ij})_h(p_{ij})^{-1}$ is diagonal for $1 \le h \le k$.

☐ **Corollary 3.** Let $(a_{ij})_1, \ldots, (a_{ij})_k$ be a set of $n \times n$ triangulable matrices over R. If each pair of the set commutes, then there exists an orthogonal matrix (p_{ij}) such that $(p_{ij})(a_{ij})_h(p_{ij})^t$ is triangular for $1 \le h \le k$.

□ **Corollary 4.** Let $(a_{ij})_1, \ldots, (a_{ij})_k$ be a set of $n \times n$ matrices over C. If each pair of the set commutes, then there exists a unitary matrix (u_{ij}) such that $(u_{ij})(a_{ij})_h(u_{ij})^*$ is triangular for $1 \leq h \leq k$.

EXERCISES

6.1. Find an orthogonal matrix (p_{ij}) such that

$$(p_{ij}) \begin{bmatrix} 1 & 0 & 0 \\ 4 & 1 & 4 \\ 1 & 0 & 2 \end{bmatrix} (p_{ij})^t$$

is triangular.

6.2. Find an orthogonal matrix (p_{ij}) such that

$$(p_{ij}) \begin{bmatrix} 2 & 1 & 0 & -1 \\ -1 & 1 & 1 & 1 \\ 0 & 1 & 1 & 0 \\ 1 & 1 & 0 & 0 \end{bmatrix} (p_{ij})^t$$

is triangular.

6.3. Find a unitary matrix (u_{ij}) such that

$$(u_{ij}) \begin{bmatrix} 0 & 1 & 0 \\ 1 & 0 & -3 \\ 0 & 1 & 0 \end{bmatrix} (u_{ij})^*$$

is triangular.

6.4. Deduce Theorem 4.1. as a corollary to Theorem 6.1.

6.5. Develop another proof of Theorem 6.1 by generalizing the reduction method employed in Example 6.2.

6.6. Show that two diagonable matrices are simultaneously diagonable if and only if they commute with each other.

6.7. Prove Corollary 1 of Theorem 6.2.

6.8. Find two matrices that are simultaneously triangularizable but do not commute.

6.9. Show that if (a_{ij}) and (b_{ij}) are 2×2 matrices over C with the same characteristic vectors, then (a_{ij}) and (b_{ij}) commute.

6.10 Show that

$$\begin{bmatrix} 1 & 0 & -1 & 0 \\ 0 & 1 & 0 & -1 \\ 1 & 0 & -1 & 0 \\ 0 & 1 & 0 & -1 \end{bmatrix}, \quad \begin{bmatrix} 4 & -1 & -1 & 0 \\ -1 & 4 & 0 & -1 \\ 1 & 0 & 2 & -1 \\ 0 & 1 & -1 & 2 \end{bmatrix}$$

commute and find a common characteristic vector of these two matrices in R^4.

6.11. Find a real matrix (p_{ij}) that simultaneously diagonalizes (a_{ij}) and (b_{ij}) if

(a) $$(a_{ij}) = \begin{bmatrix} 1 & 2 \\ 0 & 2 \end{bmatrix}, \qquad (b_{ij}) = \begin{bmatrix} 3 & -8 \\ 0 & -1 \end{bmatrix},$$

(b) $$(a_{ij}) = \begin{bmatrix} 1 & 1 \\ 1 & 1 \end{bmatrix}, \qquad (b_{ij}) = \begin{bmatrix} 1 & a \\ a & 1 \end{bmatrix}.$$

6.12. Show that Corollary 1 of Theorem 6.2 extends to a possibly infinite collection $\{A_i \mid i \in I\}$ of linear transformations.

7 Bilinear Forms and Quadratic Forms

Classically, a *homogeneous quadratic form* is a function Q of n real variables $x_1, \ldots, x_n$ such that

$$Q(x_1, \ldots, x_n) = \sum_{i,j=1}^{n} q_{ij} x_i x_j,$$

where each q_{ij} is a real number. Such functions crop up in a wide variety of mathematical contexts. For instance, a quadric surface whose center is at the origin of a rectangular coordinate system for E_3 has an equation of the form

$$ax^2 + by^2 + cz^2 + dxy + exz + fyz = g,$$

where $a, b, c, d, e, f, g \in \mathsf{R}$. The problem of simplifying this equation amounts to that of finding an orthogonal transformation (that is, a rotation) such that the quadratic form on the left is reduced to just a sum or difference of squares. In three dimensions this rotation can always be found, as we shall show. Again, the reader who has studied the differential calculus of functions of several variables may recall the role played by quadratic forms in the determination of extrema of such functions. This topic is treated in Section 2.

The principal goal of this chapter is to discuss a useful way to classify quadratic forms and to describe methods for simplifying them. Biproducts include further results about real symmetric matrices. Before we begin our study of quadratic forms, however, we discuss briefly a more general class of functions—the bilinear forms.

1. BILINEAR FORMS

Our first definition incorporates most of the features assigned to an inner-product function in a vector space over R (see Section 3.2).

☐ **Definition.** Let V be a vector space over a field F. A *bilinear form* on V is a function f on $V \times V$ into F which is linear as a function of either argument when the other is fixed. This means [designating the value of f at (α, β) by $f(\alpha, \beta)$] that for all α, β, γ in V and all scalars a

$$f(\alpha + \gamma, \beta) = f(\alpha, \beta) + f(\gamma, \beta),$$

$$f(\alpha, \beta + \gamma) = f(\alpha, \beta) + f(\alpha, \gamma),$$

$$f(a\alpha, \beta) = f(\alpha, a\beta) = af(\alpha, \beta).$$

If, in addition, $f(\alpha, \beta) = f(\beta, \alpha)$ for all α, β in V, then f is called a *symmetric bilinear form* on V.

The initial definition may be easily extended to the case of a pair of vector spaces over a common field. Since we shall have no need for this more general concept we shall not pursue the matter.

In Chapter 3 we found that values of an inner product $(\ |\ )$ in an n-dimensional euclidean space V can be computed as follows: if $(x_1, \ldots, x_n)$ and $(y_1, \ldots, y_n)$ are the n-tuples of coordinates of vectors α and β, respectively, relative to the basis $(\alpha_1, \ldots, \alpha_n)$ of V, then

$$(\alpha \mid \beta) = (x_1, \ldots, x_n)(c_{ij})(y_1, \ldots, y_n)^t,$$

where (c_{ij}) is the $n \times n$ matrix such that $c_{ij} = (\alpha_i \mid \alpha_j)$. This suggests a possible way to construct bilinear forms on a finite-dimensional space V over F. Let $(\alpha_1, \ldots, \alpha_n)$ be a basis of V, let (a_{ij}) be any $n \times n$ matrix over F, and define the function f on $V \times V$ into F as follows: If $\alpha, \beta \in V$ and $\alpha = \sum_1^n x_i \alpha_i$, $\beta = \sum_1^n y_i \beta_i$, let

$$f(\alpha, \beta) = (x_1, \ldots, x_n)(a_{ij})(y_1, \ldots, y_n)^t = \sum_{i=1}^{n} \sum_{j=1}^{n} a_{ij} x_i y_j. \tag{1}$$

It is easily verified that f is a bilinear form on V and that f is symmetric if and only if (a_{ij}) is symmetric. Further, since $a_{ij} = f(\alpha_i, \alpha_j)$, $1 \leq i, j \leq n$, and (a_{ij}) is any $n \times n$ matrix over F, we conclude that, given a basis $(\alpha_1, \ldots, \alpha_n)$ of V, a bilinear form F on V is obtained by assigning arbitrary values a_{ij} to f at the n^2 arguments (α_i, α_j) and then computing values at other arguments by the recipe given in (1).

We prove next that, conversely, every bilinear form on V is of the above type. Let f be a bilinear form on V and let $(\alpha_1, \ldots, \alpha_n)$ be a basis of V. If $\alpha = \sum_1^n x_i \alpha_1$ and $\beta = \sum_1^n y_j \alpha_j$ are vectors in V, then

$$\begin{aligned} f(\alpha, \beta) &= f\left(\sum_{i=1}^{n} x_i \alpha_i, \beta\right) \\ &= \sum_{i=1}^{n} x_i f(\alpha_i, \beta) \\ &= \sum_{i=1}^{n} x_i f\left(\alpha_i, \sum_{j=1}^{n} y_j \beta_j\right) \\ &= \sum_{i=1}^{n} \sum_{j=1}^{n} f(\alpha_i, \alpha_j) x_i y_j . \end{aligned}$$

The n^2 scalars $f(\alpha_i, \alpha_j)$ therefore completely determine the function f. Introducing the $n \times n$ matrix

$$(a_{ij}), \quad \text{where} \quad a_{ij} = f(\alpha_i, \alpha_j),$$

which is uniquely determined by f relative to the basis $\mathscr{B} = (\alpha_1, \ldots, \alpha_n)$, we have

$$\begin{aligned} f(\alpha, \beta) &= (x_1, \ldots, x_n)(a_{ij})(y_1, \ldots, y_n)^t \\ &= [\alpha]_{\mathscr{B}}(a_{ij})[\beta]_{\mathscr{B}}^t . \end{aligned}$$

□ **Definition.** Let V be a finite-dimensional vector space over F and let $\mathscr{B} = (a_1, \ldots, \alpha_n)$ be a basis of V. If f is a bilinear form on V, the matrix of f relative to $\mathscr{B}$ is the $n \times n$ matrix (a_{ij}) such that $a_{ij} = f(\alpha_i, \alpha_j)$. This matrix is often denoted by $f_{\mathscr{B}}$. If (a_{ij}) is an $n \times n$ matrix over F, the bilinear form f on V such that $f(\alpha, \beta) = [\alpha]_{\mathscr{B}}(a_{ij})[\beta]_{\mathscr{B}}$ is called the form induced by (a_{ij}) relative to $\mathscr{B}$.

Our observations may now be summarized as follows: Relative to each ordered basis $\mathscr{B}$ of an n-dimensional vector space V over F, the mapping $f \to f_{\mathscr{B}}$ is a 1-1 correspondence between the set of all bilinear forms on V and the set of all $n \times n$ matrices over F. The details are left as an exercise.

EXAMPLES

1.1. From the definition of a symmetric bilinear form it is clear that an inner product in a euclidean space is a symmetric bilinear form with the property that $(\alpha \mid \alpha) > 0$ for each nonzero vector α.

1.2. Let V be a euclidean space with inner product $(\,|\,)$ and let A be a linear transformation on V. Then the function f such that $f(\alpha, \beta) = (\alpha A \,|\, \beta)$ for all α, β in V is a bilinear form on V.

1.3. If g, h are linear functionals on a vector space V over F, then the function f on $V \times V$ into F such that $f(\alpha, \beta) = (\alpha g)(\beta h)$ for all α, β in V is a bilinear form on V.

1.4. Let k be a real-valued function of two variables which is continuous in the square $\{(x, y) \,|\, a \leq x \leq b,\ a \leq y \leq b\}$. Then the function f, defined by

$$f(\alpha, \beta) = \int_a^b dt \int_a^b k(s, t)\, \alpha(s)\, \beta(t)\, dt$$

for all α, β in $C[a, b]$, is a symmetric bilinear form on $C[a, b]$.

1.5. The function f on $\mathsf{R}^2 \times \mathsf{R}^2$ into R, defined by

$$f(\alpha, \beta) = x_1 x_2 - y_1 y_2,$$

if $\alpha = (x_1, y_1)$ and $\beta = (x_2, y_2)$, is a symmetric bilinear form whose matrix relative to $\mathscr{B} = (\epsilon_1, \epsilon_2)$ is

$$f_{\mathscr{B}} = \begin{pmatrix} 1 & 0 \\ 0 & -1 \end{pmatrix}.$$

This form is not an inner product in R^2, since, for $\alpha = (1, 1)$, $f(\alpha, \alpha) = 0$. As a symmetric bilinear form, however, it may still be used to equip the space with some metric notions. We elaborate on this matter in the next paragraph.

Let V be a vector space over a field F and let g be a symmetric bilinear form on V. We say that g introduces a *metric* in V and call the pair (V, g) a *metric space*. To capture the spirit of our present point of view we should think of $g(\alpha, \beta)$ as something like the cosine of the "angle" between α and β and of $g(\alpha, \alpha)$ as something like the "length" of α. The notions of orthogonality of vectors, orthogonal subspaces, and the orthogonal complement of a subspace can be defined as in Chapter 3. Now assume that V is finite-dimensional. Let $\mathscr{B}$ be a basis of V and let $g_{\mathscr{B}}$ be the matrix of g relative to $\mathscr{B}$. Then V is called *nonsingular* if and only if $\det g_{\mathscr{B}} \neq 0$. Shortly we shall find that this definition is independent of $\mathscr{B}$ and, hence, unambiguous. For a (finite-dimensional) nonsingular metric space (V, g) some of the results appearing in Chapter 3 can be duplicated. For example, such a space has an orthogonal basis $\{\alpha_1, \ldots, \alpha_n\}$ such that $g(\alpha_i, \alpha_i) \neq 0$, $1 \leq i \leq n$. Again, if U is a nonsingular subspace of V, the orthogonal complement $U^\perp$ of U is a subspace orthogonal to U and $V = U \oplus U^\perp$.

The metric space (R^2, f), where f is given at the beginning of this example, is called the *Lorentz plane*. The metric space (R^2, g), where

$$g(\alpha, \beta) = x_1 x_2 + y_1 y_2 \quad \text{if} \quad \alpha = (x_1, y_1), \qquad \beta = (x_2, y_2),$$

is an inner-product space and, indeed, is the familiar euclidean plane. The metric space (R^4, h), where

$$h(\alpha, \beta) = x_1 y_1 + x_2 y_2 + x_3 y_3 - x_4 y_4 \quad \text{if} \quad \begin{matrix} \alpha = (x_1, x_2, x_3, x_4), \\ \beta = (y_1, y_2, y_3, y_4), \end{matrix}$$

is called *Minkowski space*. Each of these spaces is nonsingular. The reader might attempt to show that in a Minkowski space M there is a euclidean plane E and a Lorentz plane L and that these subspaces are orthogonal and $M = E \oplus L$.

Let V be an n-dimensional vector space over F. We have found that $n \times n$ matrices over F play a role for bilinear forms on V, relative to a basis of V, similar to the one they play for linear transformations on V, relative to a basis of V. Just as for linear transformations, the question arises as to the relationship between the matrices of a bilinear form relative to different bases of V. Let $\mathscr{B} = (\alpha_1, \ldots, \alpha_n)$ and $\mathscr{B}' = (\alpha_1', \ldots, \alpha_n')$ be bases of V and let f be a bilinear form on V. If

$$\alpha_i' = \sum_{j=1}^{n} p_{ij} \alpha_j, \qquad 1 \leq i \leq n,$$

then (p_{ij}) is an invertible matrix such that, for each α in V, $[\alpha]_{\mathscr{B}} = [\alpha]_{\mathscr{B}'}(p_{ij})$. Then, for vectors α and β in V,

$$\begin{aligned} f(\alpha, \beta) &= [\alpha]_{\mathscr{B}} f_{\mathscr{B}} [\beta]_{\mathscr{B}}^t \\ &= [\alpha]_{\mathscr{B}'}(p_{ij}) f_{\mathscr{B}} (p_{ij})^t [\beta]_{\mathscr{B}'}^t \\ &= [\alpha]_{\mathscr{B}'}[(p_{ij}) f_{\mathscr{B}} (p_{ij})^t][\beta]_{\mathscr{B}'}^t. \end{aligned}$$

By the definition and uniqueness of $f_{\mathscr{B}'}$ it follows that

$$f_{\mathscr{B}'} = (p_{ij}) f_{\mathscr{B}} (p_{ij})^t.$$

☐ **Definition.** If (a_{ij}) and (b_{ij}) are $n \times n$ matrices over F, then (b_{ij}) is called *congruent* to (a_{ij}) if and only if there exists a nonsingular matrix (p_{ij}) such that $(b_{ij}) = (p_{ij})(a_{ij})(p_{ij})^t$.

The relation of congruence of $n \times n$ matrices is symmetric; that is, if (b_{ij}) is congruent to (a_{ij}), then (a_{ij}) is congruent to (b_{ij}); for if

$$(b_{ij}) = (p_{ij})(a_{ij})(p_{ij})^t,$$

then

$$(a_{ij}) = (p_{ij})^{-1}(b_{ij})[(p_{ij})^t]^{-1} = (p_{ij})^{-1}(b_{ij})[(p_{ij})^{-1}]^t.$$

Thus we may speak simply of a pair of congruent matrices. Our calculations prior to the above definition may then be summarized as the first statement in the following theorem. Its converse is left as an exercise.

☐ **Theorem 1.1.** Any two matrices of a bilinear form are congruent. Conversely, if (a_{ij}) and (b_{ij}) are congruent $n \times n$ matrices over a field F, say $(b_{ij}) = (p_{ij})(a_{ij})(p_{ij})^t$, then (a_{ij}) and (b_{ij}) induce the same bilinear form relative to a pair $\mathscr{B}$, $\mathscr{B}'$ of bases of an n-dimensional space over F such that (p_{ij}) is the matrix of $\mathscr{B}'$ to $\mathscr{B}$.

Since congruent matrices have the same rank (recall that a nonsingular matrix is a product of elementary matrices, that multiplication by an element of any matrix effects a row or column operation, and that such operations do not alter the rank of a matrix), the *rank* of a bilinear form on an n-dimensional space may be defined as the *rank* of any matrix of the form. In particular, if some matrix of a form is nonsingular, then so are all its matrices. Thus the definition of a nonsingular metric space given in Example 1.5 is justified.

EXAMPLE

1.6. Let f be the bilinear form on R^2 such that

$$f(\alpha, \beta) = x_1 y_1 - x_2 y_2 \quad \text{if} \quad \alpha = (x_1 x_2), \qquad \beta = (y_1, y_2).$$

If $\mathscr{B} = (\epsilon_1, \epsilon_2)$ and $\mathscr{B}' = ((1, 1)\,(-2, 1))$, then

$$f_{\mathscr{B}} = \begin{pmatrix} 1 & 0 \\ 0 & -1 \end{pmatrix}, \qquad f_{\mathscr{B}'} = \begin{pmatrix} 0 & -3 \\ -3 & 3 \end{pmatrix}.$$

Since the matrix of $\mathscr{B}'$ relative to $\mathscr{B}$ is

$$(p_{ij}) = \begin{pmatrix} 1 & 1 \\ -2 & 1 \end{pmatrix},$$

we may conclude that

$$\begin{pmatrix} 0 & -3 \\ -3 & 3 \end{pmatrix} = \begin{pmatrix} 1 & 1 \\ -2 & 1 \end{pmatrix} \begin{pmatrix} 1 & 0 \\ 0 & -1 \end{pmatrix} \begin{pmatrix} 1 & 1 \\ -2 & 1 \end{pmatrix}^t.$$

Now suppose that f is a bilinear form on an n-dimensional euclidean space V with orthonormal basis $\mathscr{B}$. Let A be the linear transformation on V induced by $f_{\mathscr{B}}$ relative to $\mathscr{B}$. The next theorem describes the relationship among f, A, and the inner product $(\ |\ )$ in V.

Theorem 1.2. If V is an n-dimensional euclidean space with inner-product $(\ |\ )$ and f is a bilinear form on V, then there exists a unique linear transformation A on V such that

$$f(\alpha, \beta) = (\alpha A \mid \beta) \qquad \text{all } \alpha, \beta \text{ in } V.$$

Further, f is symmetric if and only if A is self-adjoint.

PROOF. Let $\mathscr{B} = (\alpha_1, \ldots, \alpha_n)$ be an orthonormal basis of V and suppose that $(a_{ij}) = f_{\mathscr{B}}$ [so that $a_{ij} = f(\alpha_i, \alpha_j)$]. Let A be the linear transformation induced by (a_{ij}) relative to $\mathscr{B}$ [so that $\alpha_i A = \sum_1^n a_{ij}\alpha_j$]. Now

$$(\alpha_i A \mid \alpha_j) = \left(\sum_{k=1}^{n} a_{ik}\alpha_k \mid \alpha_j\right) = a_{ij} = f(\alpha_i, \alpha_j),$$

which implies that $(\alpha A \mid \beta) = f(\alpha, \beta)$ for all α, β in V.

To prove that A is unique, assume that A' is a linear transformation in V such that

$$(\alpha A' \mid \beta) = f(\alpha, \beta) = (\alpha A \mid \beta)$$

for all α, β in V. Then $(\alpha(A' - A) \mid \beta) = 0$ for all α, β in V. Hence for all α in V

$$(\alpha(A' - A) \mid \alpha(A' - A)) = 0$$

and in turn $\alpha(A' - A) = 0$. Thus $A' = A$.

To establish the final statement of the theorem observe that $f(\alpha, \beta) = f(\beta, \alpha)$ if and only if $(\alpha A \mid \beta) = (\beta A \mid \alpha)$. Since the inner product is a symmetric function, $(\beta A \mid \alpha) = (\alpha \mid \beta A)$. Thus $f(\alpha, \beta) = f(\beta, \alpha)$ if and only if $(\alpha A \mid \beta) = (\alpha \mid \beta A)$, that is if and only if A is self-adjoint. ◇

EXERCISES

1.1. Which of the following definitions of a function f yields a bilinear form on $P_\infty(\mathsf{R})$?
(a) $f(p, q) = p(1) + q(1)$.
(b) $f(p, q) = p(1)\, q(1)$.
(c) $f(p, q) = p(1)\, q'(1)$, where q' is the derivative of q.
(d) $f(p, q) = \left(\int_0^1 p(t)\, dt\right)\left(\int_0^1 q(t)\, dt\right)$.

1.2. Suppose that f is the bilinear form on R^3 whose matrix relative to the standard basis of R^3 is

$$\text{(a)} \begin{bmatrix} 1 & 0 & 2 \\ 0 & 1 & 3 \\ 2 & 0 & 3 \end{bmatrix}, \qquad \text{(b)} \begin{bmatrix} 2 & 1 & 0 \\ 1 & 3 & 0 \\ 2 & 3 & 5 \end{bmatrix}.$$

In each case find the matrix of f relative to the basis $((1, 4, 0)\ (-1, 5, 1), (3, 0, 4))$ of R^3.

1.3. Does there exist a vector space V and a bilinear form f on V such that f is not the zero function but $f(\alpha, \alpha) = 0$ for all α in V?

1.4. Let f be a symmetric bilinear form on the vector space V. The *radical* of f (in symbols, rad f) is defined by

$$\operatorname{rad} f = \{\alpha \in V \mid f(\alpha, \beta) = 0 \quad \text{for all } \beta \in V\}.$$

Show that rad f is a subspace of V and devise an example of a V and an f where rad f is a nontrivial subspace of V.

1.5. Let f be a symmetric bilinear form on a vector space V. A subspace S of V is called a *null space* of f if and only if $f(\alpha, \alpha) = 0$ for all α in S.
(a) Find all null spaces of the Lorentz plane.
(b) Decide whether rad f is equal to the intersection of all null spaces of f.

1.6. Using the familiar operations of addition of functions and multiplication of a function by a scalar, show that the set of all bilinear forms on a vector space V over a field F is a vector space W over F. If V is n-dimensional and if φ is the 1-1 correspondence of the set W with the set $M_{n\times n}(F)$ of all $n \times n$ matrices over F, show that φ is an isomorphism of the space W onto the space $M_{n\times n}(F)$.

1.7. Show in detail that congruent matrices have the same rank.

1.8. Let f be the bilinear form on R^3 such that

$$f(\alpha, \beta) = x_1x_2 + y_1y_2 + 3z_1z_2 + 2x_1y_2 + 2y_1x_2$$

if $\alpha = (x_1, y_1z_1)$ and $\beta = (x_2, y_2, z_2)$. Find a basis $\mathscr{B}$ of R^3 relative to which the matrix of f is diag $(1, 1, -1)$.

1.9. A bilinear form on a vector space V is said to be *nondegenerate* if and only if for each $\alpha \neq 0$ in V there exist β, γ in V such that

$$f(\alpha, \beta) \neq 0 \quad \text{and} \quad f(\gamma, \alpha) = 0.$$

Let f be the bilinear form on $\mathbf{R}^2$ such that

$$f(\alpha, \beta) = ax_1x_2 + \frac{b}{2}(x_1y_2 + x_2y_1) + cy_1y_2 \quad \text{if } \alpha = (x_1, y_1),\ \beta = (x_2, y_2).$$

Show that f is nondegenerate if and only if $b^2 - 4ac \neq 0$.

1.10. Let S_1 and S_2 be subspaces of a vector space V over a field F such that $V = S_1 \oplus S_2$. Let f_1 and f_2 be symmetric bilinear forms on S_1 and S_2, respectively. Show that there exists a unique symmetric bilinear form f on V such that if $\alpha, \beta \in V$ and $\alpha = \alpha_1 + \alpha_2$ and $\beta = \beta_1 + \beta_2$, where $\alpha, \beta_1 \in S_1$ and $\alpha_2, \beta_2 \in S_2$, then

$$f(\alpha, \beta) = f_1(\alpha_1, \beta_1) + f_2(\alpha_2 \beta_2).$$

If V is finite-dimensional and $\mathscr{B}_1$ and $\mathscr{B}_2$ are ordered bases of S_1 and S_2, respectively, describe the matrix of f relative to the ordered basis $\mathscr{B}_1 \cup \mathscr{B}_2$ of V.

1.11. Let V be a vector space over F and f be a bilinear form on V. For each α in V define a mapping $h_\alpha : V \to F$ by

$$\beta h_\alpha = f(\beta, \alpha), \qquad \beta \in V.$$

(a) Show that $h_\alpha \in V^*$, that is, h_α is a linear functional on V, and that the mapping $L: V \to V^*$, where $\alpha L = h_\alpha$, is a linear transformation on V into V^*.

(b) If V is finite-dimensional, show that the following conditions are equivalent.

(i) The kernel of L is $\{0\}$.

(ii) The form f is nondegenerate.

(iii) If (a_{ij}) is a matrix of f, then $\det (a_{ij}) \neq 0$.

1.12. Let V be an n-dimensional euclidean space with inner product $(\ |\ )$. Suppose that f is a bilinear form on V and A is the linear transformation on V such that $f(\alpha, \beta) = (\alpha A \mid \beta)$.

(a) Show that if f is symmetric then $f(\alpha, \alpha) > 0$ for all $\alpha \neq 0$ if and only if all characteristic values of A are positive.

(b) Does the conclusion in (a) hold if f is not necessarily symmetric?

1.13. Let (a_{ij}) be a real symmetric matrix, all of whose characteristic values are positive. Let f be the bilinear form on $\mathbf{R}^n$ induced by (a_{ij}) relative to the standard basis. Determine whether f is an inner product in $\mathbf{R}^n$.

1.14. Let V be an n-dimensional euclidean space and let f be a symmetric bilinear form on V. Is it always possible to find an orthonormal basis of V relative to which the matrix of f is diagonal?

2. QUADRATIC FORMS

Let V be a vector space over a field F. If f is a symmetric bilinear form on V, the function $q: V \to F$ defined by the rule

$$q(\alpha) = f(\alpha, \alpha), \qquad \alpha \in V,$$

has the properties numbered (1) to (4) which follow.

(1) $$q(a\alpha) = a^2 q(\alpha), \qquad a \in F, \alpha \in V,$$

since $q(a\alpha) = f(a\alpha, a\alpha) = a^2 f(\alpha, \alpha) = a^2 q(\alpha)$. In particular, $q(-\alpha) = q(\alpha)$ for all α in V.

(2) $$q(\alpha + \beta) = q(\alpha) + 2f(\alpha, \beta) + q(\beta)$$

(where 2 denotes the element $1 + 1$ of the field F), since

$$\begin{aligned} q(\alpha + \beta) = f(\alpha + \beta, \alpha + \beta) &= f(\alpha, \alpha) + f(\alpha, \beta) + f(\beta, \alpha) + f(\beta, \beta) \\ &= q(\alpha) + 2f(\alpha, \beta) + q(\beta). \end{aligned}$$

From the identity in (2) it follows that

(3) $$2f(\alpha, \beta) = [q(\alpha + \beta) - q(\alpha) - q(\beta)].$$

To derive still another identity we replace β with $-\beta$ in (2) to obtain

$$q(\alpha - \beta) = q(\alpha) - 2f(\alpha, \beta) + q(\beta).$$

Combining this result with (2) yields the *parallelogram identity*

(4) $$q(\alpha + \beta) + q(\alpha - \beta) = 2[q(\alpha) + q(\beta)].$$

□ **Definition.** Let V be a vector space over a field F. A function q on V into F is called a *quadratic form* on V if and only if there exists a symmetric bilinear function on V such that

$$q(\alpha) = f(\alpha, \alpha), \qquad \alpha \in V.$$

The function q is called the quadratic form associated with f.

According to (3), if, in F, $1 + 1 \neq 0$, the quadratic form, if associated with a bilinear form f, uniquely determines f. Thus in this case we may speak of the symmetric bilinear form associated with a given quadratic form.

To determine the explicit form of the values of a quadratic form on a finite-dimensional vector space let V be an n-dimensional vector space over

F, let $\mathscr{B} = (\alpha_1, \ldots, \alpha_n)$ be a basis of V, and let f be a symmetric bilinear form on V with $f_{\mathscr{B}} = (a_{ij})$. Thus $f(\alpha_i, \alpha_j) = a_{ij} = a_{ji}$ for $1 \leq i, j \leq n$. Let q be the quadratic form associated with f. Then, if $\alpha = \sum_1^n x_i \alpha_i \in V$,

$$q(\alpha) = f(\alpha, \alpha) = (x_1, \ldots, x_n)(a_{ij})(x_1, \ldots, x_n)^t. \tag{5}$$

Expanding the indicated product of matrices and using the fact that $a_{ji} = a_{ij}$, we obtain

$$q(\alpha) = \sum_{i=1}^{n} a_{ii} x_i^{\,2} + 2 \sum_{1 \leq i < j \leq n} a_{ij} x_i x_j .$$

The matrix (a_{ij}) of f, relative to the basis $\mathscr{B}$, is also called the matrix of q (the quadratic form associated with f), relative to the basis $\mathscr{B}$, and may be denoted by $q_{\mathscr{B}}$.

Turning matters around, let (a_{ij}) be an $n \times n$ symmetric matrix over F. Then (a_{ij}) defines a function $q{:}V \to F$ by the recipe given in (5) as well as a symmetric bilinear form on F, namely, the form f that is induced by (a_{ij}) relative to $\mathscr{B}$. Clearly q is then the quadratic form on V associated with the symmetric bilinear form f. In particular, if (a_{ij}) is an $n \times n$ symmetric matrix over R, the quadratic form q on R^n defined by

$$q(x_1, \ldots, x_n) = \sum_{i=1}^{n} a_{ii} x_i^2 + 2 \sum_{1 \leq i, j \leq n} a_{ij} x_i x_j$$

is associated with the symmetric bilinear form f induced by the symmetric matrix (a_{ij}) relative to the standard basis of R^n.

EXAMPLES

2.1. We define a function q on R^3 into R as follows: if $\alpha = (x, y, z)$, then

$$q(\alpha) = 2x^2 - y^2 + z^2 - 3xy + 5yz - 6xz.$$

Since

$$q(\alpha) = (2x^2 - y^2 + z^2) + 2(-\tfrac{3}{2}xy - 3xz + \tfrac{5}{2}yz),$$

we have

$$q(\alpha) = (x, y, z)\begin{pmatrix} 2 & -\frac{3}{2} & -3 \\ -\frac{3}{2} & -1 & \frac{5}{2} \\ -3 & \frac{5}{2} & 1 \end{pmatrix}(x, y, z)^t.$$

Thus q is the quadratic form associated with the symmetric bilinear form f whose matrix relative to the standard basis of R^3 is shown above.

2.2. Let E be an n-dimensional euclidean space and $\mathscr{B} = (\alpha_1, \ldots, \alpha_n)$ be an orthonormal basis of E. If $\alpha = \sum_1^n x_i \alpha_i$, then the function q such that $q(\alpha) = \|\alpha\|^2 = \sum_1^n x_i^2$ is a quadratic form on E. The same is true of the function q' such that

$$q'(\alpha) = \sum_{1 \le i < j \le n} x_i x_j.$$

The matrix (a_{ij}) of q' relative to $\mathscr{B}$ has the following description: $a_{ii} = 0$, $1 \le i \le n$, and $a_{ij} = \frac{1}{2}$ if $i \ne j$.

2.3. Let f denote a function on R^n into R which possesses continuous second partial derivatives in some neighborhood of the point $\alpha = (a_1, \ldots, a_n)$. We adopt the following notation:

$$f_i = \left.\frac{\partial f}{\partial x_i}\right|_\alpha, \qquad f_{ij} = \left.\frac{\partial^2 f}{\partial x_i \, \partial x_j}\right|_\alpha = f_{ji},$$

$$\varphi = (f_1, \ldots, f_n), \qquad \xi = (x_1, \ldots, x_n).$$

The Taylor series expansion of $f(\alpha + \xi)$at α is

$$f(\alpha + \xi) = f(\alpha) + \varphi \xi^t + \tfrac{1}{2}\xi(f_{ij})\xi^t + r,$$

where r is the remainder. A necessary condition for a relative extremum of f at α is that $\varphi = (0, \ldots, 0)$. If this condition is satisfied, α is called a critical point of f. Then

$$f(\alpha + \xi) - f(\alpha) = \tfrac{1}{2}\xi(f_{ij})\xi^t + r,$$

where, for small values of $x_1, \ldots, x_n$, the controlling term on the right is the quadratic form $\xi(f_{ij})\xi^t$. Thus the nature of $f(\alpha)$ at a critical point α rests with the behavior of this quadratic form near α. Later we shall develop a systematic method for classifying the values of such quadratic forms.

In order to simplify our exposition of further properties of quadratic forms, we restrict our attention in the remainder of this section to forms on finite-dimensional vector spaces over R.

Now let V be an n-dimensional euclidean space with inner product $(\ |\ )$ and let f be a symmetric bilinear form on V. Observe that if $\mathscr{B} = (\alpha_1, \ldots, \alpha_n)$ is an orthonormal basis of V such that

$$f_{\mathscr{B}} = \operatorname{diag}(c_1, \ldots, c_n),$$

which is the case if and only if $f(\alpha_i, \alpha_j) = c_i \delta_{ij}$, then the value of f at (α, β), where $\alpha = \sum_1^n x_i \alpha_i$ and $\beta = \sum_1^n y_i \alpha_i$, is

$$f(\alpha, \beta) = \sum_{i=1}^{n} c_i x_i y_i$$

and the value of the associated quadratic form is

$$q(\alpha) = \sum_{i=1}^{n} c_i x_i^2.$$

The simplicity of these formulas for function values leads us rather quickly to the question of whether such a basis of V exists. We shall show that the answer is in the affirmative. By Theorem 1.2 there exists a unique self-adjoint linear transformation A in V such that $f(\alpha, \beta) = (\alpha A \mid \beta)$ for all $\alpha, \beta \in V$. Now choose $\mathscr{B}$ to be an orthonormal basis of V whose members $\alpha_1, \ldots, \alpha_n$ are characteristic vectors of A. This is possible in view of Theorem 6.4.1. If $\alpha_i A = c_i \alpha_i$, $1 \le i \le n$, then

$$f(\alpha_i, \alpha_j) = (\alpha_i A \mid \alpha_j) = c_i(\alpha_i \mid \alpha_j) = c_i \delta_{ij}$$

for $1 \le i, j \le n$. Thus, such a basis has the features we desire. We state these results in the next theorem.

□ **Theorem 2.1.** (The principal axes theorem). If f is a symmetric bilinear form on an n-dimensional euclidean space V, then there exists an orthonormal basis $\mathscr{B} = (\alpha_1, \ldots, \alpha_n)$ of V such that if $\alpha = \sum_1^n x_i \alpha_i$ and $\beta = \sum_1^n y_i \alpha_i$, then

$$f(\alpha, \beta) = \sum_{i=1}^{n} c_i x_i y_i$$

and, for the associated quadratic form q,

$$q(\alpha) = \sum_{i=1}^{n} c_i x_i^2.$$

EXAMPLES

2.4. Let q be a quadratic form on R^2 such that if $\alpha = (x, y)$ then

$$q(\alpha) = ax^2 + bxy + cy^2.$$

Thus, the matrix of q relative to the standard basis of R^2 is

$$\begin{bmatrix} a & b/2 \\ b/2 & c \end{bmatrix}.$$

The characteristic polynomial of this matrix is

$$h = (x - a)(x - c) - \frac{b^2}{4}.$$

We are assured that h decomposes into a product of linear factors over R. If $h = (x - c_1)(x - c_2)$, then there exists an orthonormal basis (α_1, α_2) of R^2 such that

$$q(\alpha) = c_1\bar{x}^2 + c_2\bar{y}^2 \quad \text{if} \quad \alpha = \bar{x}\alpha_1 + \bar{y}\alpha_2.$$

From this result we can quickly derive the familiar classification of the graphs of equations of the form

$$ax^2 + bxy + cy^2 = d$$

(if a graph exists) by way of the discriminant

$$D = b^2 - 4ac,$$

which, in view of the definition of c_1 and c_2, is equal to $-4c_1c_2$. Indeed, if $D > 0$, then $c_1c_2 < 0$, and so the graph of $q(\alpha) = d$, which is the graph of $ax^2 + bxy + cy^2 = d$, is a hyperbola. If $D < 0$, then $c_1c_2 > 0$ and so the graph of $ax^2 + bxy + cy^2 = d$ is an ellipse. Finally, if $D = 0$, then the graph is a (degenerate) parabola.

2.5. Let us determine the graph (in E_3) of the equation

$$4x^2 + 4y^2 - 8z^2 - 10xy - 4xz - 4yz = 1.$$

The matrix of the quadratic form q such that

$$q(\alpha) = 4x^2 + 4y^2 - 8z^2 - 10xy - 4xz - yz, \quad \text{if} \quad \alpha = (x, y, z),$$

is

$$(a_{ij}) = \begin{bmatrix} 4 & -5 & -2 \\ -5 & 4 & -2 \\ -2 & -2 & -8 \end{bmatrix},$$

relative to the standard basis $(\epsilon_1, \epsilon_2, \epsilon_3)$ of R^3. Its characteristic polynomial is

$$h = (x - 9)(x + 9)x.$$

Further, we find that

$$\alpha_1 = \frac{1}{\sqrt{2}}(1, -1, 0),$$

$$\alpha_2 = \frac{1}{3\sqrt{2}}(1, 1, 4),$$

$$\alpha_3 = \frac{1}{3}(2, 2, -1)$$

are unit characteristic vectors belonging to the characteristic values 9, -9, 0, respectively. Hence $\mathscr{B} = (\alpha_1, \alpha_2, \alpha_3)$ is an orthornormal basis of R^3 (relative to the standard inner product). From the relations

$$\alpha_1 = \frac{1}{\sqrt{2}}(\epsilon_1 - \epsilon_2),$$

$$\alpha_2 = \frac{1}{3\sqrt{2}}(\epsilon_1 + \epsilon_2 + 4\epsilon_3),$$

$$\alpha_3 = \frac{1}{3}(2\epsilon_1 + 2\epsilon_2 - \epsilon_3)$$

we infer that if $\alpha = x\epsilon_1 + y\epsilon_2 + z\epsilon_3 = \bar{x}\alpha_1 + \bar{y}\alpha_2 + \bar{z}\alpha_3$ then

$$(x, y, z) = (\bar{x}, \bar{y}, \bar{z})(p_{ij})$$

and

$$(\bar{x}, \bar{y}, \bar{z}) = (x, y, z)(p_{ij})^t,$$

where

$$(p_{ij}) = \begin{bmatrix} \frac{1}{\sqrt{2}} & \frac{-1}{\sqrt{2}} & 0 \\ \frac{1}{3\sqrt{2}} & \frac{1}{3\sqrt{2}} & \frac{4}{3\sqrt{2}} \\ \frac{2}{3} & \frac{2}{3} & \frac{-1}{3} \end{bmatrix}$$

and

$$\begin{aligned} q(\alpha) &= (x, y, z)(a_{ij})(x, y, z)^t = (\bar{x}, \bar{y}, \bar{z})(p_{ij})(a_{ij})(p_{ij})^t(\bar{x}, \bar{y}, \bar{z})^t \\ &= 9\bar{x}^2 - 9\bar{y}^2. \end{aligned}$$

Hence the equation of the given quadric surface relative to the axes determined by the new basis is $9\bar{x}^2 - 9\bar{y}^2 = 1$, and so it is a hyperbolic cylinder. We can simplify this equation slightly by switching to the (orthogonal) basis $(\beta_1, \beta_2, \beta_3)$, where

$$\beta_1 = \tfrac{1}{3}\alpha_1, \qquad \beta_2 = \tfrac{1}{3}\alpha_2, \qquad \beta_3 = \alpha_3.$$

Since the matrix of this basis relative to the basis $(\alpha_1, \alpha_2, \alpha_3)$ is

$$\begin{bmatrix} \frac{1}{3} & 0 & 0 \\ 0 & \frac{1}{3} & 0 \\ 0 & 0 & 1 \end{bmatrix},$$

the matrix of q relative to the basis $(\beta_1, \beta_2, \beta_3)$ is

$$\begin{bmatrix} \frac{1}{3} & 0 & 0 \\ 0 & \frac{1}{3} & 0 \\ 0 & 0 & 1 \end{bmatrix} \begin{bmatrix} 9 & 0 & 0 \\ 0 & -9 & 0 \\ 0 & 0 & 0 \end{bmatrix} \begin{bmatrix} \frac{1}{3} & 0 & 0 \\ 0 & \frac{1}{3} & 0 \\ 0 & 0 & 1 \end{bmatrix} = \begin{bmatrix} 1 & 0 & 0 \\ 0 & -1 & 0 \\ 0 & 0 & 0 \end{bmatrix}.$$

Hence, if $\alpha = \bar{\bar{x}}\beta_1 + {}_N\beta_2 + \bar{\bar{z}}\beta_3$, then $q(\alpha) = \bar{\bar{x}}^2 - \bar{\bar{y}}^2$ and an equation of the cylinder is $\bar{\bar{x}}^2 - \bar{\bar{y}}^2 = 1$.

2.6. Consider the quadratic form of Example 2.5 on the space R^3 with the inner product given by

$$(\alpha \mid \beta) = x_1x_2 + 2y_1y_2 + z_1z_2 \quad \text{if} \quad \begin{array}{l} \alpha = (x_1, y_1, z_1), \\ \beta = (x_2, y_2, z_2). \end{array}$$

Now $\mathscr{B}_1 = ((1, 0, 0), (0, 1/\sqrt{2}, 0), (0, 0, 1))$ is an orthonormal basis of R^3 (relative to the new inner product) and the matrix of q (relative to $\mathscr{B}_1$) is

$$\begin{bmatrix} 4 & \dfrac{-5}{\sqrt{2}} & -2 \\ \dfrac{-5}{\sqrt{2}} & 2 & \dfrac{-2}{\sqrt{2}} \\ -2 & \dfrac{-2}{\sqrt{2}} & -8 \end{bmatrix}.$$

The characteristic polynomial of this matrix is found to be

$$h = x(x - d_1)(x - d_2),$$

where

$$d_1 = -\tfrac{1}{2}(2 - \sqrt{238}) > 0, \qquad d_2 = -\tfrac{1}{2}(2 + \sqrt{238}) < 0.$$

There exists an orthonormal basis $\mathscr{B}_2 = (\beta_1, \beta_2, \beta_3)$ of R^3 such that

$$q(\alpha) = d_1\bar{x}^2 + d_2\bar{y}^2 \quad \text{if} \quad \alpha = \bar{x}\beta_1 + \bar{y}\beta_2 + \bar{z}\beta_3 .$$

Further, relative to the orthogonal basis $\mathscr{B}_3 = (\gamma_1, \gamma_2, \gamma_3)$, where

$$\gamma_1 = \frac{1}{\sqrt{d_1}}\beta_1, \qquad \gamma_2 = \frac{1}{\sqrt{-d_2}}\beta_2, \qquad \gamma_3 = \beta_3,$$

we have

$$q(\alpha) = \bar{\bar{x}}^2 - \bar{\bar{y}}^2 \quad \text{if} \quad \alpha = \bar{\bar{x}}\gamma_1 + \bar{\bar{y}}\gamma_2 + \bar{\bar{z}}\gamma_3 .$$

A representation like that obtained for the quadratic form considered in the two preceding examples (as some combination of sums and differences of squares) can be obtained for every quadratic form on an n-dimensional space over R. As a preliminary to the proof, we develop a property of symmetric bilinear forms. Let f be a symmetric bilinear form on an n-dimensional vector space V over R and let $(\ |\ )$ be an inner product in V. Then there exists an orthonormal basis $\overline{\mathscr{B}} = (\alpha_1, \ldots, \alpha_n)$ of V such that

$$f_{\overline{\mathscr{B}}} = \operatorname{diag}(c_1, \ldots, c_n),$$

where, with no loss of generality, we may assume that $c_1, \ldots, c_r$ are positive numbers, that $c_{r+1}, \ldots, c_{r+s}$ are negative numbers, and $c_{r+s+1} = \cdots = c_n = 0$. Defining

$$\beta_i = \frac{1}{\sqrt{c_i}}\alpha_i, \qquad 1 \le i \le r,$$

$$\beta_i = \frac{1}{\sqrt{-c_i}}\alpha_i, \qquad r+1 \le i \le r+s,$$

and

$$\beta_i = \alpha_i, \qquad r+s+1 \le i \le n$$

yields a basis $\mathscr{B} = (\beta_1, \ldots, \beta_n)$ of V which is orthogonal but not necessarily normalized. Clearly

$$f_{\mathscr{B}} = \operatorname{diag}(\underbrace{1, \ldots, 1}_{r}, \underbrace{-1, \ldots, -1}_{s}, \underbrace{0, \ldots, 0}_{t}),$$

where $r + s + t = n$. Thus for any symmetric bilinear form f on V there exists an ordered basis $\mathscr{B}$ relative to which the matrix of f has the above form.

We show next that r, s, and t, the number of 1's, (-1)'s, and 0's, respectively, in the diagonal matrix, are uniquely determined by f. (In particular, therefore, r, s, and t are independent of the inner product introduced for the purpose of constructing such a basis). To prove this assume that relative to the basis $\mathscr{B}' = (\gamma_1, \ldots, \gamma_n)$ of V (which might accompany some other inner product in V) the matrix of f is

$$f_{\mathscr{B}'} = \operatorname{diag}(\underbrace{1, \ldots, 1}_{r'}, \underbrace{-1, \ldots, -1}_{s'}, \underbrace{0, \ldots, 0}_{t'}),$$

where $r' + s' + t' = n$. Then the ranks of $f_{\mathscr{B}}$ and $f_{\mathscr{B}'}$ are $r + s$ and $r' + s'$, respectively. Since both matrices are matrices of f, they are congruent and, hence, $r + s = r' + s'$. Further, since $n = r + s + t = r' + s' + t'$, we infer that $t = t'$. Now let $S = [\beta_{r+1}, \ldots, \beta_{r+s}]$. The dimension of S is s and, if α is a nonzero vector of S, then $f(\alpha, \alpha) < 0$; indeed, if $\alpha = a_1\beta_{r+1} + \cdots + a_s\beta_{r+s}$, then $f(\alpha, \alpha) = -(a_1{}^2 + \cdots + a_s{}^2)$. Similarly, the dimension of $U = [\gamma_1, \ldots, \gamma_{r'}, \gamma_{r'+s'+1}, \ldots, \gamma_n]$ is $n - s'$ and if $\beta \in U$ we have $f(\beta, \beta) \geq 0$. If we suppose that $s > s'$, then

$$\dim S + \dim U = s + (n - s') = n + (s - s') > n$$

and consequently there exists a nonzero vector $\alpha \in S \cap U$, which is impossible. Interchanging the roles of s and s' in the foregoing argument, we can conclude similarly that the assumption that $s' > s$ is untenable. Hence $s = s'$ and, in turn, $r = r'$. In place of the invariants r and s of f it is standard practice to use the rank of f (which is $r + s$) and the *signature* of f (which is defined to be $r - s$) to describe the matrix $f_{\mathscr{B}}$. When we summarize our findings for the quadratic form associated with the given symmetric bilinear form, we obtain the following result.

□ **Theorem 2.2.** Let q be a quadratic form on the n-dimensional euclidean space V. Then there exists an orthogonal (but not necessarily normalized) basis $\mathscr{B} = (\beta_1, \ldots, \beta_n)$ of V such that if $\alpha = \sum_1^n x_i \beta_i$ then

$$q(\alpha) = x_1{}^2 + \cdots + x_r{}^2 - x_{r+1}^2 - \cdots - x_s{}^2.$$

The sum $r + s$ is equal to the rank of q and the number r of positive squares (and, hence, the number s of negative squares) is uniquely determined by q (Sylvester's law of inertia).

In matrical language, a real symmetric matrix (a_{ij}) is congruent to a diagonal matrix of the form

$$\operatorname{diag}(1, \ldots, 1, -1, \ldots, -1, 0, \ldots, 0),$$

where the number of 1's is equal to the number of positive characteristic values and the number of (-1)'s is equal to the number of negative characteristic values of (a_{ij}).

The matrix version of the theorem follows directly from the first part of the theorem on consideration of the bilinear form induced by (a_{ij}) relative to the standard basis of R^n. The description of the number r of 1's as the number of positive characteristic values and of the number s of (-1)'s as the number of negative characteristic values of (a_{ij}) appears in the proof above. Of course, $r + s$ is the rank of (a_{ij}). Defining the *signature* of (a real symmetric matrix) (a_{ij}) to be $r - s$, that is, its number of positive characteristic values minus its number of negative characteristic values, we can establish the following result.

□ **Theorem 2.3.** Two $n \times n$ real symmetric matrices are congruent if and only if they have the same rank and signature. In particular, a real symmetric matrix is congruent to a uniquely determined matrix of the form

$$\operatorname{diag}(1, \ldots, 1, -1, \ldots, -1, 0, \ldots, 0).$$

PROOF. Consider the pair (a_{ij}) and (b_{ij}) of $n \times n$ real symmetric matrices. According to Theorem 2.2, (a_{ij}) is congruent to a diagonal matrix

$$D_a = \operatorname{diag}(1, \ldots, 1, -1, \ldots, -1, 0, \ldots, 0),$$

where the number of 1's is equal to the number of positive characteristic values of (a_{ij}) and the number of (-1)'s is equal to the number of its negative characteristic values. Also, (b_{ij}) is congruent to a diagonal matrix D_b of the same form as D_a and in which the number of 1's and (-1)'s have a similar description relative to (b_{ij}).

If (a_{ij}) and (b_{ij}) have the same rank and signature, then $D_a = D_b$ and, hence, (a_{ij}) and (b_{ij}) are congruent. Conversely, if (a_{ij}) and (b_{ij}) are congruent, then D_a and D_b are congruent. Now let f be the bilinear form on R^n induced by D_a relative to the standard basis. Then D_b is also the matrix of f relative to some basis. In view of the form of the values of f relative to these bases, Sylvester's law of inertia is applicable, and we conclude that $D_a = D_b$. Hence (a_{ij}) and (b_{ij}) have the same rank and signature. ◇

☐ **Definition.** A quadratic form q on an n-dimensional vector space V over R is called

(a) *positive (negative) definite* if and only if $q(\alpha) > 0$ (<0) for each nonzero vector α in V,

(b) *positive (negative) semidefinite* if and only if $q(\alpha) \geq 0$ (≤ 0) for each vector α in V,

(c) *indefinite* if and only if $q(\alpha) > 0$ for some vector α in V and $q(\alpha') < 0$ for some vector α' in V.

It is immediate that a quadratic form q on an n-dimensional vector space V over R is positive (negative) definite if and only if all characteristic values of any matrix of q are positive (negative). Similar statements can be formulated by the reader for positive (negative) semidefinite quadratic forms. If q is positive definite, then the $n \times n$ identity matrix is a matrix of q and conversely. In this event, the symmetric bilinear form associated with q can serve as an inner product in V.

The classification scheme for quadratic forms introduced above may be applied to real symmetric matrices by way of the quadratic forms they define. For instance, an $n \times n$ real symmetric matrix (a_{ij}) is called positive definite if and only if the quadratic form induced by (a_{ij}) on R^n relative to the standard basis is positive definite.

EXAMPLES

2.7. The quadratic form defined in Example 2.4 is found to be (a) positive definite if and only if $D < 0$ and $a + c > 0$, (b) negative definite if and only if $D < 0$ and $a + c < 0$, (c) indefinite if and only if $D > 0$, and (d) semidefinite if and only if $D = 0$.

2.8. If the quadratic form $\xi(f_{ij})\xi^t$, which appears in Example 2.3, is positive (negative) definite, then $f(\alpha)$ is a relative minimum (maximum) of f. If $\xi(f_{ij})\xi^t$ is indefinite, then $f(\alpha)$ is not an extremum of f. If $\xi(f_{ij})\xi^t$ does not fall into any of these categories, its rank is less than n and the remainder r must be studied to determine the nature of $f(\alpha)$.

2.9. Let us derive the following integral formula, which occurs in the analytic theory of probability:

$$\int_{-\infty}^{\infty} \cdots \int_{-\infty}^{\infty} e^{-q(x_1, \ldots, x_n)}\, dx_1 \cdots dx_n = \frac{\pi^{n/2}}{\sqrt{\det(a_{ij})}},$$

where (a_{ij}) is the real symmetric matrix of the positive definite quadratic form q (relative to the standard basis of R^n).

Starting with the given quadratic form, we know that there exists an orthonormal basis $\mathscr{B} = (\alpha_1, \ldots, \alpha_n)$ of R^n such that

$$q(\alpha) = c_1\bar{x}_1^2 + \cdots + c_n\bar{x}_n^2, \qquad c_i > 0, 1 \leq i \leq n$$

if $\alpha = \sum_1^n \bar{x}_i \alpha_i$. If (p_{ij}) is the matrix of $\mathscr{B}$ relative to the standard basis of R^n (thus $\alpha_i = \sum_1^n p_{ij} \epsilon_j$), then (p_{ij}) is orthogonal and

$$(p_{ij})(a_{ij})(p_{ij})^t = \operatorname{diag}(c_1, \ldots, c_n).$$

Since $\det(p_{ij}) = \pm 1$, the foregoing equation implies that

$$\det(a_{ij}) = c_1 \cdots c_n.$$

In particular, since $c_i > 0$, $1 \leq i \leq n$, $\det(a_{ij}) > 0$. Designating the given integral by I, we transform it to the new coordinate system $(\bar{x}_1, \ldots, \bar{x}_n)$, where

$$(x_1, \ldots, x_n) = (\bar{x}_1, \ldots, \bar{x}_n)(p_{ij}).$$

Since $\bar{x}_i = \sum_1^n p_{ij} x_j$, $1 \leq i \leq n$, the partial derivative of $\bar{x}_i$ with respect to x_j is equal to p_{ij}. It follows that the Jacobian of the transformation is simply $\det(p_{ij})$. Thus, since the absolute value of the Jacobian is equal to 1, we have

$$\begin{aligned} I &= \int_{-\infty}^{\infty} \cdots \int_{-\infty}^{\infty} \exp[-(c_1\bar{x}_1{}^2 + \cdots + c_n\bar{x}_n{}^2)]\, d\bar{x}_1 \cdots d\bar{x}_n \\ &= \int_{-\infty}^{\infty} e^{-c_1\bar{x}_1{}^2}\, d\bar{x}_1 \cdots \int_{-\infty}^{\infty} e^{-c_n\bar{x}_n{}^2}\, d\bar{x}_n = \prod_{i=1}^{n} \int_{-\infty}^{\infty} e^{-c_i\bar{x}_i{}^2}\, d\bar{x}_i. \end{aligned}$$

Using the well-known formula (which is derived in most calculus texts),

$$\int_{-\infty}^{\infty} e^{-ct^2}\, dt = \left(\frac{\pi}{c}\right)^{1/2}, \qquad c > 0,$$

we conclude that

$$I = \prod_{i=1}^{n} \left(\frac{\pi}{c_i}\right)^{1/2} = \frac{\pi^{n/2}}{\sqrt{c_1 \cdots c_n}} = \frac{\pi^{n/2}}{\sqrt{\det(a_{ij})}}.$$

We conclude this section with a proof of the result that it is possible to diagonalize a pair of quadratic forms simultaneously, provided one of them is positive definite. This theorem has applications in mechanics.

The proof is an abstraction of the following geometric argument, which is adequate for the case of two quadratic forms of rank 3 on R^3. Let q_1 and q_2 be two such forms and suppose that q_1 is positive definite. Then the graph of $q_1(\alpha) = c$ (a positive constant) is an ellipsoid with its center at the origin and that of $q_2(\alpha) = d$ (a constant) is a central quadric surface whose center is also at the origin. Let us now rotate the axes of the initial coordinate system so that they are aligned with those of the ellipsoid. Next, we change the scale on the new set of axes so as to deform the ellipsoid into a sphere. Finally, we rotate these axes to align them with the principal axes of the second quadric surface. Referred to this last set of axes, the equation of the ellipsoid is $\sum_1^3 \bar{x}_i{}^2 = c$ and that of the other surface has the form $\sum_1^3 c_i \bar{x}_1{}^2 = d$; in particular, the desired reduction of q_1 and q_2 is achieved. We turn now to the general result.

□ **Theorem 2.4.** Let q_1 and q_2 be quadratic forms on an n-dimensional vector space V over R. If q_1 is positive definite, there exists a basis $\mathscr{B} = (\beta_1, \ldots, \beta_n)$ of V such that if $\alpha = \sum_1^n \bar{x}_i \beta_i$ then

$$q_1(\alpha) = \sum_{i=1}^{n} \bar{x}_i{}^2 \quad \text{and} \quad q_2(\alpha) = \sum_{i=1}^{n} c_i \bar{x}_i{}^2.$$

PROOF. Suppose that the positive definite form q_1 is associated with the (symmetric) bilinear form f_1. Then f_1 is an inner product in V. With f_1 as inner product, V is a euclidean space. According to Theorem 2.1, there is an orthonormal basis $\mathscr{B} = (\beta_1, \ldots, \beta_n)$ of the euclidean space V (with inner product f_1) such that if $\alpha = \sum_1^n \bar{x}_i \beta_i$, then $q_2(\alpha) = \sum_1^n c_i \bar{x}_i{}^2$. Simultaneously, $q_1(\alpha) = f_1(\alpha, \alpha) = \sum_1^n \bar{x}_i{}^2$. ◇

EXAMPLE

3.10. Let q_1 and q_2 be the quadratic forms on the vector space R^2 defined as follows: if $\alpha = (x, y) \in \mathsf{R}^2$, then

$$q_1(\alpha) = x^2 - 2xy + 2y^2 \quad \text{and} \quad q_2(\alpha) = 2xy.$$

Since $q_1(\alpha) = (x - y)^2 + y^2$, q_1 is positive definite and we use its associated bilinear form f_1 as the inner product in R^2. Thus

$$f_1(\alpha, \beta) = x_1x_2 - (x_1y_1 + x_2y_2) + 2y_1y_2 \qquad \text{if} \quad \alpha = (x_1, y_1), \quad \beta = (x_2, y_2).$$

Now $\mathscr{B}_1 = (\alpha_1, \alpha_2)$, where $\alpha_1 = (1, 0)$ and $\alpha_2 = (1, 1)$ is an orthonormal basis of the euclidean space R^2 with inner product f_1 and the matrix of q_2 relative to $\mathscr{B}_1$ is

$$\begin{pmatrix} 0 & 1 \\ 1 & 2 \end{pmatrix}.$$

The characteristic values of this matrix of q_2 are $c_1 = 1 + \sqrt{2}$ and $c_2 = 1 - \sqrt{2}$. Now

$$\beta_1 = \frac{\alpha_1 + (1 + \sqrt{2})\alpha_2}{\sqrt{4 + 2\sqrt{2}}}, \qquad \beta_2 = \frac{(1 + \sqrt{2})\alpha_1 - \alpha_2}{\sqrt{4 + 2\sqrt{2}}}$$

are unit characteristic vectors associated with c_1 and c_2, respectively. Hence $\mathscr{B}_2 = (\beta_1, \beta_2)$ is an orthonormal basis of R^2 (relative to the inner product f_1) and, if $\alpha = \bar{x}\beta_1 + \bar{y}\beta_2$, then

$$q_1(\alpha) = \bar{x}^2 + \bar{y}^2, \qquad q_2(\alpha) = (1 + \sqrt{2})\bar{x}^2 + (1 - \sqrt{2})\bar{y}^2.$$

EXERCISES

2.1. If q is the quadratic form on R^3 such that

$$q(\alpha) = 2x^2 - y^2 + xz + yz \quad \text{if} \quad \alpha = (x, y, z),$$

find the matrix of q relative to the standard basis of R^3 and relative to the basis $((1, 0, 1), (2, 1, 0), (0, -1, 1))$. Classify q in accordance with the definitions given after Theorem 2.3.

2.2. Reduce each of the following quadratic forms on R^3 to a sum of squares by referring them to suitable orthonormal bases:

(a) $q(\alpha) = 6x^2 + 5y^2 + 7z^2 - 4xy + 4xz$;

(b) $q(\alpha) = x^2 - 5y^2 + z^2 + 4xy + 2xz + 4yz, \qquad \alpha = (x, y, z)$

2.3. Show that the pair of quadratic forms in R^2 given by

$$q_1(\alpha) = x^2 + 2xy + 2y^2, \qquad q_2(\alpha) = 2x^2 - xy, \qquad \alpha = (x, y),$$

satisfy the hypothesis of Theorem 2.4 and find a basis relative to which the values of both are given by a sum of squares.

2.4. Let q be a nonzero quadratic form on a vector space V. Show that $q(\alpha)$ can be written in the form

$$q(\alpha) = a(\alpha f)^2, \qquad |a| = 1,$$

where f is in the dual space of V, if and only if the rank of q is 1.

2.5. Find the values of k for which the following quadratic forms on $\mathbf{R}^3$ are positive definite.
(a) $q(\alpha) = 5x^2 + y^2 + kz^2 + 4xy - 2xz - 2yz$;
(b) $q(\alpha) = 2x^2 + 2y^2 + z^2 + 2kxy + 6xz + 2yz, \qquad \alpha = (x, y, z)$.

2.6. Let q_1 and q_2 be quadratic forms on $\mathbf{R}^3$ such that

$$[q_1(\alpha)]^2 + [q_2(\alpha)]^2 > 0,$$

for all nonzero vectors α in $\mathbf{R}^3$. Show that there exists an ordered basis $\mathscr{B}$ of $\mathbf{R}^3$ such that the matrices of q_1 and q_2 relative to $\mathscr{B}$ are both diagonal. *Note.* This result holds true for $\mathbf{R}^n$ with $n \geqslant 3$ but is false if $n < 3$.

2.7. Let (a_{ij}) be a 3×3 real symmetric matrix. Let

$$D_1 = (a_{11}), \qquad D_2 = \begin{bmatrix} a_{11} & a_{12} \\ a_{21} & a_{22} \end{bmatrix}, \qquad D_3 = (a_{ij}).$$

Show that (a_{ij}) is positive definite if and only if $\det D_i > 0$, $i = 1, 2, 3$.

2.8. Generalize the foregoing result to the case of $n \times n$ matrices by showing that an $n \times n$ real symmetric matrix (a_{ij}) is positive definite if and only if $\det D_i > 0$, $1 \leq i \leq n$, where

$$D_i = \begin{bmatrix} a_{11} & \cdots & a_{1i} \\ \vdots & & \\ a_{i1} & \cdots & a_{ii} \end{bmatrix}, \qquad 1 \leq i \leq n.$$

2.9. Let V be a vector space over a field F such that $1 + 1 \neq 0$. Let q_i be the quadratic form on V that is associated with the symmetric bilinear form f_i on V, $i = 1, 2$. Show that $q_1 + q_2$ is the quadratic form associated with the symmetric bilinear form $f_1 + f_2$, where $(f_1 + f_2)(\alpha, \beta)$ is defined to be $f_1(\alpha, \beta) + f_2(\alpha, \beta)$.

2.10. If (a_{ij}) and (b_{ij}) are $n \times n$ real symmetric matrices, let us define (a_{ij}) to be less than (b_{ij}) if and only if $(b_{ij}) - (a_{ij})$ is positive definite. Show that if (a_{ij}) is less than (b_{ij}) and (b_{ij}) is less than (c_{ij}) then (a_{ij}) is less than (c_{ij}).

2.11. Let (a_{ij}) be an $n \times n$ real symmetric matrix. Show that there exists a real number c such that $(a_{ij}) + cI$ (where I is the $n \times n$ identity matrix) is positive definite.

2.12. Let q be a quadratic form on an n-dimensional vector space V over $\mathbf{R}$. Show that there exist subspaces S_0, S_1, and S_2 of V such that (a) $V = S_0 \oplus S_1 \oplus S_2$ and (b) $q(\alpha) = 0$ if and only if $\alpha \in S_0$, q is positive definite on S_1 and negative definite on S_2.

3. EXTREMAL PROPERTIES OF CHARACTERISTIC VALUES OF A SYMMETRIC MATRIX

In this section we shall obtain a characterization of the characteristic values of a real symmetric matrix as extrema of suitably defined functions. In addition to its theoretical interest, the result provides the basis for an algorithm to determine a set of principal axes for a real quadratic form if there is available some method to approximate the extreme values of a quadratic form subject to a constraint.

To motivate the presentation we begin with an example. Let A be a 3×3 positive definite matrix. Then the graph E of the quadric surface with equation

$$(x, y, z)A(x, y, z)^t = 1$$

relative to a cartesian coordinate system for E_3 is an ellipsoid. When referred to a set of principal axes, the equation of E has the form

$$UDU^t = \frac{x^2}{a^2} + \frac{y^2}{b^2} + \frac{z^2}{c^2} = 1,$$

where $c_1 = 1/a^2$, $c_2 = 1/b^2$, and $c_3 = 1/c^2$ are the characteristic values of A. We may and shall assume that $c_1 \geq c_2 \geq c_3 > 0$ and, consequently, that $a^2 \leq b^2 \leq c^2$. Then it is clear that a is the least and c the greatest distance from the origin O to E. Our observation concerning a can be described as follows: a^2 is the minimum value of UU^t subject to the constraint $UDU^t = 1$; then 1 is the maximum value of UDU^t subject to the constraint $UU^t = a^2$. Normalizing the side condition, we may conclude that $1/a^2 = c_1$, the greatest characteristic value of A, is the maximum value of $UDU^t = 1$ for those U's such that $UU^t = 1$. In turn, this maximum value may be determined as that of the function ρ defined by

$$\rho(U) = \frac{UDU^t}{UU^t}, \qquad U \neq 0.$$

In a similar manner it can be shown that c_3 is the least value of ρ.

Armed with these facts we turn our attention to an $n \times n$ real symmetric matrix. For such a matrix $A = (a_{ij})$, define the function ρ on the set N of nonzero vectors $X = (X_1, \ldots, X_n)$ of R^n by

$$\rho(X) = \frac{XAX^t}{XX^t}.$$

Assuming that R^n is equipped with the standard inner product, we have

$$\rho(X) = \frac{(XA \mid X)}{\|X\|^2} = \frac{1}{\|X\|^2}\left(\sum_{i=1}^{n} a_{ii}x_i^2 + 2 \sum_{i<j} a_{ij}x_i x_j\right).$$

Clearly ρ is continuous when regarded either as a function of the single variable X or as a function of the n variable $x_1, \ldots, x_n$. Also, if c is a characteristic value of A and X_0 is a characteristic vector of A belonging to c (that is, $X_0 A = cX_0$), then $\rho(X_0) = c$.

We intend to investigate the extrema of ρ. That ρ has a maximum and a minimum value can be shown as follows. Let S denote the unit sphere in R^n; thus, $S = \{X \in \mathsf{R}^n \mid \|X\| = 1\}$. If $X \in N$, then $X = \|X\| U$ where $U \in S$ and

$$\rho(X) = \frac{(\|X\| UA \mid \|X\| U)}{(\|X\| U \mid \|X\| U)} = \frac{(UA \mid A)}{(U \mid U)} = \rho(U).$$

This means that every value of ρ is assumed at an argument in S. Since the restriction of ρ to S is continuous and S is a closed bounded set, this restriction has a maximum and a minimum value. By the observation made above, these extrema are the maximum and the minimum values of ρ. We shall prove that if $c_1 = \rho(X_1)$ is the maximum value of ρ and $c_2 = \rho(X_2)$ is the minimum value of ρ, then c_1 and c_2 are characteristic values and the accompanying arguments are corresponding characteristic vectors.

We begin by introducing the auxiliary function u on $N \times \mathsf{R}$ defined as follows:

$$u(X, \varphi) = XAX^t - XX^t\varphi = XAX^t - \|X\|^2\varphi.$$

Two immediate properties of u are

(1) if $u(X, \rho) = 0$, then $\varphi = \rho(X)$, and

(2) $u(cX, \varphi) = c^2 u(X, \varphi)$ for all $c \in \mathsf{R}$.

The essential properties of u for our purposes are those stated next.

□ **Lemma 3.1.** If $u(X_0, \varphi_0) \geq u(X, \varphi_0)$ for all X in N, then $\varphi_0 = \rho(X_0)$ and $\rho(X_0) \geq \rho(X)$ for all X in N. Conversely, if $\rho_0 = \rho(X_0) \geq \rho(X)$ for all X in N, then $u(X_0, \rho_0) \geq u(X, \rho_0)$ for all X in N. Corresponding statements hold when all inequalities are reversed.

PROOF. Suppose $u(X_0, \varphi_0) \geq u(X, \varphi_0)$ for all X in N. In particular, therefore, $u(X_0, \varphi_0) \geq u(cX_0, \varphi_0)$ for all c in R. Using (2) it follows that

$(1 - c^2)u(X_0, \varphi_0) \geq 0$ for all c. This implies that $u(X_0, \varphi_0) = 0$ and, in turn, using (1) that $\varphi_0 = \rho(X_0)$. Thus for each $X \in N$,

$$\begin{aligned} u(X, \varphi_0) &= XAX^t - \|X\|^2\varphi. \\ &= \|X\|^2\rho(X) - \|X\|^2\rho(X_0) \\ &= \|X\|^2(\rho(X) - \rho(X_0)). \end{aligned}$$

Since $u(X, \varphi_0) \leq 0$ and $\|X\|^2 > 0$, we conclude that $\rho(X_0) \geq \rho(X)$ for all X.

For the converse, assume that $\rho_0 = \rho(X_0) \geq \rho(X)$ for all X in N. Then

$$\begin{aligned} u(X_0, \rho_0) - u(X, \rho_0) &= -XAX^t + \|X\|^2\rho(X_0) \\ &= \|X\|^2(\rho(X_0) - \rho(X)) \geq 0 \end{aligned}$$

for all X in N. Hence, $u(X_0, \rho_0) \geq u(X, \rho_0)$ for all X in N. ◇

The importance of this lemma rests on the fact that it converts the problem of maximizing (minimizing) the function ρ under consideration to that of maximizing (minimizing) the corresponding function u, treating φ as a constant. The steps involved in the conversion are of the type taken when establishing the rule known as *Lagrange's method of undetermined multiples* for finding the stationary points of a function whose variables are constrained by a side condition. In effect, we have validated this rule for the case at hand. By restricting ourselves to this particular problem we have obtained more precise information (in terms of extreme rather than stationary values) than the Lagrange method gives in general.

To determine extrema of u, regarded as a function of $x_1, \ldots, x_n$, we locate its critical points, that is, points $(x_1, \ldots, x_n)$ such that

$$\frac{\partial u}{\partial x_r} = 0, \qquad 1 \leq r \leq n.$$

Since

$$u(X, \varphi) = \sum_{i=1}^{n} a_{ii} x_i^2 + 2 \sum_{i<j} a_{ij} x_i x_j - \varphi \sum_{i=1}^{n} x_i^2,$$

the above system of equations has the form

$$a_{rr} x_r + \sum_{i \neq r} a_{ir} x_r - \varphi x_r = 0, \qquad 1 \leq r \leq n$$

or

$$\sum_{i=1}^{n} a_{ir} x_r = \varphi x_r, \qquad 1 \le r \le n$$

or

$$XA = \varphi X.\dagger$$

Thus, a critical point of u is a characteristic vector of A. It follows that if $c_1 = \rho(X_1)$ is the maximum (minimum) value of ρ, then, in turn, $u(X_1, c_1)$ is the maximum (minimum) value of u, c_1 is a characteristic value of A, and X_1 is a characteristic vector belonging to c_1.

EXAMPLES

3.1. For the 2×2 symmetric matrix

$$A = \begin{bmatrix} 2 & 1 \\ 1 & 2 \end{bmatrix},$$

the function ρ is given by

$$\rho(x, y) = \frac{2x^2 + 2y^2 + 2xy}{x^2 + y^2}.$$

The equations $\partial f/\partial x = 0$ and $\partial f/\partial y = 0$, which determine the critical points of ρ are found to be

$$y(y^2 - x^2) = 0$$
$$x(x^2 - y^2) = 0.$$

For $(x, y) \neq (0, 0)$ it follows that $y^2 - x^2 = 0$; that is, $y = x$ or $y = -x$. Since $\rho(x, x) = 3$ and $\rho(x, -x) = 1$, we conclude that 3 is a characteristic value with $(1, 1)$ as an accompanying characteristic vector and 1 is a characteristic value with $(1, -1)$ as an accompanying characteristic vector. Observe that an advantage of this method of determining characteristic values is the fact that a corresponding characteristic vector is determined simultaneously.

3.2. Let A be the matrix

$$\begin{bmatrix} 4 & -5 & -2 \\ -5 & 4 & -2 \\ -2 & -2 & -8 \end{bmatrix}$$

† We call attention to the fact that the above calculation is the only point in our discussion where the symmetry of A is used.

of the quadratic form discussed in Example 2.5. The corresponding function ρ is given by

$$\rho(x, y, z) = \frac{N}{D}$$

where

$$N = 4x^2 + 4y^2 - 8z^2 - 10xy - 4xz - 4yz,$$
$$D = x^2 + y^2 + z^2.$$

Setting the partial derivatives of ρ with respect to x, y, and z equal to 0, we obtain the three equations

$$D(8x - 10y - 4z) - N(2x) = 0 \tag{3}$$

$$D(8y - 10x - 4z) - N(2y) = 0 \tag{4}$$

$$D(-16z - 4x - 4y) - N(2z) = 0 \tag{5}$$

Equations (3) and (4) imply that

$$(4x - 5y - 2z)y = (4y - 5x - 2z)x,$$

which is equivalent to

$$y = x \qquad \text{or} \qquad y = \frac{2z - 5x}{5}.$$

Equations (4) and (5) imply that

$$(4y - 5x - 2z)z = (-8z - 2x - 2y)y. \tag{6}$$

With $y = x$, equation (6) reduces to $(4x - z)(x + 2z) = 0$. If $y = x$ and $z = 4x$, then ρ has the value -9. Thus, -9 is a possible characteristic value having $(1, 1, 4)$ as an accompanying characteristic vector. Since $(1, 1, 4)A = -9(1, 1, 4)$, we have indeed found a characteristic value and a corresponding characteristic vector. With $y = x$ and $z = -x/2$, the value of ρ is 0. This is another characteristic value. An accompanying characteristic vector is $(2, 2, -1)$.

We continue by choosing $y = (2z - 5x)/5$ in (6). It follows that $z(78z - 445x) = 0$. Choosing $z = 0$, we obtain the remaining characteristic value 9. The vector $(1, -1, 0)$ is a characteristic vector belonging to 9.

There are available† a variety of computational methods for approximating the maximum and the minimum value of the quotient $\rho(X) = XAX^t/XX^t$ where A is an $n \times n$ real symmetric matrix. Once the maximum value c_1 of ρ has been determined, we have the largest characteristic value of A and a characteristic vector X_1 belonging to c_1. The next characteristic value c_2 (in order of decreasing magnitude) may then be described as the maximum of ρ when X is restricted to the orthogonal complement in R^n of the one-dimensional space $[X_1]$. In general, the kth characteristic value c_k is obtained after $c_1, \ldots, c_{k-1}$ (together with associated characteristic vectors $X_1, \ldots, X_{k-1}$) have been found, as the maximum value of ρ under the side condition of orthogonality to $[X_1, \ldots, X_{k-1}]$.

For quadratic forms the foregoing algorithm amounts to the following. Let q be a quadratic form on R^n whose matrix relative to the standard basis is A. If $\alpha = \sum_1^n x_i \epsilon_i$, then $q(\alpha) = XAX^t$, where $X = (x_1, \ldots, x_n)$. If α_1 maximizes q on the unit sphere, then $q(\alpha_1) = c_1$ is the largest characteristic value of A and α_1 is a unit characteristic vector belonging to c_1. Extend α_1 to an ordered basis $(\alpha_1, \ldots, \alpha_n)$ of R^n. Relative to this basis, $q(\alpha)$ has the form

$$q(\alpha) = YBY^t = c_1 {y_1}^2 + q_1(\alpha)$$

where $\alpha = \sum_1^n y_k \alpha_i$, $Y = (y_1, \ldots, y_n)$ and q_1 is a quadratic form on the $(n-1)$-dimensional orthogonal complement of $[\alpha_1]$. Upon repeating the foregoing procedure with q_1 we will obtain an orthonormal basis $(\beta_2, \ldots, \beta_n)$ of the orthogonal complement such that β_2 maximizes q_1 on a unit sphere. This splits off another diagonal term. Continuing in this manner, we finally obtain a bases of principal axes $(\alpha_1, \beta_2, \gamma_3, \ldots)$ for q along with a representation of $q(\alpha)$ in the form mentioned in Theorem 2.1.

† See, for example, Faddeev, D. K. and Faddeeva, V. N., *Computational Methods of Linear Algebra*. Freeman, San Francisco, 1963.

8 Decomposition Theorems for Normal Transformations

This chapter is an introduction to techniques for analyzing individual linear transformations on a finite-dimensional vector space. Since our most decisive results are obtained for normal transformations (defined in Section 3), we have elected to mention them in the title of the chapter. The methods that we discuss in this connection serve to decompose a linear transformation A on a vector space V into linear transformations on certain subspaces of V in such a way that the behavior of A can be deduced from that of its "component" linear transformations. An illustration is provided by the result appearing in Theorem 5.4.1. Turning to that theorem, it will be noted that accompanying the decomposition of V into the direct sum of subspaces $S_1, \ldots, S_n$, each invariant under A, there is a "decomposition" of A into components $A_1, \ldots, A_n$, where A_i is the restriction of A to S_i, $1 \leq i \leq n$. Clearly the nature of A can be inferred from that of $A_1, \ldots, A_n$. The circumstances present in this example involve an extreme, namely, that of a linear transformation which has a diagonal matrix. Such an extreme is highly desirable, however. From this point of view the current chapter may be described by saying that it presents methods designed to obtain a matrix of a given linear transformation which is as "close as possible" to a diagonal matrix.

1. DIRECT SUMS AND PROJECTIONS

We recall that if $S_1, \ldots, S_k$ are subspaces of a vector space V, then the sum $S = S_1 + \cdots + S_k$ is said to be *direct*, or S is said to be the *direct sum* of $S_1, \ldots, S_k$ if and only if α in S can be expressed uniquely in the form $\alpha = \alpha_1 + \cdots + \alpha_k$ with $\alpha_i \in S_i$, $1 \leq i \leq k$. In this event we write $S = S_1 \oplus \cdots \oplus S_k$.

Next we recall that a linear transformation E on a vector space V is called idempotent if and only if $E^2 = E$. In Exercise 4.3.12 it is stated that if E is an idempotent linear transformation on V, then V is the direct sum of R, the range of E, and N, the null space of E. [This result follows easily from the representation of each vector α of V in the form $\alpha = \alpha E + \alpha(1 - E)$.] Since the range and the null space of a given linear transformation are invariant under the transformation, the restriction E_R of E to R is a linear transformation on R and the restriction E_N of E to N is a linear transformation on N. Now E_R is the identity transformation on R [since, if $\alpha \in R$, then $\alpha = \beta E$ for some β and, hence, $\alpha E = (\beta E)E = \beta E^2 = \beta E = \alpha$] and E_N is the zero transformation on N. Because of these properties, an idempotent transformation E is often called a *projection.* If R is the range of E and N is the null space of E, then we call E the *projection on R along N.*

Projections can be used to describe direct-sum decompositions of a vector space V. Suppose that

$$V = S_1 \oplus \cdots \oplus S_k.$$

Since the representation of a vector α in V in the form $\alpha = \alpha_1 + \cdots + \alpha_k$, $\alpha_i \in S_i$, $1 \leq i \leq k$, is unique, a function E_i on V into S_i is defined on setting $\alpha E_i = \alpha_i$. This function is a linear transformation and has S_i as its range and $S_1 + \cdots + S_{i-1} + S_{i+1} + \cdots S_k$ as its null space. Furthermore, it is immediate that

(a) $E_i^2 = E_i, \qquad 1 \leq i \leq k,$
(b) $E_i E_j = 0, \qquad \text{if } i \neq j,$
(c) $1 = E_1 + \cdots + E_k.$

Two projections, A and B, in V are said to be *orthogonal* if and only if $AB = BA = 0$ and a set $\{A_1, \ldots, A_k\}$ of linear transformation on V is called a *complete set of projections for* V if and only if (a) each A_i is a projection, (b) distinct pairs are orthogonal, and (c) $A_1 + \cdots + A_k = 1$.

Thus our observations may be summarized by the statement that a direct-sum decomposition of a vector space V determines a complete set of projections for V. Conversely, a complete set $\{E_1, \ldots, E_k\}$ of projections for V determines a direct-sum decomposition of V. Indeed, setting

$$S_i = \{\alpha E_i | \alpha \in V\}, \qquad 1 \leq i \leq k$$

(so that S_i is the range of E_i), we contend that $V = S_1 \oplus \cdots \oplus S_k$. That V is the sum of $S_1, \ldots, S_k$ follows from the fact that each α in V has the representation

$$\alpha = \alpha E_1 + \cdots + \alpha E_k.$$

The uniqueness of this representation is demonstrated as follows: if

$$\alpha = \alpha_1 + \cdots + \alpha_k, \qquad \alpha_j \in S_j,\ 1 \le j \le k,$$

then $\alpha_j = \beta_j E_j$ for some β_j in V and, consequently,

$$\alpha E_i = \left(\sum_{j=1}^{k} \alpha_j\right) E_i = \sum_{j=1}^{k} \beta_j E_j E_i = \beta_i E_i^2 = \beta_i E_i = \alpha_i.$$

Hence V is the direct sum of $S_1, \ldots, S_k$. It follows that there exists a $1-1$ correspondence between all collections of subspaces $\{S_1, \ldots, S_k\}$ of V such that $V = S_1 \oplus \cdots \oplus S_k$ and the collection of all complete sets of projections for V.

Another version of the result in Theorem 5.4.1 (see also Theorem 6.3.2) can be given in terms of projections. This is our next theorem.

□ **Theorem 1.1.** A linear transformation A on a finite dimensional vector space V is diagonable if and only if

$$A = c_1 E_1 + \cdots + c_k E_k,$$

where $c_1, \ldots, c_k$ are scalars and $\{E_1, \ldots, E_k\}$ is a complete set of projections for V.

PROOF. Assume that $\{E_1, \ldots, E_k\}$ is a complete set of projections for V and let S_i be the range of E_i, $1 \le i \le k$. If B_i is a basis of S_i, then $\mathscr{B} = \cup \mathscr{B}_i$ is a basis of V (see Exercise 1.3.17). We show next that the matrix of $A = c_1 E_1 + \cdots + c_k E_k$ is diagonal relative to $\mathscr{B}$. Let $\beta \in \mathscr{B}$; then, for some i, $\beta \in \mathscr{B}_i$. Hence

$$\beta A = \beta(c_1 E_1 + \cdots + c_k E_k) = c_i \beta E_i = c_i \beta,$$

and the proof is complete.

For the converse, assume that there exists a basis $\mathscr{B} = (\beta_1, \ldots, \beta_n)$ of V such that

$$\beta_i A = c_i \beta_i, \qquad 1 \le i \le n.$$

Without loss of generality, we may assume that $c_1, \ldots, c_k$ exhaust the distinct scalars among $c_1, \ldots, c_n$. Now define S_{c_i} to be the space spanned by

$$\{\beta_j \in \mathscr{B} | \beta_j A = c_i \beta_j\}, \qquad 1 \leq i \leq k.$$

Then it is clear that

$$V = S_{c_1} \oplus \cdots \oplus S_{c_k}$$

and $\alpha_i A = c_i \alpha_i$ if $\alpha_i \in S_{c_i}$. Let $\{E_1, \ldots, E_k\}$ be the complete set of projections for V corresponding to this direct-sum decomposition of V. Since $1 = E_1 + \cdots + E_k$, it follows that for α in V

$$\alpha = \alpha_1 + \cdots + \alpha_k,$$

where $\alpha_i = \alpha E_i \in S_{c_i}$, $1 \leq i \leq k$. Hence

$$\begin{aligned} \alpha A = \alpha_1 A + \cdots + \alpha_k A &= c_1 \alpha_1 + \cdots + c_k \alpha_k \\ &= c_1(\alpha E_1) + \cdots + c_k(\alpha E_k) \\ &= \alpha(c_1 E_1 + \cdots + c_k E_k) \end{aligned}$$

and so $A = c_1 E_1 + \cdots + c_k E_k$. ◇

Our next theorem is a consequence of the foregoing result.

☐ **Theorem 1.2.** A linear transformation A on a finite-dimensional inner-product space V is orthogonally diagonable if and only if $A = c_1 E_1 + \cdots + c_k E_k$, where $\{E_1, \ldots, E_k\}$ is a complete set of projections for V such that each member is self-adjoint.

Proof. Let $\{E_1, \ldots, E_k\}$ be a complete set of projections for V and $V = S_1 \oplus \cdots \oplus S_k$ be the associated representation of V as a direct sum of subspaces. Since an orthonormal basis can be constructed for each S_i, the theorem will follow from Theorem 1.1 if we prove that $S_1, \ldots, S_k$ are mutually orthogonal if and only if each of $E_1, \ldots, E_k$ is self-adjoint.

Assume that $S_1, \ldots, S_k$ are mutually orthogonal subspaces. Let $\mathscr{B}_i$ be an orthonormal basis of S_i, $1 \leq i \leq k$; then $\mathscr{B} = \cup \mathscr{B}_i$ is an orthonormal basis of V. The matrix of E_i relative to $\mathscr{B}$ is a diagonal matrix with 1's and 0's as diagonal elements. Hence E_i is self-adjoint for $1 \leq i \leq k$.

Conversely, if $E_i = E_i^*$, $1 \leq i \leq k$, then, for $\alpha_i \in S_i$ and $\alpha_j \in S_j$ with $i \neq j$,

$$\begin{aligned}(\alpha_i \mid \alpha_j) = (\alpha_i E_i \mid \alpha_j E_j) &= (\alpha_i \mid \alpha_j E_j E_i^*) \\ &= (\alpha_i \mid \alpha_j E_j E_i) \\ &= (\alpha_i \mid 0) \\ &= 0.\end{aligned}$$

Hence $S_1, \ldots, S_k$ are mutually orthogonal. ◇

EXERCISES

1.1. Show that the following three statements about a vector space V and subspaces $S_1, \ldots, S_k$ are equivalent.
(a) $V = S_1 \oplus \cdots \oplus S_k$.
(b) $V = S_1 + \cdots + S_k$ and a representation of the zero vector in the form $0 = \alpha_1 + \cdots + \alpha_k$, where $\alpha_i \in S_i$, $1 \leq i \leq k$, implies that each $\alpha_i = 0$.
(c) $V = S_1 + \cdots + S_k$ and the intersection of S_i and the sum of $S_1, \ldots, S_{i-1}, S_{i+1}, \ldots, S_k$ is $\{0\}$, $1 \leq i \leq k$.

1.2. Let V be a finite-dimensional inner-product space and let E be a projection on V. Show that E is self-adjoint if and only if $EE^* = E^*E$.

1.3. Determine two 2×2 matrices (a_{ij}) and (b_{ij}) over R such that

$$(a_{ij})^2 = (a_{ij}), \qquad (b_{ij})^2 = (b_{ij}),$$
$$(a_{ij})(b_{ij}) = O \qquad (b_{ij})(a_{ij}) \neq O.$$

1.4. Find a projection on R^2 (with the standard inner product) that is not self-adjoint.

1.5. Show that if A is a diagonable linear transformation on a finite-dimensional vector space V over F and p is a polynomial over F then $p(A)$ is a diagonable linear transformation on V. Further, show that if $A = \sum_1^k c_i E_i$, then $p(A) = p(c_1)E_1 + \cdots + p(c_k)E_k$.

1.6. Suppose that a vector space V is the direct sum of the subspaces $S_1, \ldots, S_k$ and that $\{E_1, \ldots, E_k\}$ is the associated complete set of projections for V. Let A be a linear transformation on V. Show that each S_i is A-invariant if and only if A commutes with each E_i, $1 \leq i \leq k$.

1.7. Let A be a linear transformation on the vector space V. Show that there exists a linear transformation B on V such that AB and BA are projections and $ABA = A$.

1.8. Let A be a linear transformation on the finite dimensional vector space V. Show that there exist nonsingular linear transformations T_1 and T_2 on V

such that T_1A and AT_2 are projections. Derive from this result the fact that for any $n \times n$ matrix (a_{ij}) over a field F there exist nonsingular $n \times n$ matrices (p_{ij}) and (q_{ij}) over F such that $(p_{ij})(a_{ij})(q_{ij})$ is a diagonal matrix.

1.9. A projection E on an inner-product space V is called a *perpendicular projection* if and only if the range of E and the null space of E are orthogonal subspaces. Show that
(a) A projection E is perpendicular if and only if $E = E^*$;
(b) A perpendicular projection E has the property that $(\alpha E|\alpha) > 0$ for $\alpha \neq 0$ and $\|\alpha E\| \leq \|\alpha\|$ for all $\alpha \in V$;
(c) If E is an idempotent linear transformation in an inner-product space V and $\|\alpha E\| \leq \|\alpha\|$ for all $\alpha \in V$, then E is a perpendicular projection.

2. A DECOMPOSITION THEOREM

The result that follows is of the sort promised in the introduction to this chapter.

☐ **Theorem 2.1.** Let A be a linear transformation on a vector space V over the field F. Suppose that A has a minimal polynomial m and that $p_1^{r1} \cdots p_k^{r_k}$ is a factorization over F of m as a product of powers of distinct, irreducible, monic polynomials $p_1, \ldots, p_k$. Let S_i be the null space of $[p_i(A)]^{r_i}$, $1 \leq i \leq k$. Then

(a) S_i is an A-invariant subspace, $1 \leq i \leq k$;
(b) $V = S_1 \oplus \cdots \oplus S_k$;
(c) the minimal polynomial of the restriction A_i of A to S_i is $p_i^{r_i}$, $1 \leq i \leq k$;
(d) if $\{E_1, \ldots, E_k\}$ is the complete set of projections for V associated with the direct-sum decomposition of V indicated in (b), then A commutes with each E_i.

PROOF. To establish (a), consider a vector α_i in S_i; we will show that $\alpha_i A \in S_i$. Since A and $[p_i(A)]^{r_i}$ commute,

$$(\alpha_i A)[p_i(A)]^{r_i} = (\alpha_i[p_i(A)]^r)A = 0,$$

which yields the desired conclusion.

To prove (b) we introduce the polynomials

$$q_i = \frac{m}{p_i^{r_i}}, \qquad 1 \leq i \leq k,$$

whose only common divisors are units. According to Corollary 1 of Theorem 6.1.5 there exist polynomials $s_1, \ldots, s_k$ such that $1 = s_1 q_1 + \cdots + s_k q_k$. It follows that the identity transformation in V may be written as

$$1 = s_1(A)\, q_1(A) + \cdots + s_k(A)\, q_k(A),$$

and so for α in V

$$\alpha = \alpha s_1(A)\, q_1(A) + \cdots + \alpha s_k(A)\, q_k(A).$$

Now $\alpha s_i(A)\, q_i(A) \in S_i$, $1 \leq i \leq k$, because

$$\alpha s_i(A)\, q_i(A)[p_i(A)]^{r_i} = \alpha s_i(A)\, m(A) = 0.$$

Hence $V = S_1 + \cdots + S_k$. To prove that this sum is direct (see Exercise 1.1) we show that if

$$0 = \alpha_1 + \cdots + \alpha_k, \qquad \alpha_i \in S_i, \qquad 1 \leq i \leq k,$$

then each $\alpha_i = 0$. Since $p_i^{r_i}$ and q_i are relatively prime, there exist polynomials u_i and v_i such that

$$1 = u_i p_i^{r_i} + v_i q_i, \qquad 1 \leq i \leq k.$$

Hence

$$\begin{aligned} \alpha_i &= \alpha_i v_i(A)\, q_i(A) = \alpha_i q_i(A)\, v_i(A) \\ &= -(\alpha_1 + \cdots + \alpha_{i-1} + \alpha_{i+1} + \cdots + \alpha_k)\, q_i(A)\, v_i(A) \\ &= 0, \qquad 1 \leq i \leq k, \end{aligned}$$

as desired.

To prove (c) let m_i be the minimal polynomial of A_i, $1 \leq i \leq k$. (The existence of m_i follows from the existence of a minimal polynomial for A.) Then, since $[p_i(A)]^{r_i}$ is the zero transformation on S_i, $p_i^{r_i}$ is a multiple of m_i. Since p_i is irreducible, this means that $m_i = p_i^{r'_i}$ for $r'_i \leq r_i$. It follows that

$$[p_1(A)]^{r_1} \cdots [p_i(A)]^{r'_i} \cdots [p_k(A)]^{r_k} = 0,$$

and, hence, that $p_i^{r_i} \cdots p_i^{r'_i} \cdots p_k^{r_k}$ is a multiple of $p_1^{r_1} \cdots p_i^{r_i} \cdots p_k^{r_k}$, the minimal polynomial of A. Thus $r'_i \geq r_i$ and so $r'_i = r_i$.

To complete the proof let $E_1 \ldots, E_k$ be defined as in (d). Since S_i is A-invariant, $1 \leq i \leq k$,

$$\alpha(AE_i) = (\alpha_1 + \cdots + \alpha_k)AE_i = (\alpha_i A)E_i = \alpha_i A = (\alpha_i E_i)A = \alpha(E_i A)$$

for each α in V. Hence A and E_i commute, $1 \leq i \leq k$. ◇

☐ **Corollary.** Referring to part (d) of the theorem, each E_i is a polynomial in A and, hence, if a linear transformation B on V commutes with A, then B commutes with each E_i; that is, each subspace S_i is invariant under B.

PROOF. This is left as an exercise. ◇

Several comments about this theorem are in order. Suppose that V is finite-dimensional, that $\dim S_i = d_i$, and that $\mathscr{B}_i$ is a basis of S_i, $1 \leq i \leq k$. Then $\mathscr{B} = \bigcup \mathscr{B}_i$ is a basis of V and the matrix of A relative to $\mathscr{B}$ has the "block" form

$$\begin{bmatrix} M_1 & O & \cdots & O \\ O & M_2 & \cdots & O \\ \vdots & & & \\ O & O & \cdots & M_k \end{bmatrix},$$

where M_i is the $d_i \times d_i$ matrix of A_i relative to $\mathscr{B}_i$ and the O's designate rectangular matrices whose entries are the zero scalar; that is, the matrix of A is the *direct sum* of the matrices $M_1, \ldots, M_k$. Thus our decomposition theorem yields a "generalized" diagonal matrix of any linear transformation A. In particular, if the minimal polynomial of A is the product of k distinct linear factors, then each M_i, and consequently the matrix of A, is diagonal.

The theorem has an interesting implication in another special case. Suppose that each p_i in the factorization of the minimal polynomial is of the form $p_i = x - c_i$. Then the range of E_i is the null space S_i of $(A - c_i)^{r_i}$. Now define the linear transformation D on V by

$$D = c_1 E_1 + \cdots + c_k E_k.$$

By Theorem 1.1 D is diagonable if V is finite-dimensional. In any case we call it the *diagonable part* of A. Next consider the linear transformation $N = A - D$ in V. Since we may write A in the form

$$A = AE_1 + \cdots + AE_k,$$

we have

$$N = (A - c_1)E_1 + \cdots + (A - c_k)E_k.$$

Using the distinguishing properties of the E_i's, we can prove easily by induction that

$$N^r = (A - c_1)^r E_1 + \cdots + (A - c_k)^r E_k.$$

If $r \geq \max\{r_1, \ldots, r_k\}$, it follows that $N^r = 0$, since $(A - c_i)^r$ is the zero transformation on the range of E_i, $1 \leq i \leq k$. Thus N is nilpotent, and we have proved that if the minimal polynomial of a linear transformation is the product of linear polynomials over the scalar field at hand then it is the sum of a diagonable and a nilpotent linear transformation. Further, since each E_i can be expressed as a polynomial in A, so can D and N. It follows that $DN = ND$. This result is also a consequence of the form of the representation of D and N in terms of $E_1, \ldots, E_k$ and the fact that A commutes with each E_i. Finally, we mention without proof that the representation of A in the form $D + N$ is unique. Our next theorem summarizes the foregoing results.

□ **Theorem 2.2.** Let A be a linear transformation on a finite-dimensional vector space over a field F. Suppose that the minimal polynomial of A decomposes over F into a product of linear polynomials. Then there is a diagonable linear transformation D on V and a nilpotent linear transformation N on V such that $A = D + N$ and $DN = ND$. The transformations D and N are uniquely determined by A and each is a polynomial in A.

□ **Corollary.** Let A be a linear transformation on a finite-dimensional unitary space V. Then A is the sum of a diagonable linear transformation D on V and a nilpotent linear transformation N on V. Further, D and N are unique commuting transformations and each is a polynomial in A.

EXAMPLE

2.1. Consider the linear transformation A on R^3 induced by the matrix

$$\begin{bmatrix} 3 & 3 & -2 \\ 0 & 1 & 0 \\ 1 & 2 & 0 \end{bmatrix},$$

relative to the standard basis of R^3. The characteristic polynomial and the minimal polynomial of A are each equal to $(x - 1)^2\,(x - 2)$. The null space S_1 of $(A - 1)^2$ is spanned by $\{(0, 1, 0), (1, 0, -2)\}$ and the null space S_2 of $A - 2$ is spanned by $\{(1, 1, -1)\}$. Thus $\mathsf{R}^3 = S_1 \oplus S_2$ and, if $\alpha = (x, y, z) \in \mathsf{R}^3$, then $\alpha = \alpha_1 + \alpha_2$, where

$$\alpha_1 = (-x - z,\ y - 2x - z,\ 2x + 2z),$$
$$\alpha_2 = (2x + z,\ 2x + z,\ -2x - z).$$

Further, if $\{E_1, E_2\}$ is the complete set of projections for R^3 associated with the direct-sum decomposition of R^3 indicated above, D is the diagonable part of A, and $N = A - D$, then

$$D = E_1 + 2E_2, \qquad N = (A - 1)E_1 + (A - 2)E_2,$$

and the matrices of D and N relative to the standard basis of R^3 are, respectively,

$$\begin{bmatrix} 3 & 2 & -2 \\ 0 & 1 & 0 \\ 1 & 1 & 0 \end{bmatrix}, \qquad \begin{bmatrix} 0 & 1 & 0 \\ 0 & 0 & 0 \\ 0 & 1 & 0 \end{bmatrix}.$$

We conclude this section with a further observation about Theorem 2.1, namely, that parts (a) and (b) do not require that m be the minimal polynomial of A; that is, if p is any polynomial such that $p(A) = 0$ and $p_1^{r_1} \cdots p_k^{r_k}$ is its factorization over F into powers of distinct irreducible polynomials and S_i is defined to be the null space of $[p_i(A)]^{r_i}$, then S_i is an A-invariant subspace and $V = S_1 \oplus \cdots \oplus S_k$. The earlier proofs apply. This result has an interesting application to the theory of linear differential equations discussed in Chapter 9.

EXERCISES

2.1. Verify each part of Theorem 2.1 for the linear transformation A in R^3 whose matrix relative to the standard basis of R^3 is

$$(a_{ij}) = \begin{bmatrix} 6 & -3 & -2 \\ 4 & -1 & -2 \\ 10 & -5 & -3 \end{bmatrix}.$$

Note. The minimal polynomial of (a_{ij}) is $(x^2 + 1)(x - 2)$.

2.2. Prove the result stated in the corollary to Theorem 2.1.

2.3. Decompose the linear transformation defined in Example 6.6.1 into the sum of a diagonable linear transformation and a nilpotent linear transformation on R^4.

2.4. Let A be the linear transformation on R^3 induced by the matrix

$$\begin{bmatrix} 3 & 1 & -1 \\ 2 & 2 & -1 \\ 2 & 2 & 0 \end{bmatrix}$$

relative to the standard basis of R^3. Show that there is a diagonable linear transformation D on R^3 and a nilpotent linear transformation N on R^3 such that $A = D + N$ and $DN = ND$. Find the matrices of D and of N relative to the standard basis of R^3.

2.5. Let V be a finite-dimensional vector space over a field F and let A be a linear transformation on V of rank 1. Prove that A is either diagonable or nilpotent, but not both.

2.6. Let A be a linear transformation on a finite-dimensional vector space V. Suppose that $A = D + N$, where D is a diagonable linear transformation, N is a nilpotent linear transformation and $DN = ND$. Show that D and N are uniquely determined by A.

2.7. Let A be a linear transformation on the finite-dimensional vector space V with characteristic polynomial $f = (x - c_1)^{d_1} \cdots (x - c_k)^{d_k}$ and minimal polynomial $m = (x - c_1)^{r_1} \cdots (x - c_k)^{r_k}$. Let S_i be the null space of $(A - c_i)^{r_i}$, $1 \leq i \leq k$. Prove that

(a) $S_i = \{\alpha \in V \mid \alpha(A - c_i)^n = 0$ for some positive integer $n\}$,

(b) $\dim S_i = d_i, \quad 1 \leq i \leq k$.

3. NORMAL TRANSFORMATIONS

Theorem 1.2 characterizes an orthogonally diagonable linear transformation on a finite-dimensional inner product space V as follows: A is orthogonally diagonable if and only if $A = \sum_1^k c_i E_i$, where $\{E_1, \ldots, E_k\}$ is a complete set of projections for V such that each E_1 is self-adjoint. If A has this form, then $A^* = \sum_1^k \bar{c}_i E_i$ and consequently

$$AA^* = \sum_{i=1}^{r} |c_i|^2 E_i = A^*A$$

Other types of linear transformations on finite-dimensional inner-product space which have the property that they commute with their adjoints (as does A above) include self-adjoint transformations, unitary transformations, as well as orthogonal transformations. It is the class of such linear transformations on finite-dimensional inner-product spaces that we analyze in this section.

□ **Definition.** A linear transformation A on an inner-product space V is called *normal* if and only if its adjoint A^* exists and $AA^* = A^*A$. A square matrix (a_{ij}) over R is called *normal* if and only if $(a_{ij})(a_{ij})^t = (a_{ij})^t(a_{ij})$, and a square matrix (c_{ij}) over C is called *normal* if and only if

$$(c_{ij})(c_{ij})^* = (c_{ij})^*(c_{ij}).$$

EXAMPLE

3.1. We have already mentioned four types of normal linear transformations. Normal matrices include unitary matrices, orthogonal matrices, real symmetric matrices, and hermitian matrices. Further, a matrix of the form

$$\begin{bmatrix} a & -b \\ b & a \end{bmatrix}, \qquad a, b \in \mathsf{R},$$

is normal. The matrix

$$\begin{bmatrix} 1+i & -1+i \\ -1+i & 1+i \end{bmatrix}$$

is an example of a normal matrix that is not hermitian.

□ **Theorem 3.1.** Let A be a normal linear transformation on an inner-product space V. If α is a characteristic vector of A belonging to the characteristic value c of A, then α is a characteristic vector of A^* belonging to the characteristic value $\bar{c}$ of A^*.

PROOF. Since

$$(\alpha A \mid \alpha A) = (\alpha \mid \alpha A A^*) = (\alpha \mid \alpha A^* A) = (\alpha A^* \mid \alpha A^*),$$

we conclude that $\|\alpha A\| = \|\alpha A^*\|$ for each α in V. Hence $\alpha A = 0$ if and only if $\alpha A^* = 0$ for a normal linear transformation A. With A normal, so is $A - c$ for any scalar c, and $(A - c)^* = A^* - \bar{c}$. Thus $\alpha A = c\alpha$ if and only if $\alpha A^* = \bar{c}\alpha$ for a scalar c and a vector α. The conclusions of the theorem then follow. ◇

□ **Lemma.** Let A be a normal linear transformation in an inner-product space V and let h be a polynomial over the scalar field. Then the null space S of $h(A)$ and $S^\perp$ are A-invariant (hence A^*-invariant).

PROOF. Clearly, S is A-invariant, hence $S^\perp$ is A^*-invariant, regardless of whether or not A is normal. With A normal, it follows that $A^* h(A) = h(A)A^*$ and that S is A^*-invariant. To show that $S^\perp$ is A-invariant let $\beta \in S^\perp$ and $\alpha \in S$. Then

$$(\alpha \mid \beta A) = \overline{(\beta A \mid \alpha)} = \overline{(\beta \mid \alpha A^*)} = \bar{0} = 0,$$

which implies that $\beta A \in S^\perp$ is A-invariant. ◇

With this lemma we can prove that a normal linear transformation on a finite-dimensional unitary space is orthogonally diagonable (how?). We choose, however, to derive this result as a corollary of the following theorem about normal linear transformations.

□ **Theorem 3.2.** Let A be a normal linear transformation on a finite-dimensional inner-product space V. Let $p_1^{r_1} \cdots p_k^{r_k}$ be a factorization of the minimal polynomial m of A as a product of powers of distinct irreducible monic polynomials $p_1, \ldots, p_k$. Let S_i be the null space of $[p_i(A)]^{r_i}$, $1 \leq i \leq k$. Then

(a) $S_1, \ldots, S_k$ are mutually orthogonal subspaces of V,

(b) $r_1 = \cdots = r_k = 1$.

PROOF. To prove the orthogonality of S_i and S_j, $i \neq j$, it suffices to show that $S_j \subseteq S_i^\perp$. For this consider a vector $\alpha_j \in S_j$. Since $V = S_i \oplus S_i^\perp$, $a_j = \alpha_i + \alpha_i^\perp$, where $\alpha_i \in S_i$ and $\alpha_i^{\perp} \in S_i^\perp$. Hence

$$\alpha_j[p_j(A)]^{r_j} = \alpha_i[p_j(A)]^{r_j} + \alpha_i^\perp[p_j(A)]^{r_j} = 0.$$

By the above lemma S_i and $S_i^\perp$ are A-invariant and, hence, are $[p_j(A)]^{r_j}$-invariant. Therefore $\alpha_i[p_j(A)]^{r_j} \in S_i$ and $\alpha_i^\perp[p_j(A)]^{r_j} \in S_i^\perp$. Since $\{0\} = S_i \cap S_i^\perp$, it follows in turn that $\alpha_i[p_j(A)]^{r_j} = 0$, $\alpha_i \in S_j$, and $\alpha_i = 0$. Hence $\alpha = \alpha_i^\perp \in S_i^\perp$ and so $S_j \subseteq S_i^\perp$.

We begin the proof of (b) by recalling (Theorem 2.1) that the minimal polynomial of A_i, the restriction of A to S_i, is $p_i^{r_i}$. If $r_i > 1$, there exists an element α_i of S_i such that $\alpha_i \neq 0$ and $\alpha_i p_i(A) = 0$. Hence $T_i = \{\alpha \in V \mid \alpha p_i(A) = 0\} \subset S_i$. Then S_i is the direct sum of T_i and its orthogonal complement $T_i^\perp$ in S_i. Again by the lemma, T_i and $T_i^\perp$ are A_i-invariant and, hence, $[p_i(A_i)]^{r_i-1}$-invariant. Since $T_i^\perp[p_i(A_i)]^{r_i-1}\, p_i(A_i) = 0$, we may conclude that $T_i^\perp[p_i(A_i)]^{r_i-1} = 0$ because $T_i \cap T_i^\perp = \{0\}$. Of course, $T_i[p_i(A_i)]^{r_i-1} = 0$, and it follows that $S_i[p_i(A_i)]^{r_i-1} = 0$, contrary to the fact that $p_i^{r_i}$ is the minimal polynomial of A_i. Since the assumption that $r_i > 1$ yields a contradiction, we conclude that $r_i = 1$. ◇

Now consider the case of a normal linear transformation A on a finite-dimensional *unitary* space V. Since the only irreducible polynomials over C have degree 1, the minimal polynomial m of A is a product of factors of the form $x - c$. In view of Theorem 3.2, none of these factors is repeated and so m has the form

$$m = (x - c_1) \cdots (x - c_k),$$

where the c_i's are distinct complex numbers. Moreover, V is the direct sum of the mutually orthogonal subspaces $S_1, \ldots, S_k$, where

$$S_i = \{\alpha \in V \mid \alpha A = c_i \alpha\}, \qquad 1 \leq i \leq k,$$

and each member of the complete set of projections $\{E_1, \ldots, E_k\}$ associated with this direct sum is self-adjoint. We summarize these results in the next theorem.

□ **Theorem 3.3.** If A is a normal linear transformation on a finite-dimensional unitary space V, then $A = c_1 E_1 + \cdots + c_k E_k$, where $\{E_1, \ldots, E_k\}$ is a complete set of self-adjoint projections and $c_1, \ldots, c_k$ are the distinct characteristic values.

□ **Corollary 1.** A linear transformation A on a finite-dimensional unitary space is orthogonally diagonable if and only is A is normal.

PROOF. This follows from Theorem 1.2 and a remark at the beginning of this section. ◇

□ **Corollary 2.** If (c_{ij}) is a normal matrix over C, then there exists a unitary matrix (u_{ij}) such that $(u_{ij})(c_{ij})(u_{ij})^*$ is diagonal.

According to Corollary 1 the normality condition merely serves to characterize those linear transformations on finite-dimensional unitary spaces that are orthogonally diagonable. The state of affairs is different, however, in finite-dimensional euclidean spaces, as we proceed to show.

According to Theorem 3.2, the minimal polynomial m of a normal linear transformation A on a finite-dimensional euclidean space V is a product $p_1 \cdots p_k$ of distinct factors, each of which is either of the form $x - c$ or $(x - a)^2 + b^2$, where $b \neq 0$. Moreover, if S_i is the null space of $p_i(A)$, $1 \leq i \leq k$, then the S_i's are mutually orthogonal. Our concern is with the choice of a basis for a null space S_i determined by a quadratic factor such that the matrix of the restriction of A to S_i is "close to" diagonal form relative to this basis. In view of the decomposition of A which accompanies the decomposition of V as the direct sum of the subspaces $S_1 \ldots, S_k$, it is sufficient to investigate the following case: A is a normal linear transformation on V having minimal polynomial

$$m = (x - a)^2 + b^2, \qquad a, b \in \text{R} \quad \text{and} \quad b \neq 0.$$

Then (see Theorem 6.5.1) the characteristic polynomial h of A has the form

$$h = m^r.$$

To begin our analysis we note that $A + A^*$ is self-adjoint; hence, according to Theorem 6.4.1, there exists a real number c and a vector α in V such that $\alpha \neq 0$ and

$$\alpha(A + A^*) = c\alpha.$$

The subspace V_1 of V spanned by α and αA has dimension 2 and is A-invariant, since from $(A - a)^2 + b^2 = 0$ it follows that $A^2 = 2aA - (a^2 + b^2)$ and so

$$(\alpha A)A = \alpha A^2 = -(a^2 + b^2)\alpha + 2a\alpha A.$$

Moreover, V_1 is A^*-invariant, since both

$$\alpha A^* = c\alpha - \alpha A$$

and

$$(\alpha A)A^* = (\alpha A^*)A = (c\alpha - \alpha A)A$$

are in V_1. For reference we mention the further conclusion that $V_1^{\perp}$ is also both A and A^*-invariant.

We prove next that the number c introduced above is equal to $2a$. Since $A^2 - 2aA + (a^2 + b^2) = 0$ and $A^{*2} - 2aA^* + (a^2 + b^2) = 0$, $A^2 - A^{*2} = 2a(A - A^*)$. Hence, in turn,

$$\alpha(A^2 - A^{*2}) = \alpha(A + A^*)(A - A^*) = \alpha[2a(A - A^*)].$$

Since $c\alpha = \alpha(A + A^*)$, we have also that

$$\alpha(A^2 - A^{*2}) = c\alpha(A - A^*).$$

Equating these two values for $\alpha(A^2 - A^{*2})$ gives

$$c\alpha(A - A^*) = (2a)\, \alpha(A - A^*)$$

and its equivalent

$$(c - 2a)\, \alpha(A - A^*) = 0.$$

If $\alpha(A - A^*) = 0$, then $\alpha A = \alpha A^*$, which implies that $c\alpha = \alpha(A + A^*) = 2\alpha A$ and so $\alpha A = (c/2)\alpha$. Since this conclusion is impossible, we infer that $c - 2a = 0$ or $c = 2a$.

Next, set $\delta_1 = \alpha/\|\alpha\|$ and let δ_2 be a unit vector in V_1 that is orthogonal to δ_1 to obtain the orthonormal basis (δ_1, δ_2) of V_1. If d, e, f, and g are real numbers such that

$$\delta_1 A = d\delta_1 + e\delta_2,$$
$$\delta_2 A = f\delta_1 + g\delta_2,$$

then

$$\delta_1 A^* = d\delta_1 + f\delta_2,$$
$$\delta_2 A^* = e\delta_1 + g\delta_2,$$

and so

$$\delta_1(A + A^*) = \frac{\alpha}{\|\alpha\|}(A + A^*) = \frac{2a\alpha}{\|\alpha\|} = 2a\delta_1.$$

On the other hand,

$$\delta_1(A + A^*) = 2d\delta_1 + (e + f)\delta_2.$$

Thus $d = a$ and $e = -f$. Still further calculations are required to obtain values for e and f. On the one hand

$$\begin{aligned}\delta_1 A^2 &= a\delta_1 A - f\delta_2 A = a(a\delta_1 - f\delta_2) - f(f\delta_1 + g\delta_2)\\ &= (a^2 - f^2)\delta_1 - (af + fg)\delta_2\end{aligned}$$

and on the other

$$\begin{aligned}\delta_1 A^2 &= \delta_1[2aA - (a^2 + b^2)] = 2a\delta_1 A - (a^2 + b^2)\delta_1\\ &= 2a(a\delta_1 - f\delta_2) - (a^2 + b^2)\delta_1\\ &= (a^2 - b^2)\delta_1 - 2af\delta_2.\end{aligned}$$

Hence $a^2 - f^2 = a^2 - b^2$ and $af + fg = 2af$; that is, $f^2 = b^2$ and $fg = af$ From the latter equation it follows that $g = a$, since $f \neq 0$. Further, there is no loss of generality to infer from $f^2 = b^2$ that $f = b$, since the alternative $(f = -b)$ may be eliminated by replacing δ_2 by $-\delta_2$. Thus, in summary,

$$d = g = a, \qquad f = b, \qquad e = -b.$$

It follows that the matrix of the restriction A to V_1, relative to (δ_1, δ_2), is

$$\begin{bmatrix} a & -b \\ b & a \end{bmatrix}.$$

We now turn our attention to $V_1^\perp$ and the restriction A_1 of A to this subspace (recall that $V_1^\perp$ is both A and A^*-invariant). The minimal polynomial of A_1 is m and its characteristic polynomial is m^{r-1}. Thus (assuming that $r > 1$) the circumstances concerning A_1 on $V_1^\perp$ are precisely the same for A on V except that r has been reduced by 1. It is clear that repeating the foregoing sequence of computations r times yields an orthonormal basis of V relative to which the matrix of A is the direct sum of r matrices, each of which is equal to the 2×2 matrix displayed above.

From the result just established, together with Theorem 3.2 and Corollary 2 of Theorem 3.3, we have the following definitive theorem for normal linear transformations on finite-dimensional euclidean vector spaces.

□ **Theorem 3.4.** Suppose that the normal linear transformation A on an n-dimensional euclidean vector space V has characteristic polynomial

$$h = (x - c_1)^{m_1} \cdots (x - c_k)^{m_k}[(x - a_1)^2 + b_1{}^2]^{m_{k+1}} \cdots [(x - a_p)^2 + b_p{}^2]^{m_{k+p}}$$

and (therefore) minimal polynomial

$$m = (x - c_1)\cdots(x - c_k)[(x - a_1)^2 + b_1{}^2]\cdots[x - a_p)^2 + b_p{}^2],$$

where $c_1, \ldots, c_k$ are distinct real numbers, $(a_1, b_1), \ldots, (a_p, b_p)$ are distinct pairs of real numbers, and each $b_i \neq 0$. Let S_i be the null space $A - c_i$, $1 \leq i \leq k$, and let S_{k+j} be the null space of $[(A - a_j)^2 + b_j{}^2]$, $1 \leq j \leq p$. Then an orthonormal basis $\mathscr{B}$ of V can be constructed such that the matrix of A relative to $\mathscr{B}$ is the direct sum of matrices $M_1, \ldots, M_k$, $M_{k+1}, \ldots, M_{k+p}$, where

$$M_i = \operatorname{diag}(c_i, \ldots, c_i) \quad \text{is} \quad m_i \times m_i, \qquad 1 \leq i \leq k,$$

and

$$M_{k+j} = \begin{bmatrix} N_j & O & \cdots & O \\ O & N_j & \cdots & O \\ \vdots & & & \\ O & O & \cdots & N_j \end{bmatrix}$$

is a $2m_{k+j} \times 2m_{k+j}$ matrix such that

$$N_j = \begin{bmatrix} a_j & -b_j \\ b_j & a_j \end{bmatrix}, \quad 1 \leq j \leq p.$$

□ **Corollary 1.** If (a_{ij}) is a normal real matrix, then there exists an orthogonal matrix (p_{ij}) such that $(p_{ij})(a_{ij})(p_{ij})^t$ is a matrix of the form described in the theorem.

□ **Corollary 2.** If A is an orthogonal linear transformation on V, then each c_i is either 1 or -1 and $a_j^2 + b_j^2 = 1$, $1 \leq j \leq p$. Hence there exists a real number θ_j such that $0 \leq \theta_j \leq \pi$, $a_j = \cos\theta_j$, $b_j = \sin\theta_j$, and N_j has the form

$$N_j = \begin{bmatrix} \cos\theta_j & -\sin\theta_j \\ \sin\theta_j & \cos\theta_j \end{bmatrix}, \qquad 1 \leq j \leq p.$$

EXAMPLES

3.1. The matrix

$$(a_{ij}) = \begin{bmatrix} -1 & 0 & -2 & 0 \\ 0 & -1 & 0 & -2 \\ 2 & 0 & -1 & 0 \\ 0 & 2 & 0 & -1 \end{bmatrix}$$

is normal. Its minimal polynomial is $m = (x+1)^2 + 2^2$ and its characteristic polynomial is m^2. If

$$(p_{ij}) = \begin{bmatrix} 1 & 0 & 0 & 0 \\ 0 & 0 & 1 & 0 \\ 0 & 1 & 0 & 0 \\ 0 & 0 & 0 & 1 \end{bmatrix},$$

then

$$(p_{ij})(a_{ij})(p_{ij})^t = \begin{bmatrix} -1 & -2 & 0 & 0 \\ 2 & -1 & 0 & 0 \\ 0 & 0 & -1 & -2 \\ 0 & 0 & 2 & -1 \end{bmatrix},$$

which is in the form predicted by the theorem.

3.2. The normal matrix

$$(a_{ij}) = \begin{bmatrix} 1 & 0 & 0 & 0 \\ 0 & 0 & -1 & 0 \\ 0 & 1 & 0 & 0 \\ 0 & 0 & 0 & -1 \end{bmatrix}$$

has minimal polynomial $(x-1)(x^2+1)$ and characteristic polynomial $(x-1)^2(x^2+1)$. Hence there exists (p_{ij}) such that

$$(p_{ij})(a_{ij})(p_{ij})^t = \begin{bmatrix} 1 & 0 & 0 & 0 \\ 0 & 1 & 0 & 0 \\ 0 & 0 & 0 & -1 \\ 0 & 0 & 1 & 0 \end{bmatrix}.$$

EXERCISES

3.1. Analyze all orthogonal linear transformations on $\mathbf{R}^3$.

3.2. Find two normal linear transformations whose sum is not normal.

3.3. Show that if A and B are commuting normal linear transformations then $A+B$ and AB are normal.

3.4. Let A be a normal linear transformation on a finite-dimensional unitary space. Show that
(a) A is self-adjoint if and only if all characteristic values of A are real,
(b) A is unitary if and only if all characteristic values of A have absolute value 1.

3.5. Let A be a linear transformation on a finite-dimensional unitary space. Show that A is normal if and only if $A^* = p(A)$ for some polynomial p over $\mathbf{C}$.

3.6. Let A be an orthogonal linear transformation on $\mathbf{R}^3$ such that $\det A = 1$. Show that there exists an α in $\mathbf{R}^3$ such that $\alpha A = \alpha$.

3.7. Let A be a normal linear transformation on a finite-dimensional unitary space. If all coefficients of the characteristic polynomial of A are real numbers, what is the characteristic polynomial of A^*?

3.8. Show that two normal linear transformations on a finite-dimensional inner-product space are similar if and only if they have the same characteristic polynomial. What is the matrix version of this result?

3.9. Let A be a linear transformation on a finite-dimensional unitary space. Show that there exist self-adjoint linear transformations B and C such that

$$A = B + iC.$$

Also show that A is normal if and only if $BC = CB$.

3.10 Let A be a linear transformation on a finite-dimensional unitary space V. Show that
(a) if c_1 and c_2 are complex numbers such that $|c_1| = |c_2| = 1$ then $c_1 A + c_2 A^*$ is normal;
(b) if $\|\alpha A\| = \|\alpha A^*\|$ for all $\alpha \in V$, then A is normal.

3.11. Let A and B be normal linear transformations on an inner-product space. If $AB = 0$, what can be said about BA?

3.12. Can a nonzero, normal linear transformation on a finite-dimensional inner-product space be nilpotent?

4. THE JORDAN NORMAL FORM†

It was established in Section 2 that if the minimal polynomial m of a linear transformation A on a finite-dimensional vector space V over a field F is a product of linear polynomials over F, say

$$m = (x - c_1)^{r_1} \cdots (x - c_k)^{r_k},$$

then $A = N + D$, where D is a diagonable linear transformation on V, N is a nilpotent linear transformation on V, and $DN = ND$. As long as we choose *any* basis $\mathscr{B}_i$ of the null space S_i of $(A - c_i)^{r_i}$, $1 \leq i \leq k$, their union $\mathscr{B}$ is a basis of V relative to which the matrix of D is

$$\operatorname{diag}(c_1, \ldots, c_1, \ldots, c_k, \ldots, c_k).$$

Here $c_1, \ldots, c_k$ are the characteristic values of A and c_i appears as many times as the linear factor $x - c_i$ appears in a factorization of the characteristic polynomial of A into irreducible factors. Thus, apart from the ordering selected for the characteristic values of A, this matrix is completely determined by the characteristic polynomial of A. On the other hand, the matrix of N depends on the particular choice of the bases $\mathscr{B}_1, \ldots, \mathscr{B}_k$. It is possible to choose these bases in such a way that the matrix of N relative to the resulting basis $\mathscr{B}$ of V is close to diagonal form. Consequently, the sum of this matrix of N and the diagonal matrix of D, which is the matrix of A relative to $\mathscr{B}$, is particularly simple.

To describe with some precision what may be achieved in this connection, let us define the *Jordan matrix* $J_m(c)$ to be the $m \times m$ matrix having c's appearing along the principal diagonal, 1's appearing along the next diagonal above, and 0's elsewhere. Thus

$$J_m(c) = \begin{bmatrix} c & 1 & 0 & \cdots & 0 \\ 0 & c & 1 & \cdots & 0 \\ 0 & 0 & c & \ddots & \\ \vdots & & & \ddots & \\ \vdots & & & \ddots & 1 \\ 0 & & \cdots & & c \end{bmatrix}.$$

† The material in this section is optional.

Then it may be shown that there exists a basis of V relative to which the matrix of A is a direct sum of Jordan matrices $J_{m_i}(c_i)$, where c_i is a characteristic value of A. A matrix of A in this form is unique (once the c_i's have been ordered) and is called the *Jordan normal form* among the matrices of A.

An exhaustive treatment of the results stated above would include the derivation of an algorithm for obtaining the Jordan normal form of A and a proof of its uniqueness. We forego this and restrict our attention to the proof that if A is a linear transformation on a finite-dimensional unitary space then there exists a matrix of A that is a direct sum of Jordan matrices. (By adopting $\mathbf{C}$ as the field of scalars, the decomposition of the minimal polynomial of A as a product of linear factors is assured and the assumption that the space is unitary makes possible a much shorter proof.) Our theorem follows easily from the next theorem.

□ **Theorem 4.1.** Let N be a nonzero nilpotent linear transformation on a finite-dimensional unitary space V. Then V can be expressed as the direct sum of subspaces that are cyclic with respect to N.

PROOF. With N and V as stated, suppose that N has index k and $\dim V = n$. Then $1 < k \leq n$, and so there exists a nonzero vector α in V such that $\alpha N^{k-1} \neq 0$. It follows that k is the value of the least positive integer s such that $\{\alpha, \alpha N, \ldots, \alpha N^s\}$ is linearly dependent. Hence (see Theorems 6.2.4 and 6.2.5) $C_{\alpha,N}$, the (N-invariant) cyclic subspace of α relative to N, has $(\alpha, \alpha N, \ldots, \alpha N^{k-1})$ as a basis and the matrix of N relative to this basis is $J_k(0)$.

If $k = n$, the proof is complete. If $k < n$, we continue by proving the existence of an N-invariant subspace V_1 of V such that $V = C_{\alpha,N} \oplus V_1$. Clearly, the adjoint N^* of N is nilpotent of index k. Since $\alpha N^{k-1} \neq 0$, there exists a vector β such that $(\alpha N^{k-1} | \beta) \neq 0$. Thus

$$\{\beta, \beta N^*, \ldots, \beta(N^*)^{k-1}\}$$

is linearly independent and the dimension of W, the space spanned by this set, is k. Further, W is N^*-invariant and, in turn, $W^{\perp}$ is N-invariant. We now prove that $V = C_{\alpha,N} \oplus W^{\perp}$. Since

$$\dim W^{\perp} = \dim V - \dim W = \dim V - \dim C_{\alpha,N},$$

the proof will be complete when we show that $C_{\alpha,N} \cap W^{\perp} = \{0\}$. Let γ

be a member of the intersection. Then there exists scalars $a_0, \ldots, a_{k-1}$ such that

$$\gamma = a_0\alpha + \cdots + a_{k-1}\,(\alpha N^{k-1}),$$

since $\gamma \in C_{\alpha,A}$ and $(\gamma \mid \beta(N^*)^{k-1}) = 0$, since $\gamma \in W^\perp$. Thus

$$0 = (\gamma \mid \beta(N^*)^{k-1}) = (\gamma N^{k-1} \mid \beta) = a_0\,(\alpha N^{k-1} \mid \beta),$$

and, hence, $a_0 = 0$. Similarly, the orthogonality of γ and $\beta(N^*)^{k-2}$ implies that $\alpha_1 = 0$. Continuing in this way, we deduce that every $a_i = 0$, hence $\gamma = 0$. Thus $V = C_{\alpha,N} \oplus W^\perp$.

Letting $V_1 = W^\perp$ and $k_1 = \dim V_1$, we may summarize our conclusions at this point: Accompanying a nilpotent transformation on a finite-dimensional unitary space V is a decomposition of V as a direct sum $C_{\alpha,N} \oplus V_1$ of a cyclic subspace and an N-invariant subspace V_1 of dimension $n_1 = n - k$, where $k = \dim C_{\alpha,N}$.

Continuing, we note that the restriction of N to V_1 is a nilpotent transformation on V_1 whose index k_1, let us say, is less than or equal to k. Thus the above result implies that $V_1 = c_{\alpha_1,N} \oplus V_2$, where V_2 is an N-invariant subspace of V_2 and, hence, of V, of dimension $n_2 = n_1 - \dim C_{\alpha_1,N}$. Thus

$$V = C_{\alpha,N} \oplus (C_{\alpha_1,N} \oplus V_2) = (C_{\alpha,N} \oplus C_{\alpha_1,N}) \oplus V_2.$$

If $n_2 > 0$, the foregoing argument may be repeated. Since V is finite-dimensional and with each application the dimension of the "residual" subspace is reduced, in a finite number of steps we reach a decomposition of V that has the desired form. ◇

As a consequence of Theorem 4.1 we have the result that if N is a nilpotent transformation on a finite-dimensional unitary space, there exists a basis of V relative to which the matrix of N is a direct sum of Jordan matrices of the form $J_{k_j}(0)$. Our principal theorem now follows easily.

□ **Theorem 4.2.** Let A be a linear transformation on a finite-dimensional unitary space V whose minimal polynomial has the form $(x - c_1)^{r_1} \cdots (x - c_k)^{r_k}$. Then there exists a basis of V such that the matrix of A relative to this basis is the direct sum of Jordan matrices $J_{k_j}(c_i)$, where each c_i is a characteristic value of A.

PROOF. According to Theorem 2.1, V is the direct sum of the null spaces S_i of $(A - c_i)^{r_i}$, $1 \leq i \leq k$; $A - c_i$ is a nilpotent transformation on the space S_i. Thus there exists a basis $\mathscr{B}_i$ of S_i such that the matrix (b_{pq}) of $A - c_i$ is the direct sum of matrices of the form $J_{k_j}(0)$. If the matrix of the restriction of A to S_i is (a_{pq}) relative to $\mathscr{B}_i$, then

$$(b_{pq}) = (a_{pq}) - c_i I \quad \text{and} \quad (a_{pq}) = (b_{pq}) + c_i I;$$

that is, (a_{pq}) is a direct sum of Jordan matrices $J_{k_j}(c_i)$. Combining the bases $\mathscr{B}_1, \ldots, \mathscr{B}_k$, we obtain a basis of V relative to which the matrix of A has the desired form. ◇

EXAMPLE

4.1. The matrix

$$\begin{bmatrix} 2 & 1 & & & & & \\ 0 & 2 & & & & & \\ & & 2 & 1 & 0 & & \\ & & 0 & 2 & 1 & & \\ & & 0 & 0 & 2 & & \\ & & & & & 3 & 1 \\ & & & & & 0 & 3 \end{bmatrix}$$

is the direct sum of three Jordan matrices. It is the Jordan normal form of a linear transformation A on a unitary space V of dimension 7. The space V is the direct sum of the null spaces S_1 and S_2 of $(A - 2)^3$ and $(A - 3)^2$, respectively. The null space S_1, in turn, is the direct sum of two subspaces, one of dimension 2 and one of dimension 3. These subspaces are cyclic with respect to the restriction of $A - 2$ to S_1.

9 Several Applications of Linear Algebra

Both physical and social scientists, when idealizing their problems so that they can be formulated mathematically, rely heavily on linear approximations for variables. Thus the idealized mathematical models they create often exhibit instances of linearity. Since the theory of linear algebra is the development of the concept of linearity, it should come as no surprise to find that the theory we have expounded has a wide variety of applications. In this chapter we present a sampling. The first is to the theory of systems of linear differential equations with constant coefficients. The second is concerned with Leontief's interindustry (linear) model of an economy. In Section 3 we illustrate how systems of linear equations arise in mixture problems of interest to chemists. Section 4 is concerned with a problem in classical mechanics, and in the final section we treat a problem in quantum mechanics.

A list of references appears at the end of each section and at the end of the chapter there is a more comprehensive bibliography of books in which applications of our subject matter can be found.

1. LINEAR DIFFERENTIAL EQUATIONS

The notation, the techniques, and some theorems of linear algebra are extremely useful in the study of linear differential equations. As evidence to support this statement we discuss two problems in this section. The first is that of finding solutions of a system of first-order linear homogeneous differential equations. The second problem, that of finding solutions of an nth-order linear homogeneous equation is a special case of the first, but we have chosen to treat it separately.

Consider the system

$$
(1) \qquad \begin{aligned} \frac{dy_1}{dt} &= a_{11}y_1 + \cdots + a_{1n}y_n \\ &\vdots \\ \frac{dy_n}{dt} &= a_{n1}y_1 + \cdots + a_{nn}y_n \end{aligned}
$$

of n linear homogeneous linear differential equations determined by the matrix $A = (a_{ij})$ over C. We wish to determine solutions $y_1, \ldots, y_n$ which assume prescribed initial values $y_1(0), \ldots, y_n(0)$, respectively, at $t = 0$. In order to launch our investigation, several definitions are required.

Suppose that (a_{ij}) is an $n \times n$ matrix where each a_{ij} is a complex-valued function having an interval I of real numbers as its domain. Then for each t in I,

$$
(a_{ij}(t)) = \begin{bmatrix} a_{11}(t) \cdots a_{1n}(t) \\ \vdots \\ a_{n1}(t) \cdots a_{nn}(t) \end{bmatrix}
$$

is a matrix over C. If we denote (a_{ij}) by A, then we shall write $A(t)$ for the displayed matrix. It is possible to define a useful calculus of matrices of functions by applying the familiar definitions to each entry. Two such definitions which we shall need are the following:

$$
\lim_{t \to c} A(t) = \left(\lim_{t \to c} a_{ij}(t)\right),
$$

provided all the limits on the right hand side exist, and

$$
\frac{dA}{dt} = \left(\frac{d}{dt} a_{ij}(t)\right)
$$

provided the derivatives of all a_{ij}'s exist. Two further generalizations of familiar definitions are given next. Let $A^{(0)}, A^{(1)}, \ldots, A^{(k)}, \ldots$ be a sequence of $n \times n$ matrices over C. If $A^{(k)} = (a_{ij}{}^{(k)})$, then the sequence is said to have the matrix $A = (a_{ij})$ as limit if and only if, for all i and j, $\lim_{k \to \infty} a_{ij}{}^{(k)} = a_{ij}$. In this event we write

$$
\lim_{k \to \infty} A^{(k)} = A.
$$

An infinite series $\sum_0^\infty B_r$ of $n \times n$ matrices over C is said to converge to the matrix B if and only if its sequence of partial sums $A^{(0)}$, $A^{(1)}$, ..., $A^{(k)}$, ..., where $A^{(k)} = B_0 + \cdots + B_k$, has B as limit.

Employing the second of the above definitions, we may write the system (1) as

$$\frac{dY}{dt} = AY \tag{2}$$

upon defining $Y = (y_1, \ldots, y_n)^t$. If the vector function Y has just one component (that is, $n = 1$), the above equation (after dropping subscripts) is the simple differential equation.

$$\frac{dy}{dt} = ay \tag{3}$$

for which the solution that takes the value $y(0)$ at $t = 0$ is $y(t) = y(0)e^{at}$. It is a remarkable fact that it is possible to define the exponential function for arguments of the form tA, where A is an $n \times n$ matrix over C and t is a scalar, so that the solution of (2) can be written in the same form as that of (3).

The inspiration for a suitable definition in this connection comes from the power series representation of the exponential function. If $A = (a_{ij})$ is an $n \times n$ matrix over C, consider the sequence of matrices A_m, $m = 0, 1, 2, \ldots$, where

$$A_m = I + A + \frac{1}{2!}A^2 + \cdots + \frac{1}{m!}A^m.$$

We shall prove that $\lim_{m\to\infty} A_m$ exists. Let b be an upper bound for $\{|a_{ij}| \mid 1 \le i, j \le n\}$; then $|a_{ij}| \le b$ for all i and j. In addition, let

$$A^k = (a_{ij}^{(k)}), \qquad k = 0, 1, 2, \ldots$$

Then the (i, j)th entry of A_m is

$$a_{ij}^{(0)} + a_{ij}^{(1)} + \frac{1}{2!}a_{ij}^{(2)} + \cdots + \frac{1}{m!}a_{ij}^{(m)} \tag{4}$$

(of course, $a_{ij}^{(0)} = \delta_{ij}$ and $a_{ij}^{(1)} = a_{ij}$) and our assertion will be established if we can show that all such sequences have limits. The existence of the limit of a sequence whose generic term if (4) is equivalent to the convergence of the infinite series

$$\sum_{k=0}^{\infty} \frac{1}{k!} a_{ij}^{(k)}. \tag{5}$$

The convergence of this series may be shown as follows. Using induction it is easily seen that for $k = 1, 2, \ldots$ and all i and j,

$$|a_{ij}^{(k)}| \leq n^{k-1}b^k.$$

Thus, each term of (5) is dominated in absolute value by the corresponding term in the positive term series

$$\sum_{k=0}^{\infty} \frac{1}{k!} n^{k-1}b^k,$$

and this series converges for all b by the ratio test.

With the proof of the existence of the limit of A_m completed, we continue by defining

$$e^A = \lim_{m \to \infty} A_m = \sum_{k=0}^{\infty} \frac{1}{k!} A^k,$$

Straightforward calculations establish the following properties of such matrices.

$$\frac{d}{dt} e^{tA} = Ae^{tA}. \tag{6}$$

$$e^{A+B} = e^A e^B, \qquad \text{if } A \text{ and } B \text{ commute.} \tag{7}$$

$$P^{-1}e^A P = e^{P^{-1}AP} \qquad \text{for an invertible matrix } P. \tag{8}$$

We are now in a position to obtain the solution of the differential equation (4).

Theorem 1.1. The differential equation

$$\frac{dY}{dt} = AY,$$

where $Y = (y_1, \ldots, y_n)^t$ and A is an $n \times n$ matrix over C, has the solution

$$Y(t) = e^{tA}Y(0)$$

which takes on the initial value $(y_1(0), \ldots, y_n(0))^t$ at $t = 0$.

PROOF. It is clear that Y assumes the prescribed initial values at $t = 0$. It remains to verify that $Y(t)$ is actually a solution of the differential equation. With the definition of the derivative of a matrix of functions given earlier it is easy to prove the familiar rule for the derivative of a product. In particular, if $A(t)$ is a matrix of functions and B is a matrix of constants,

$$\frac{d}{dt}[A(t)B] = \frac{dA}{dt}B.$$

It follows that

$$\frac{dY}{dt} = \frac{d}{dt}[e^{tA}Y(0)] = Ae^{tA}Y(0) = AY(t)$$

using (6). ◇

For the solution of (4) given in the theorem to have practical importance, a method for computing e^{tA} must be available. If the minimal polynomial can be found, then Theorem 8.2.2 can be used for this purpose as we shall now show. Let

$$m = (x - c_1)^{r_1} \cdots (x - c_k)^{r_k}$$

be the minimal polynomial of A. From the results established in Sections 8.1 and 8.2 it is clear that there exists a nonsingular matrix P over $\mathbf{C}$ such that $P^{-1}AP$ is the sum of a diagonal matrix D and a nilpotent matrix N which commutes with D. Hence

$$A = P(D + N)P^{-1}$$

and so

$$e^{tA} = e^{tP(D+N)P^{-1}} = Pe^{tD+tN}P^{-1} = Pe^{tD}e^{tN}P^{-1}$$

using (7) and (8). Now D is the diagonal matrix

$$\mathrm{diag}(c_1, \ldots, c_1, \ldots, c_k, \ldots, c_k)$$

where the number of occurrences of c_i is d_i, $1 \leq i \leq k$, if the characteristic polynomial of A is $(x - c_1)^{d_1} \cdots (x - c_k)^{d_k}$ (see Exercise 8.2.7). Thus, by the definition of e^{tD}, we deduce that

$$e^{tD} = \mathrm{diag}(e^{c_1 t}, \ldots, e^{c_1 t}, \ldots, e^{c_k t}, \ldots, e^{c_k t}).$$

Moreover, e^{tN} is not difficult to calculate; for if N is of index r, then

$$e^{tN} = I + tN + \frac{t^2N^2}{2!} + \cdots + \frac{t^{r-1}N^{r-1}}{(r-1)!}$$

EXAMPLE

9.1. Let us find the solution of the system of linear differential equations

$$\frac{dy_1}{dt} = -y_1 + 3y_2$$

$$\frac{dy_2}{dt} = 2y_2$$

$$\frac{dy_3}{dt} = 2y_1 + y_2 - y_3$$

such that $y_1(0) = 1$, $y_2(0) = -1$, and $y_3(0) = 0$. When written in the form (2), the matrix of the system is

$$A = \begin{bmatrix} -1 & 3 & 0 \\ 0 & 2 & 0 \\ 2 & 1 & -1 \end{bmatrix}.$$

The minimal and characteristic polynomial of A is $(x+1)^2(x-2)$. Using the technique illustrated in Example 8.2.1 we find that

$$P^{-1}AP = \begin{bmatrix} -1 & 2 & 0 \\ 0 & -1 & 0 \\ 0 & 0 & 2 \end{bmatrix} \quad \text{if } P = \begin{bmatrix} 0 & 1 & 1 \\ 0 & 0 & 1 \\ 1 & 0 & 1 \end{bmatrix}.$$

Hence

$$D = \begin{bmatrix} -1 & 0 & 0 \\ 0 & -1 & 0 \\ 0 & 0 & 2 \end{bmatrix} \quad \text{and} \quad N = \begin{bmatrix} 0 & 2 & 0 \\ 0 & 0 & 0 \\ 0 & 0 & 0 \end{bmatrix},$$

where $N^2 = O$. It follows that

$$e^{tD} = \begin{bmatrix} e^{-t} & 0 & 0 \\ 0 & e^{-t} & 0 \\ 0 & 0 & e^{2t} \end{bmatrix} \quad \text{and} \quad e^{tN} = I + tN = \begin{bmatrix} 1 & 2t & 0 \\ 0 & 1 & 0 \\ 0 & 0 & 1 \end{bmatrix}.$$

Hence

$$\begin{bmatrix} y_1(t) \\ y_2(t) \\ y_3(t) \end{bmatrix} = Pe^{tD}e^{tN}P^{-1}\begin{bmatrix} 1 \\ -1 \\ 0 \end{bmatrix} = \begin{bmatrix} 2e^{-t} - e^{2t} \\ -e^{2t} \\ (1+4t)e^{-t} - e^{2t} \end{bmatrix}$$

which means that

$$y_1(t) = 2e^{-t} - e^{2t}$$
$$y_2(t) = -e^{2t}$$
$$y_3(t) = (1 + 4t)e^{-t} - e^{2t}$$

The foregoing method of solving the system of equations (1) is applicable to an nth-order linear homogeneous differential equation

$$\frac{d^n y}{dt^n} + a_1 \frac{d^{n-1} y}{dt^{n-1}} + \cdots + a_{n-1} \frac{dy}{dt} + a_n y = 0, \tag{9}$$

where each a_i is a complex number. Indeed, if we set $y_1 = y$ and rename the ith derivative of y as y_{i+1}, $1 \leq i \leq n - 1$, then the functions $y_1, \ldots, y_n$ satisfy the system

$$\begin{aligned} \frac{dy_1}{dt} &= y_2 \\ \frac{dy_2}{dt} &= y_3 \\ &\vdots \\ \frac{dy_n}{dt} &= -a_n y_1 - \cdots - a_1 y_n \end{aligned} \tag{10}$$

and, conversely, any set of solutions of (10) yields a solution $y = y_1$ of (9). The initial conditions amount to specifying the values of y and its first $n - 1$ derivatives at $t = 0$.

It is also of interest to tackle equation (9) directly using methods of Chapter 8 (in particular, Theorem 8.2.1 and the remark made in the final paragraph of Section 8.2 abouts parts (a) and (b) of that theorem). We begin with some definitions. For a positive integer n let W be the space of all complex-valued, n-fold continuously differentiable functions having R as domain. Then the subset V of W consisting of all y that satisfy (9) is a subspace of W. Let D denote the derivative operator on W, and let p be the polynomial

$$p = x^n + a_1 x^{n-1} + \cdots + a_{n-1} x + a_n .$$

From its definition it is clear that V is the null space of the linear transformation $p(D)$ on W and, hence, regarding D as restricted to V, $p(D)$ is the zero transformation on V.

The polynomial p has a factorization

$$p = (x - c_1)^{r_1} \cdots (x - c_k)^{r_k} \tag{11}$$

where $c_1, \ldots, c_k$ are distinct complex numbers. If S_i is the null space of $(D - c_i)^{r_i}$, then according to Theorem 8.2.1

$$V = S_1 \oplus \cdots \oplus S_k.$$

Thus each solution y of (9) is uniquely expressible in the form

$$y = y_1 + \cdots + y_k$$

where y_i satisfies the differential equation $(D - c_i)^{r_i} y_i = 0$. Observe that the problem of finding solutions of (9) is thereby reduced to a study of the solution space of a differential equation of the form

$$(D - c)^r y = 0 \tag{12}$$

and that this reduction is a sole consequence of the linearity of the operator D.

To determine the solution space to (12) we must use additional properties of D. It is easy to prove by induction on r that if y is an r-fold continuously differentiable function, then

$$(D - c)^r y = e^{ct} D^r (e^{-ct} y)$$

and, hence, $(D - c)^r y = 0$ if and only if $D^r(e^{-ct})y = 0$. A function f such that $D^r f = 0$ is a polynomial function of degree less than or equal to $r - 1$:

$$f(t) = b_0 t^{r-1} + \cdots + b_{r-2} t + b_{r-1}.$$

Thus y satisfies (12) if and only if it has the form

$$y(t) = e^{ct}(b_0 t^{r-1} + \cdots + b_{r-2} t + b_{r-1}).$$

It follows that the set of functions

$$\{e^{ct}, te^{ct}, \ldots, t^{r-1} e^{ct}\}$$

(committing the common sin of denoting functions by their values) span the solution space of (12). Clearly this set of r functions is linearly independent and consequently forms a basis of the solution space.

Returning to the original equation (9) and the related polynomial p in (11), we conclude that the n functions $t^m e^{c_i t}$, $0 \leq m \leq r_i - 1$, $1 \leq i \leq k$, form a basis of the solution space to (9). In particular, the solution space has dimension equal to the order of the differential equation.

REFERENCES

1. Coddington, E. A., and Levinson, N., *Theory of Ordinary Differential Equations.* McGraw-Hill, New York, 1955.
2. Ince, E. L., *Ordinary Differential Equations.* 4th ed. Dover, New York, 1953.
3. Kaplan, W., *Ordinary Differential Equations.* Addison-Wesley, Reading, Mass., 1962.

2. ECONOMICS: INTERACTIONS AMONG INDUSTRIES AND CONSUMERS†

The discovery of linear algebra by economists has had a profound influence on the development of modern economics. Linear programming, game theory, the analysis of systems of interrelated markets, and econometrics, for example, now employ this branch of mathematics extensively. The single example that we shall discuss in this section is one of the first significant applications of linear algebra to economics. To describe it we introduce the following notation. Consider an economy in which the producers are n industries $I_1, \ldots, I_n$. Let x_i denote the value of the output of industry I_i, $1 \leq i \leq n$, and let x_{ij}, $1 \leq i, j \leq n$, denote the value of the product produced by I_i and purchased by I_j. In general, the x_i's and x_{ij}'s are functions of the time t. For the mathematical model (a so-called "Leontief" model) of this economy that we shall consider, the assumption is made that x_{ij} is proportional to x_j; that is,

$$x_{ij} = a_{ij} x_j, \quad 1 \leq i, j \leq n,$$

where each a_{ij} is a constant. In defence of this assumption we remark merely that by virtue of its definition, x_{ij} should have some relationship to the output x_j of I_j; that which we have selected is a simple one.

In terms of the notation introduced, the output of industry I_i that is available for sale to ultimate consumers has the value

$$x_i - \sum_{j=1}^{n} x_{ij} = x_i - \sum_{j=1}^{n} a_{ij} x_j, \qquad 1 \leq i \leq n.$$

† For an understanding of this section, familiarity with Section 1 is a prerequisite.

Let y_i denote the value of the consumer demand for the output of industry I_i, $1 \le i \le n$. We assume that each y_i is nonnegative and note that, in general, it is a function of t. Clearly an ideal state for the economy exists when the consumer demand for each product is equal to the available amount of that product. Defining

$$X = \begin{bmatrix} x_1 \\ \vdots \\ x_n \end{bmatrix}, \qquad Y = \begin{bmatrix} y_1 \\ \vdots \\ y_n \end{bmatrix}, \qquad A = \begin{bmatrix} a_{11} & \cdots & a_{1n} \\ \vdots & & \\ a_{n1} & \cdots & a_{nn} \end{bmatrix},$$

the ideal state of affairs may be summarized by the matrix equation

$$(I - A)X = Y. \tag{1}$$

When regarded as known vectors (matrices), X and Y are called the *output vector* and *consumer vector*, respectively, and A is called the *production matrix*. We emphasize that a given X and Y may not satisfy (1). If in (1) we regard A and Y as known and X as an unknown, then any solution of (1) with nonnegative components is called an *equilibrium solution* of our model.

Questions an economist might ask about such a mathematical model are concerned with conditions on the production matrix A that insure (a) the existence of an equilibrium solution for a given consumer vector Y, (b) the existence of an equilibrium solution for all permissible choices of Y (that is, Y's such that every $y_i \ge 0$), and (c) the stability of the model, by which is meant that if $X(t)$ and $\overline{X}(t)$ are the actual output vector and equilibrium solution vector, respectively, at time t, then $\lim_{t\to\infty}[X(t) - \overline{X}(t)] = 0$. We shall supply some partial answers to these questions on the basis of some additional assumptions.

We consider first the trivial case where $I - A$ is nonsingular. Then (1) has the unique solution $(I - A)^{-1}Y$ for a given Y. Clearly if every entry in the matrix $(I - A)^{-1}$ is nonnegative, then this solution is an equilibrium solution for every Y. At the other extreme if every entry of $(I - A)^{-1}$ is negative, then no equilibrium solution exists for any Y that has at least one positive component. If $(I - A)^{-1}$ has both positive and negative entries, then it is easy to construct examples which show that for some but not all Y's an equilibrium solution exists.

Next we turn to an investigation of consequences of the assumption that A is so restricted that an equilibrium solution exists for each consumer vector Y. Then a solution $\overline{X}$ of (1) depends upon Y and since, in general, Y depends upon t, it follows that the same is true of $\overline{X}$. Let X be the production vector at time t. It is realistic to suppose that X is predetermined—hopefully in such a way that $X = \overline{X}$; it is not to be expected, however, that this is the case. The best that can reasonably be hoped for is the continuous adjustment of X in such a way that $\lim_{t\to\infty}(X - \overline{X}) = 0$. We study the difference $X - \overline{X}$ under the assumptions that $\overline{X}$ is independent of t and that X satisfies the following differential equation:

$$\frac{dX}{dt} = K(A - I)X + KY, \tag{2}$$

where Y is a consumer vector and

$$K = \operatorname{diag}(k_1, \ldots, k_n), \qquad k_i > 0, \qquad 1 \leq i \leq n.$$

In expanded form, (2) is the system of differential equations

$$\frac{dx_i}{dt} = k_i\left[\sum_{j=1}^{n} a_{ij}x_j + y_i - x_i\right], \qquad 1 \leq i \leq n.$$

That these conditions constitute a set of plausible assumptions may be argued as follows. If each industry is continuously adjusting its output with the goal of having supply equal demand, then certainly dx_i/dt will depend on the difference between total demands and the amount produced. That dx_i/dt be proportional to this difference is a manageable example of such a dependence. Having made the assumption that dx_i/dt is proportional to this difference, the further assumption that each $k_i > 0$ is easily justified. For if $\sum_j a_{ij}x_j + y_i - x_i > 0$ (<0), then for supply to approach demand, industry I_i should increase (decrease) production which means that $dx_i/dt > 0$ (<0). Either circumstance requires that $k_i > 0$.

To solve equation (2) we introduce a new variable $U = X - \overline{X}$, where $\overline{X}$ is the equilibrium solution corresponding to Y and $I - A$, that is, $(I - A)\overline{X} = Y$. Substituting $U + \overline{X}$ for X in (2) we obtain, assuming that $\overline{X}$ is constant,

$$\begin{aligned}\frac{dU}{dt} &= K(A - I)(U + \overline{X}) + KY \\ &= K(A - I)U + K(A - I)\overline{X} + KY \qquad (3) \\ &= K(A - I)U\end{aligned}$$

According to the results obtained prior to Example 9.1 in Section 1, if P is a matrix such that

$$P^{-1}[K(A-I)]P = D + N,$$

where D is diagonal and N is a nilpotent matrix that commutes with D, then the solution of (3) can be written in the form

$$U(t) = Pe^{tD}e^{tN}P^{-1}U_0,$$

where, if $c_1, \ldots, c_n$ are the characteristic values of $K(A-I)$,

$$e^{tD} = \text{diag}(e^{c_1 t}, \ldots, e^{c_n t}),$$

e^{tN} is an $n \times n$ matrix whose entries are polynomials in t, and U_0 is a column vector (that is, an $n \times 1$ matrix) of constants. It is left to the reader to convince himself that

$$U(t) = P\begin{bmatrix} p_1(t)e^{c_1 t} \\ \vdots \\ p_n(t)e^{c_n t} \end{bmatrix}$$

where $p_1, \ldots, p_n$ are polynomials in t. Hence

$$X(t) - \overline{X} = \begin{bmatrix} p_1(t)e^{c_1 t} \\ \vdots \\ p_n(t)e^{c_n t} \end{bmatrix}$$

We call a model of the economy stable if and only if $\lim_{t\to\infty}(X(t) - \overline{X}) = 0$. To any one who has studied the calculus and is familiar with the fact that if a and b are real numbers,

$$|e^{a+ib}| = e^a,$$

it should be clear that we have stability if and only if the real part of each characteristic value is negative. Interestingly enough, necessary and sufficient conditions that a matrix of real numbers have this property are known. E. J. Reuth proved that if

$$x^n + a_1x^{n-1} + \cdots + a_{n-1}x + a_n$$

is the characteristic polynomial of an $n \times n$ matrix over R, then the real part of each characteristic value is negative if and only if each of the following n determinants is positive:

$$|a_1| = a_1, \quad \begin{vmatrix} a_1 & a_3 \\ 1 & a_2 \end{vmatrix}, \quad \begin{vmatrix} a_1 & a_3 & a_5 \\ 1 & a_2 & a_4 \\ 0 & a_1 & a_3 \end{vmatrix}, \ldots, \quad \begin{vmatrix} a_1 & a_3 & \cdots & a_{2s-1} \\ 1 & a_2 & \cdots & a_{2s-2} \\ 0 & a_1 & \cdots & a_{2s-3} \\ 0 & 1 & \cdots & \\ \vdots & \vdots & & \\ 0 & 0 & \cdots & a_s \end{vmatrix}, \ldots$$

The rule of formation of the matrices involved is as follows. The first row of the sth consists of the first s coefficients of the characteristic polynomial with an odd subscript. If the uppermost entry in a column is a_j, then below it appear, in turn, the coefficients of successively higher powers of x in the characteristic polynomial, followed by the necessary number of zeros to fill out the matrix. Further, any a with a subscript greater than n is set equal to zero.

The magnitude of the computations required to evaluate these determinants has forced economists to search for simpler, sufficient conditions for stability in the case of particular classes of models. We shall consider one such—namely, the class of models for which all entries a_{ij} of the production matrix satisfy the conditions

$$0 < a_{ij} < 1. \tag{4}$$

When thought of in economic terms, this is a reasonable assumption. With (4) in force both of the matrices $A - I$ and $K(A - I)$ have positive elements on the principal diagonal and negative elements elsewhere. According to a theorem of O. Perron it follows that the characteristic value of $K(A - I)$ with largest real part is a real number. Thus, if this characteristic value is negative, the real part of each characteristic value of $K(A - I)$ is negative. So we are assured of stability if no root of the characteristic equation of $K(A - I)$ is zero or a positive number. Clearly the latter condition holds if all coefficients of the characteristic polynomial of $K(A - I)$ are positive.

There is a simple recipe for computing the coefficients of the characteristic polynomial of a matrix. To state it requires a definition. Let (b_{ij}) be an $n \times n$ matrix. If the rows and columns that include a specified set of m diagonal elements of (b_{ij}) are discarded, then the determinant of the remaining matrix of $n - m$ rows and columns is called a principal minor of (b_{ij}) of degree $n - m$. It can be proved that the coefficient of x^m in the characteristic polynomial of (b_{ij}) is equal to $(-1)^{n-m}$ times the sum of all its principal minors of degree $n - m$. Since the characteristic polynomial of $K(A - I)$ is

$$\det[xI - K(A - I)] = \det[xI + K(I - A)]$$
$$= x^n + a_1 x^{n-1} + \cdots + a_{n-1}x + a_n,$$

let us say, it follows that

$$a_1 = (-1)^1 \sum_{i=1}^{n} k_i(a_{ii} - 1) = \sum_{i=1}^{n} k_i(1 - a_{ii}),$$

$$a_2 = (-1)^2 \sum_{i \neq j} k_i k_j \det\begin{bmatrix} a_{ii}-1 & a_{ij} \\ a_{ji} & a_{jj}-1 \end{bmatrix} = \sum_{i \neq j} k_i k_j \det\begin{bmatrix} 1 - a_{ii} & -a_{ij} \\ -a_{ji} & 1 - a_{jj} \end{bmatrix}.$$

$$\vdots$$

$$a_n = k_1 \cdots k_n \det(I - A).$$

Since each $k_i > 0$, it is obvious that each $c_i > 0$ if all principal minors of $I - A$ are positive. The condition that all entries of A and all principal minors of $I - A$ are positive is known as the Hick's condition (after J. R. Hicks). Thus we have proved that the Hick's condition implies stability. It can also be shown that if all entries of A are positive and the corresponding model is stable, then all principal minors of $I - A$ are positive.

In passing we state and solve another problem that is of interest to economists. In mathematical terms it is this: If the consumer vector Y is positive, that is, each of its components is positive, what is a condition which will insure that X, the production vector, is positive? Assume that the Hick's condition is satisfied by the production matrix A. Then $\det(I - A) > 0$ and, moreover, via a result due to J. Mosak, it follows that the cofactors of each element of $I - A$ is positive. Since $(I - A)^{-1}$ is equal to $1/\det(I - A)$ times the matrix whose (i, j)th entry is the cofactor of the (j, i)th entry of $I - A$ (see Exercise 5.5.12), we deduce that all elements of $(I - A)^{-1}$ are positive. Therefore, since $\overline{X} = (I - A)^{-1} Y$, we conclude that if Y is positive, then X is positive. This result illustrates an important concept of economics known as the correspondence principle.

We have proved that if a production matrix $A = (a_{ij})$ satisfies (4) and the Hick's condition, then the corresponding model of our economy is stable. Another set of sufficient conditions for stability consists of (4) and the so-called column sum rule—the sum of the elements in each column of A is less than 1 or, equivalently, the sum of the elements in each column of $I - A$ is positive.† To establish this result it suffices to show that if they hold for A, then each principal minor of $I - A$ is positive, for then $I - A$ satisfies the Hick's condition. That each principal minor of $I - A$ is positive in the light of our assumptions is an immediate consequence of the following result. If

$$B = \begin{bmatrix} b_{11} & -b_{12} & \cdots & -b_{1n} \\ -b_{21} & b_{22} & \cdots & -b_{2n} \\ \vdots & & & \\ -b_{n1} & -b_{n2} & \cdots & b_{nn} \end{bmatrix}$$

is an $n \times n$ matrix over R such that (a) every $b_{ij} > 0$ and (b) $\Sigma_i b_{ij} > 0$ for all j, then every principal minor of B is positive.

To prove this we observe first that every principal submatrix of B (that is, those submatrices that give rise to the principal minors of B) has properties (a) and (b). Next we note that the determinant of every 1×1 principal submatrix, that is, every principal minor of degree 1, is positive. Now assume that every principal minor of degree less than n, of all $n \times n$ matrices satisfying (a) and (b), is positive. If it can be inferred that $\det B > 0$, the proof will be complete by the principle of mathematical induction. Turning to $\det B$, we have

$$\det B = \det \begin{bmatrix} b_{11} & b_{12} & \cdots & -b_{1n} \\ 0 & b_{22} - \dfrac{b_{12} b_{21}}{b_{11}} & \cdots & -b_{2n} - \dfrac{b_{1n} b_{21}}{b_{11}} \\ \vdots & & & \\ 0 & -b_{n2} - \dfrac{b_{12} b_{n1}}{b_{11}} & \cdots & b_{nn} - \dfrac{b_{1n} b_{n1}}{b_{11}} \end{bmatrix} = b_{11} \det B_1,$$

where B_1 is the $(n-1) \times (n-1)$ matrix obtained from the displayed matrix by deleting the first row and the first column. Since $b_{11} > 0$ we can conclude that $\det B > 0$ if $\det B_1 > 0$. To show that $\det B_1 > 0$ it is sufficient to verify that B_1 satisfies (a) and (b) in view of the induction hypothesis. Now each diagonal element of B_1 is positive because it is a principal minor of degree 2 of B divided by a positive number. Each non-diagonal element is negative since every b_{ij} is positive. Thus B_1 has property

† Since $x_{ij} = a_{ij} x_j$ for all i and j, $\Sigma_i x_{ij} = (\Sigma_i a_{ij}) x_j$. Thus the column sum rule means that the value of all purchases of industry I_j is less than the value of its production.

(a). Next, let us compute the sum of the jth column of B_1. It is

$$\left(-b_{2j}-\frac{b_{ij}b_{21}}{b_{11}}\right)+\left(-b_{3j}-\frac{b_{ij}b_{31}}{b_{11}}\right)+\cdots+\left(b_{ij}-\frac{b_{ij}b_{ji}}{b_{11}}\right)+\cdots+\left(-b_{nj}-\frac{b_{ij}b_{n1}}{b_{11}}\right)$$

$$=[b_{jj}-(b_{2j}+\cdots+b_{j-1,j}+b_{j+1,j}+\cdots+b_{nj}]-\frac{b_{ij}}{b_{11}}(b_{21}+\cdots+b_{n1})$$

$$>b_{jj}-(b_{ij}+b_{2j}+\cdots+b_{j-1,j}+b_{j+1,j}+\cdots+b_{nj}),$$

which is the sum of the jth column of B and, hence, positive. Thus B_1 satisfies (b) and the proof is complete.

REFERENCES

1. Allen, R. G. D. *Mathematical Economics*, 2nd ed. Macmillan, London, 1959
2. Bellman, Richard, *Introduction to Matrix Analysis*. McGraw-Hill, New York, 1960.
3. Boot, John C. G , *Mathematical Reasoning in Economics and Management Science: Twelve Topics*. Prentice-Hall, Englewood Cliffs, New Jersey, 1967.
4. Karlin, Samuel, *Mathematical Methods and Theory in Games, Programming, and Economics*, Vol. 1. Addison-Wesley, Reading, Mass., 1959.
5. Leontief, Wassily W., *The Structure of American Economy 1919–1939*, 2nd ed. Oxford Univ. Press, New York, 1953.
6. Samuelson, Paul Anthony, *Foundations of Economic Analysis*, Vol. LXXX. Harvard Univ. Press, Cambridge, Mass., 1958.
7. Samuelson, Paul Anthony, *The Collected Scientific Papers of Paul A. Samuelson*, Vol. 1. The M.I.T. Press, Cambridge, Mass., 1966.

3. CHEMISTRY: ANALYSIS OF MULTICOMPONENT MIXTURES

Linear algebra is a useful tool to both the practicing chemist and the theoretical chemist. In this section we discuss two examples of problems that fall under its heading and are of interest to the practicing chemist. Mathematically they are trivial but nonetheless of considerable importance. In Section 5 we consider a problem that is of interest to the theoretical chemist and has considerable mathematical sophistication.

Instrumental methods of analysis of mixtures of several components usually make use of the linearity of response of the instrument to the amount of each substance present. By reading the response of an n-component system at n values of an appropriate property of the applied "stimulus" (for example, frequency or wavelength of absorbed light, intensity of electric and magnetic fields), a system of n linear equations results that may be solved for the concentrations of the n components. Our first example of this technique is a problem in absorption spectroscopy.

If a solution of the ith component in an inert solvent (one that does not absorb light in the spectral region of interest) is placed in the path of a beam of light of wavelength γ and intensity I_0, the intensity of transmitted light is attenuated to the value I. The absorbance A of the solution is defined by the equation

$$A(\gamma) = \log_{10}\left|\frac{I_0(\gamma)}{I(\gamma)}\right|$$

and is proportional to the length of the light path l, through the solution, and to the concentration of the solute C_i in moles liter^{-1}; thus

$$A(\gamma) = l\epsilon_i(\gamma)C_i,$$

where the constant of proportionality $\epsilon_i(\gamma)$ is called the molar absorptivity of component i at wavelength γ, and exhibits maxima and minima at values of γ that are characteristic of component i. By measuring $A(\gamma)$ for binary solutions of each of the n components in an inert solvent, we can determine $\epsilon_i(\gamma)$ for $1 \le i \le n$.

The absorbance of ϵ_j of an unknown solution containing all n components is given by

$$A(\gamma_j) = l\sum_{i=1}^{n} C_i\epsilon_i(\gamma_j), \qquad 1 \le j \le n. \tag{1}$$

Thus, the determination of A at $\gamma_1, \ldots, \gamma_n$ gives a system of linear equations that can be solved for $C_1, \ldots, C_n$. Defining α, β, and E by

$$\alpha = (A(\gamma_1), \ldots, A(\gamma_n)), \qquad \beta = (C_1, \ldots, C_n)$$

$$E = \begin{bmatrix} \epsilon_1(\gamma_1) \cdots \epsilon_1(\gamma_n) \\ \vdots \\ \epsilon_n(\gamma_1) \cdots \epsilon_n(\gamma_n) \end{bmatrix},$$

the system of equations (1) can be written as the matrix equation

$$\alpha = l\beta E,$$

whose solution is

$$\beta = (l^{-1}\alpha)E^{-1},$$

provided E is nonsingular. This equation gives the chemical analysis of the mixture. Incidentally, in practice the wavelengths $\gamma_1, \ldots, \gamma_n$ can usually be chosen so that the off-diagonal elements of E are small, in which case an accurate numerical inversion of E is easy.

If several components of a mixture have large values of ϵ in the same ranges of γ (as in the important case of the hydrocarbon mixtures found in crude oil, for example), the numerical inversion of the matrix E involves the subtraction of pairs of numbers with experimental uncertainties of the same order of magnitude as their differences. In these cases, absorption spectroscopy gives a rather poor analysis. An alternative technique (our second example) which often gives good results for mixtures whose components are easily volatilized is that of mass spectrometry. In the mass spectrometer the sample is vaporized and the component molecules bombarded by electrons to produce molecular ions whose deflections in an applied magnetic and electric field are determined by the ratio of their mass m to charge q. This ratio, $M = m/q$, is called the *mass number*. The ions are constrained within an analyzer tube which is so constructed that for a given choice of intensity of the applied magnetic and electric fields only ions with a known value of M are mechanically allowed to reach the metal plate that serves as the ion detector. We may accordingly treat M as the independent variable analogous to the wavelength γ in the previous example.

Preliminary to performing a multicomponent analysis, the mass spectrometrist introduces a pure sample of each possible component i into the spectrometer at partial pressure P_i. In addition to the primary ions formed by the initial bombardment of the neutral molecules by electrons some of these ions usually break up very rapidly into smaller ionic and neutral fragments. Measurement of the intensity of the ion beam as a function of mass number $I_i(M_j)$ thus gives a "mass spectrum" characteristic of the substance i. The value of M for which I_i has its maximum value is called the mass number of the "base peak" for substance i and will be designated B_i. The relative intensities of R are defined by

$$R_i(M_j) = \frac{I_i(M_j)}{I_i(B_i)}.$$

The intensity of the base peak is proportional to the partial pressure P_i of the component; that is,

$$I_i(B_i) = s_i P_i,$$

where the proportionality constant s_i is called the *sensitivity factor*.

After s_i and $R_i(M_j)$ have been determined separately for all possible components, the mass spectrum of the unknown mixture is recorded. The intensities at each mass number are related to the partial pressures of each component by the equations

$$I(M_j) = \sum_{i=1}^{n} s_i P_i R_i(M_j), \qquad 1 \leq j \leq n. \tag{2}$$

Defining μ, σ, and R by

$$\mu = (I(M_1), \ldots, I(M_n)), \qquad \sigma = (s_1 P_1, \ldots, s_n P_n)$$

$$R = \begin{bmatrix} R_1(M_1) & \cdots & R_1(M_n) \\ \vdots & & \\ R_n(M_1) & \cdots & R_n(M_n) \end{bmatrix}$$

the system of equations (2) may be written as the matrix equation

$$\mu = \sigma R$$

with solution $\sigma = \mu R^{-1}$. Then the partial pressures P_i and the mole fractions x_i of the n components are computed from the relations $P_i = \sigma_i / s_i$, where σ_i is the ith component of σ, and $x_i = P_i / \sum_k P_k$.

REFERENCE

1. H. H. Willard, L. L. Merritt, and J. A. Dean, *Instrumental Methods of Analysis*, 4th ed. Van Nostrand, New York, 1965.

4. PHYSICS: COUPLED OSCILLATIONS AND NORMAL MODES

The example we discuss is from classical mechanics. When a mechanical system is displaced slightly from its equilibrium position, it will execute periodic motion about this equilibrium. Familiar examples of this are the response of a violin string when it is plucked and a drum head when struck. An example where the periodicity is more complicated is the sloshing of water in a container. Even in this case, however, there are modes of motion where the water moves in phase with the applied force. Such a mode, where the system oscillates with a single frequency, is called a *normal mode* of the system and the description of the response of the system to an arbitrary disturbance is most conveniently referred to these modes.

In order to expedite the formulation of the general problem in terms of the concepts of linear algebra (whenever this theory is applicable) a word about coordinate systems for euclidean space is in order. The reader is well aware of the fact that in some physical problems a cylindrical or spherical coordinate system has marked advantages over a cartesian coordinate system. For problems in mechanics it is desirable to have available forms of Newton's laws of motion that are suitable for use in any coordinate system. This can be achieved via the notion of generalized coordinates for a physical system. Consider a system of n particles in space and let (x_i, y_i, z_i) be the coordinates of the ith at time t, relative to some cartesian coordinate system. Suppose that $q_1, \ldots, q_{3n}$ are $3n$ real valued functions of t such that

$$\begin{aligned} x_i &= x_i(q_1, \ldots, \alpha_{3n}, t) \\ y_i &= y_i(q_1, \ldots, q_{3n}, t), \qquad 1 \leq i \leq n. \\ z_i &= z_i(q_1, \ldots, q_{3n}, t) \end{aligned}$$

These equations determine the location of each particle, relative to the cartesian coordinate system, at time t. The functions $q_1, \ldots, q_{3n}$ are called the *generalized coordinates* of the system. In terms of generalized coordinates, the equations of motion of the system may be written as the system of $3n$ so-called Lagrange equations

$$\frac{\partial L}{\partial q_j} - \frac{d}{dt}\frac{\partial L}{\partial \dot{q}_j} = 0 \tag{1}$$

where $L = T - V$, T is the total kinetic energy, and V is the total potential energy of the system. Further, $\dot{q}_j = dq/dt$.

To illustrate the usage of generalized coordinates in connection with a problem in coupled oscillations, we offer an example. Consider three identical masses and four identical springs strung in a line between two supports as indicated in Fig. 1. Subscripts are employed only for identification purposes. Constraining the masses to move along the line of the springs, we need merely one generalized coordinate to specify the position

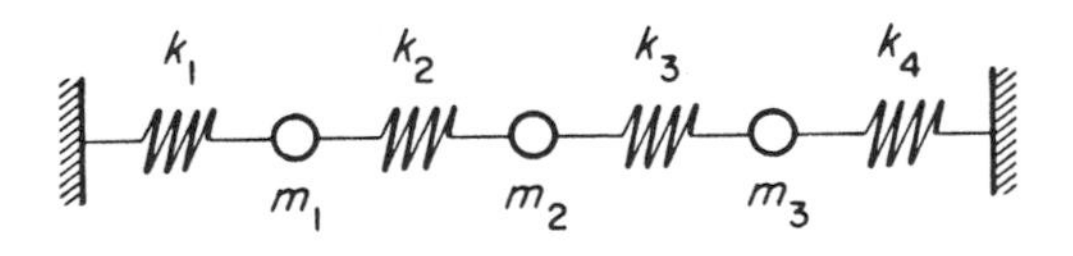

Fig. 1

of each mass. Let $q_i(t)$ be the displacement of the ith mass from its equilibrium position at time t. The velocity of the ith mass is then $\dot{q}_i$. The total kinetic energy of the system is given by

$$T = \sum_{i=1}^{3} \frac{m_i}{2} \dot{q}_i^{\,2} = \frac{m}{2} \sum_{i=1}^{3} \dot{q}_i^{\,2}$$

where $m = m_1 = m_2 = m_3$. The potential energy of the jth spring is given by $V_j = \frac{1}{2} k_j x_j^{\,2}$, where k_j is the spring constant and x_j is the displacement of the spring from its original length. In this case, $x_1 = q_1$, $x_2 = q_2 - q_1$, $x_3 = q_3 - q_2$, and $x_4 = -q_3$. Thus the total potential energy is

$$V = \sum_{j=1}^{4} V_j = \frac{k}{2} \sum_{j=1}^{4} x_j^{\,2}$$

where $k = k_1 = k_2 = k_3 = k_4$ and so

$$(2) \quad L = \frac{m}{2}(\dot{q}_1^2 + \dot{q}_2^2 + \dot{q}_3^2) - \frac{k}{2}[q_1^2 + (q_2 - q_1)^2 + (q_3 - q_2)^2 + q_2^3].$$

From (1) and (2) we obtain the three coupled equations

$$(3) \quad \begin{aligned} -m\ddot{q}_1 - 2kq_1 + kq_2 \qquad\quad &= 0 \\ -m\ddot{q}_2 - 2kq_2 + kq_1 + kq_3 &= 0 \\ -m\ddot{q}_3 - 2kq_3 + kq_2 \qquad\quad &= 0. \end{aligned}$$

Since we are attempting to find a solution in which all masses move with the same periodicity we investigate solutions of the form

$$q_j(t) = a_j \exp(i(\omega t + \delta)), \qquad 1 \le j \le 3.$$

If each q_j has this form then the system (3) may be written as

$$\begin{aligned} 2\frac{k}{m} q_1 - \frac{k}{m} q_2 \qquad\qquad &= \omega^2 q_1 \\ -\frac{k}{m} q_1 + 2\frac{k}{m} q_2 - \frac{k}{m} q_3 &= \omega^2 q_2 \\ - \frac{k}{m} q_2 + 2\frac{k}{m} q_3 &= \omega^2 q_3. \end{aligned}$$

Setting $k/m = \omega_0^2$ and $\omega^2/\omega_0^2 = \lambda$, this system of equations may be written as the matrix equation

$$\begin{bmatrix} 2 & -1 & 0 \\ -1 & 2 & -1 \\ 0 & -1 & 2 \end{bmatrix} \begin{bmatrix} q_1 \\ q_2 \\ q_3 \end{bmatrix} = \begin{bmatrix} q_1 \\ q_2 \\ q_3 \end{bmatrix}.$$

Thus the permissible values of λ are simply the characteristic values of the symmetric matrix that appears in the equation. These characteristic values are found to be $\lambda_1 = 2$, $\lambda_2 = 2 + \sqrt{2}$, $\lambda_3 = 2 - \sqrt{2}$. Corresponding normalized characteristic vectors are, respectively,

$$\alpha_1 = \begin{bmatrix} \frac{1}{\sqrt{2}} \\ 0 \\ \frac{-1}{\sqrt{2}} \end{bmatrix}, \qquad \alpha_2 = \begin{bmatrix} \frac{1}{2} \\ \frac{-\sqrt{2}}{2} \\ \frac{1}{2} \end{bmatrix}, \qquad \alpha_3 = \begin{bmatrix} \frac{1}{2} \\ \frac{\sqrt{2}}{2} \\ \frac{1}{2} \end{bmatrix}.$$

Since the α_j's are mutually orthogonal, they determine a basis of the coordinate space.

For the problem at hand, the first characteristic vector corresponds to the outer two masses oscillating out of phase and the center mass remaining at rest, and the second corresponds to the outer masses oscillating in phase and the center mass oscillating out of phase and with an amplitude which is $\sqrt{2}$ times that of the others. Finally, α_3 corresponds to all three masses oscillating in phase and the center mass moving with an amplitude $\sqrt{2}$ times that of the other masses. In each case the frequency of the oscillation is given by $\omega_j = \sqrt{\lambda_j}\,\omega_0$.

An arbitrary motion of the system will consist of a linear combination of motions in the three normal modes described above and in each of these the oscillation frequency will be the characteristic frequency previously found. Thus, writing a displacement d in terms of the basis $(\alpha_1, \alpha_2, \alpha_3)$,

$$d = \sum_{j=1}^{3} a_j \alpha_j,$$

it follows that the time dependence of the system is given by

$$d(t) = \sum_{j=1}^{3} a_j \alpha_j \exp(i(\omega_j t + \delta))$$

The approach exemplified above can be applied to a much more general class of problems. Here, the existence of three degrees of freedom (the motion of each mass along the line) led to three independent, or normal, modes of motion. In the general case with n degrees of freedom there will exist n normal modes and the specification of these modes and their characteristic frequencies will allow us to specify the response of the system to any arbitrary linear disturbance.

Consider such a mechanical system with n degrees of freedom at equilibrium. The system may be specified by n generalized coordinates and the time t. We assume that the basis to which the coordinates are referred is not time dependent. If we choose the generalized coordinates to describe departures from equilibrium, the equilibrium state may be specified by the conditions $q_j = 0$, $\dot{q}_j = 0$, and $\ddot{q}_i = 0$ for all i. Lagrange's equations (1) hold and, in particular, hold at equilibrium. For normal systems and small excursions from equilibrium the function $d(\partial L/\partial \dot{q}_j)/dt$ is a function only of $\dot{q}_j$ and $\ddot{q}_j$ and vanishes at equilibrium, which implies that $(\partial L/\partial q_j)_{eq} = 0$ for all j. The kinetic energy is generally a function of the $\dot{q}_j$ only and also vanishes at equilibrium. This implies that $(\partial V/\partial q_j)_{eq} = 0$ for all j.

With just small excursions from equilibrium anticipated, we introduce now a Taylor series expansion of the potential energy function V:

$$V(q_1, \ldots, q_n) = V(0, \ldots, 0) + \sum_{j=1}^{n} \left(\frac{\partial V}{\partial q_j}\right)_{eq} q_j + \tfrac{1}{2} \sum_{j,k=1}^{n} \left(\frac{\partial^2 V}{\partial q_j \, \partial q_k}\right)_{eq} q_j q_k + \cdots.$$

Since the constant term has no consequence for equation (1), we may make a scale transformation and set it equal to zero. We observed that each of the first partial derivatives is zero, so that the leading terms in the expansion of V are

$$\tfrac{1}{2} \sum_{j,k=1}^{n} A_{jk} q_j q_k,$$

where

$$A_{jk} = \left(\frac{\partial^2 V}{\partial q_j \, \partial q_k}\right)_{eq}.$$

We adopt this quadratic form with (symmetric) matrix $A = (A_{ij})$ as the value of V:

(4) $$V = \tfrac{1}{2} \sum_{j,k=1}^{n} A_{jk} q_j q_k .$$

A similar series expansion can be found for the kinetic energy T. Since the generalized coordinates do not involve time explicitly, T is (approximated by) a quadratic form in the velocities $\dot{q}_j$:

(5) $$T = \tfrac{1}{2} \sum_{j,k=1}^{n} m_{jk} \dot{q}_j \dot{q}_k .$$

The coefficients m_{jk} are functions of the q's in general and may be expressed in a Taylor series about the equilibrium configuration:

$$m_{jk}(q_1, \ldots, q_n) = m_{jk}(0, \ldots, 0) + \sum_{l=1}^{n} \left(\frac{\partial m_{jk}}{\partial q_l} \right)_{eq} q_l + \cdots .$$

For the sake of simplicity we retain only the constant terms in these series, which means that we take the m_{jk}'s in (5) to be constants. Since a kinetic energy by its nature must be positive definite, we may classify T, as given in (5), a positive definite quadratic form. Hence the $n \times n$ matrix M such that $M = (m_{jk})$ is positive definite.

It follows from (4) and (5) that the Lagrangian of the system is

$$L = \tfrac{1}{2} \sum_{j,k=1}^{n} (m_{jk} \dot{q}_j \dot{q}_k - A_{jk} q_j q_k).$$

With the q's as the generalized coordinates of the system, the equations of motion (1) have the form

(6) $$\sum_{j=1}^{n} A_{jk} q_j + \sum_{j=1}^{n} m_{jk} \ddot{q}_j = 0, \qquad 1 \le k \le n.$$

Again assuming that the q's have the form

(7) $$q_j(t) = B_j \exp(i(\omega t + \delta)), \qquad 1 \le j \le n,$$

we deduce from (6) that

$$\sum_{j=1}^{n} (A_{kj} - \omega^2 m_{kj}) q_j = 0, \qquad 1 \le k \le n.$$

In terms of the matrices A and M defined earlier, this set of defining equations for the q_j's can be written as the matrix equation

(8) $$(A - \omega^2 M) q = 0$$

where $q = (q_1, \ldots, q_n)^t$. In particular we note that if $M = mI$, where I is the $n \times n$ identity matrix then, after defining λ to be $\omega^2 m$, equation (8) may be written $Aq = \lambda q$ and the frequencies of the oscillations can be calculated from the characteristic values of A. This situation is illustrated by the example discussed earlier.

To cope with (8) in general we need the matrix version of Theorem 7.2.5, which we derive next. Since M is positive definite, there exists a nonsingular matrix Q such that $Q^tMQ = I$. Then for any orthogonal matrix R,

$$R^tQ^tMQR = R^tIR = R^{-1}R = I.$$

Since A is symmetric, Q^tAQ is also symmetric and so there exists an orthogonal matrix R such that $R^tQ^tAQR = \mathrm{diag}(d_1, \ldots, d_n)$. Let $P = QR$. Then

$$P^tMP = I \qquad \text{and} \qquad P^tAP = \mathrm{diag}(d_1, \ldots, d_n).$$

Here the d_i's are the characteristic values of Q^tAQ and, hence, the roots of the polynomial equation

$$\begin{aligned} 0 = \det(Q^tAQ - \lambda I) &= \det(Q^tAQ - \lambda Q^tMQ) \\ &= \det Q^t(A - \lambda M)Q. \end{aligned}$$

Equivalently, they are the roots of the equation

$$\det(A - \lambda M) = 0. \tag{9}$$

It can be shown that all solutions of (9) are positive and so the equations of motion will be satisfied by an oscillatory motion of the form (7) not merely for one frequency but in general for a set of n frequencies $\omega_1, \ldots, \omega_n$.

In the example at the beginning of this section, after finding the characteristic values of the 3×3 matrix, we derived an orthonormal basis of the coordinate space such that the motion of the system could be described as a linear combination of motions in the normal modes of vibration. There is a similar basis of the coordinate space for the system now under consideration. We begin our demonstration of this result by introducing into the coordinate space the inner product which is determined by the matrix P which simultaneously diagonalizes M and A. This means that if $a = (a_1, \ldots, a_n)^t$ and $b = (b_1, \ldots, b_n)^t$ are elements of the coordinate space then we define $(a \mid b)$, the inner product of a and b, by

$$(a \mid b) = a^tMb.$$

Then, relative to this inner product, the equation $P^tMP = I$ means that P is an orthogonal matrix. Hence, the columns of P, which we denote by $\rho_1, \ldots, \rho_n$, provide us with an orthonormal basis of the coordinate space. Each ρ_j corresponds to a vibration of the system with only one frequency and the component oscillations are called the *normal modes of vibration*. All of the particles in each mode vibrate with the same frequency and with the same phase, the relative amplitudes being determined by the elements of ρ_j. If an arbitrary displacement of the system from equilibrium is given by $d = \sum_j c_j \rho_j$, then the ensuing time dependence of the system is obtained by following each of the normal coordinates through its normal mode of motion,

$$d(t) = \sum_j c_j \rho_j \exp(i(\omega_j t + \delta)).$$

REFERENCES

Goldstein, H. J., *Classical Mechanics*. Addison-Wesley, Reading, Mass., 1950.

Landau, L. D., and Lifschitz, E. M., *Mechanics*. Translated from Russian by J. B. Sykes and J. S. Bell (Vol. 1 in *Course of Theoretical Physics*). Addison-Wesley, Reading, Mass., 1958.

Synge, J. L., and Griffith, B. A., *Principles of Mechanics*, 3rd ed. McGraw-Hill, New York, 1959.

5. CHEMICAL PHYSICS: THE HARMONIC OSCILLATOR

Besides serving as a means of analysis of mixtures, the observation of absorption and emission of radiation can be used to determine many important properties of molecules. In a large number of instances, the quantum-theoretical connection between the property of interest and the spectrum of observed energy levels is facilitated by the use of linear algebra. In this section, we derive the energy spectrum of a diatomic molecule, the vibrations of which are assumed to be adequately approximated by the motions of two atoms connected by a spring—that is, a single harmonic oscillator. The force constant (constant of proportionality between the displacement from the equilibrium interatomic distance and the restoring force) which gives the best agreement between the observed spectrum and that predicted by the simpler model is taken to be a measure of the strength of the chemical bond in the actual molecule.

The classical Hamiltonian function H for a harmonic oscillator of mass m and force constant k is defined as follows:

$$H(p, q) = \frac{1}{2m} p^2 + \frac{k}{2} q^2 \tag{1}$$

where p is the momentum and q is the displacement of the position coordinate system from equilibrium. Although they may be given in more general form, it suffices for our purposes to state the axioms of quantum mechanics in the following form.

(a) In (1) the functions p and q are self-adjoint linear transformations on a separable Hilbert space V (that is, an inner-product space with a countable basis over the field of complex numbers).

(b) The linear transformations p and q satisfy the commutation rule

$$pq - qp = -i\hbar I, \tag{2}$$

where $i = \sqrt{-1}$, I is the identity transformation on V, and $\hbar$ is a scalar (Planck's constant divided by 2π).

(c) The energy spectrum of the oscillator is the set of characteristic values of the transformation H.

Since p and q are self-adjoint, H is a self-adjoint linear transformation on V. Therefore there exists an orthonormal basis $\mathscr{B} = \{\alpha_1, \alpha_2, \ldots\}$ of V where α_i's are characteristic vectors of H. Thus the characteristic values $\epsilon_1, \epsilon_2, \ldots$ which we wish to compute are determined by the equations $\alpha_n H = \epsilon_n \alpha_n$, $n = 1, 2, \ldots$

Defining P and Q by

$$P = (2m)^{-1/2} p, \qquad Q = (k/2)^{1/2} q$$

simplifies the form of H to

$$H = P^2 + Q^2. \tag{3}$$

Simultaneously, rule (2) takes the form

$$PQ - QP = -\frac{i}{2} \hbar\omega I, \tag{4}$$

where the scalar ω is defined by $\omega = (k/m)^{1/2}$. Now let us define transformations L and R by

$$L = P - iQ \tag{5}$$

$$R = P + iQ \tag{6}$$

We may infer from equations (3)–(6) that

(7) $$RL = P^2 + Q^2 + i(QP - PQ) = H - \tfrac{1}{2}\bar{h}\omega I$$

and

(8) $$LR = P^2 + Q^2 - i(QP - PQ) = H + \tfrac{1}{2}\bar{h}\omega I.$$

Left multiplication of (7) by L and right multiplication of (8) by L gives

(9) $$LRL = L(H - \tfrac{1}{2}\bar{h}\omega I) = (H + \tfrac{1}{2}\bar{h}\omega I)L.$$

If $(\ |\ )$ is the inner product function in V, define L_{ij} for $i, j = 1, 2, \ldots$ by

$$L_{ij} = (\alpha_i L \mid \alpha_j).$$

Then

$$\alpha_i L = \sum_{j=1}^{\infty} L_{ij} \alpha_j$$

and this series converges because the α_i's form an orthonormal basis of the space V. Since from (9) we have

$$\alpha_i L(H - \tfrac{1}{2}\bar{h}\omega I) = \alpha_i(H + \tfrac{1}{2}\bar{h}\omega I)L$$

it follows, in turn that

$$\sum_{j=1}^{\infty} L_{ij} \alpha_j(H - \tfrac{1}{2}\bar{h}\omega I) = (\epsilon_i \alpha_i + \tfrac{1}{2}\bar{h}\omega\alpha_i)L,$$

$$\sum_{j=1}^{\infty} L_{ij}(\epsilon_j - \tfrac{1}{2}\bar{h}\omega)\alpha_j = (\epsilon_i + \tfrac{1}{2}\bar{h}\omega) \sum_{j=1}^{\infty} L_{ij} \alpha_j$$

$$\sum_{j=1}^{\infty} (\epsilon_j - \bar{h}\omega - \epsilon_i)L_{ij} \alpha_j = 0.$$

This implies that

(10) $$L_{ij}(\epsilon_j - \epsilon_i - \bar{h}\omega) = 0, \qquad i, j = 1, 2, \ldots.$$

Defining R_{ij} by

$$R_{ij} = (\alpha_i \mid \alpha_j R)$$

a sequence of operations similar to the above yields

(11) $$R_{ij}(\epsilon_j - \epsilon_i + \bar{h}\omega) = 0, \qquad i, j = 1, 2, \ldots.$$

Next, it follows from (7) that

(12) $$(\alpha_i RL \mid \alpha_i) = (\epsilon_i \alpha_i - \tfrac{1}{2}\bar{h}\omega\alpha_i \mid \alpha_i) = \epsilon_i - \tfrac{1}{2}\bar{h}\omega.$$

On the other hand

$$(\alpha_i RL \mid \alpha_i) = \left(\left(\sum_{k=1}^{\infty} R_{ik}\alpha_k\right) L \,\middle|\, \alpha_i\right)$$

$$= \sum_{k=1}^{\infty} R_{ik}(\alpha_k L \mid \alpha_i) = \sum_{k=1}^{\infty} R_{ik}\left(\sum_{j=1}^{\infty} L_{kj}\alpha_j \,\middle|\, \alpha_i\right)$$

$$= \sum_{k=1}^{\infty} R_{ik} L_{ki}.$$

Equating these two values for $(\alpha_i RL \mid \alpha_i)$ gives

$$\epsilon_i - \tfrac{1}{2}\hbar\omega = \sum_{k=1}^{\infty} R_{ik} L_{ki}. \tag{13}$$

From (10) and (11) it follows that

$$R_{ik} L_{ki}(\epsilon_i - \epsilon_k - \hbar\omega)^2 = 0. \tag{14}$$

For a fixed i, either $\sum_k R_{ik} L_{ki} = 0$, in which case

$$\epsilon_i = \tfrac{1}{2}\hbar\omega$$

in accordance with (13), or $\sum_k R_{ik} L_{ki} \neq 0$, in which case

$$\epsilon_i = \epsilon_k + \hbar\omega$$

for some k, in accordance with (14). Now the set of characteristic values of H has a least member since the minimum value of the continuous function G, where $G(\alpha) = (\alpha H \mid \alpha)$ for α restricted to the closed, bounded subset of V where $\|\alpha\| = 1$, is a characteristic value of H. Thus, the least characteristic value of H must be $\epsilon_1 = \frac{1}{2}\hbar\omega$. If we arrange the characteristic values of H in order of increasing magnitude $\epsilon_1 < \epsilon_2 < \cdots < \epsilon_n < \cdots$, then

$$\begin{aligned}
\epsilon_2 \quad &= (1 + \tfrac{1}{2})\hbar\omega \\
\epsilon_3 \quad &= \epsilon_2 + \hbar\omega = (2 + \tfrac{1}{2})\hbar\omega \\
&\vdots \\
\epsilon_{n+1} &= (n + \tfrac{1}{2})\hbar\omega \\
&\vdots
\end{aligned}$$

The value of $k = m\omega^2$ that gives the best correlation between these formulas for characteristic values and the observed vibrational energy spectrum of a diatomic molecule is taken to be the force constant of the molecule in the harmonic oscillator approximation.

REFERENCES

1. H. Margenau and G. M. Murphy, *The Mathematics of Physics and Chemistry*, 2nd ed. van Nostrand, New York, 1956.
2. L. Pauling and E. B. Wilson, Jr., *Introduction to Quantum Mechanics*. McGraw-Hill, New York, 1935.
3. H. Eyring, J. Walter, and G. E. Kimball, *Quantum Chemistry*. Wiley, New York, 1944.
4. R. G. Parr, *Quantum Theory of Molecular Electronic Structure*. Benjamin, New York, 1963.
5. F. A. Matsen and P. L. M. Plummer, *Operator Quantum Chemistry*. Academic Press, New York, to be published.

SUGGESTIONS FOR FURTHER READING

Engineering and Physical Sciences

Amundson, N. R., *Mathematical Methods in Chemical Engineering*. Prentice-Hall, Englewood Cliffs, New Jersey, 1966.

Aris, R., *Vectors, Tensors, and the Basic Equations of Fluid Mechanics*. Prentice-Hall, Englewood Cliffs, New Jersey, 1962.

Asplund, S. O., *Structural Mechanics: Classical and Matrix Methods*. Prentice-Hall, Englewood Cliffs, New Jersey, 1966.

Athans, M., and Falb, P. L., *Optimal Control*. McGraw-Hill, New York, 1966.

Borg, S. F., *Matrix-Tensor Methods in Continuum Mechanics*. Van Nostrand, Princeton, New Jersey, 1963.

Brouwer, W., *Matrix Methods in Optical Instrument Design*. Benjamin, New York, 1964.

Frazer, R. A., Duncan, W. J., and Collar, A. R., *Elementary Matrices*. Cambridge Univ. Press, London and New York, 1947.

Goldstein, H., *Classical Mechanics*. Addison-Wesley, Reading, Mass., 1951.

Huelsman, L. P., *Circuits, Matrices, and Linear Vector Spaces*. McGraw-Hill, New York, 1963.

Langhaar, H. L., *Dimensional Analysis and Theory of Models*. Wiley, New York, 1951.

Laursen, H. I., *Matrix Analysis of Structures*. McGraw-Hill, New York, 1966.

Martin, H. C., *Introduction to Matrix Methods of Structural Analysis*. McGraw-Hill, New York, 1966.

Ogata, K., *State Space Analysis of Control Systems*. Prentice-Hall, Englewood Cliffs, New Jersey, 1967.

Pestel, E. C., and Leckie, F. A., *Matrix Methods in Elastomechanics*. McGraw-Hill, New York, 1963.

Robinson, J., *Structural Matrix Analysis for the Engineer*. Wiley, New York, 1966.

Tou, J. T., *Modern Control Theory*. McGraw-Hill, New York, 1964.

Behavioral, Biological, Management, and Social Sciences

Gale, D., *The Theory of Linear Economic Models*. McGraw-Hill, New York, 1960.
Horst, P., *Matrix Algebra for Social Scientists*. Holt, Rinehart and Winston, New York, 1963.
Johnston, J. B., Price, G. B., and Van Vleck, F. S., *Linear Equations and Matrices*. Addison-Wesley, Reading, Mass., 1966.
Kemeny, J. G., Schleifer, A., Snell, J. L., and Thompson, G. L., *Finite Mathematics with Business Applications*. Prentice-Hall, Englewood Cliffs, New Jersey, 1962.
Kemeny, J. G., and Snell, J. L., *Mathematical Models in the Social Sciences*. Ginn, Boston, 1962.

Systems

DeRusso, P. M., Roy, R. J., and Close, C. M., *State Variables for Engineers*. Wiley, New York, 1965.
Elgerd, O. I., *Control Systems Theory*. McGraw-Hill, New York, 1967.
Gupta, S. C., *Transform and State Variable Methods in Linear Systems*. Wiley, New York, 1966.
Koenig, H E., Tokad, Y., and Kesavan, H. K., *Analysis of Discrete Physical Systems*. McGraw-Hill, New York, 1966.
Kuo, B. C., *Linear Networks and Systems*. McGraw-Hill, New York, 1967.
MacFarlane, A. G. J., *Engineering System Analysis*. Addison-Wesley, Reading, Mass., 1964.
Peschon, J., *Disciplines and Techniques of Systems Control*. Random House (Blaisdell), New York, 1965.
Roe, P. H. O'N., *Networks and Systems*. Addison-Wesley, Reading Mass., 1966.
Schwarz, R. J., and Friedland, B., *Linear Systems*. McGraw-Hill, New York, 1965.
Zadeh, L. A., and Desoer, C. A., *Linear System Theory*. McGraw-Hill, New York, 1963.

Computation

Bodewig, E., Matrix Calculus, 2nd ed. Wiley (Interscience), New York, 1959.
Crandall, S. H., *Engineering Analysis*. McGraw-Hill, New York, 1956.
Faddeev, D. K., and Faddeeva, V. N., *Numerical Methods in Linear Algebra*. Freeman, San Francisco, 1963.
Fox, L., *An Introduction to Numerical Linear Algebra*. Oxford Univ. Press, London and New York, 1965.
Freeman, H., *Discrete-Time Systems*. Wiley, New York, 1965.
Householder, A. S., *Principles of Numerical Analysis*. McGraw-Hill, New York, 1953.
Householder, A. S., *The Theory of Matrices in Numerical Analysis*. Random House (Blaisdell), New York, 1964.

Kuo, F. F., and Kaiser, J. F., *System Analysis by Digital Computers*. Wiley, New York, 1966.
Lanczos, C., *Applied Analysis*. Prentice-Hall, Englewood Cliffs, New Jersey, 1956.
Varga, R. S., *Matrix Iterative Analysis*. Prentice-Hall, Englewood Cliffs, New Jersey, 1962.

Linear Analysis

Bellman, R., *Introduction to Matrix Analysis*. McGraw-Hill, New York, 1960.
Block, H. D., Cranch, E. T., Hilton, P. J., and Walker, R. J., *Engineering Mathematics*, Vol. 1, 2. Cornell, Univ. Press, Ithaca, New York, 1965.
Gantmacher, F. R., *Applications of the Theory of Matrices*. Wiley (Interscience), New York, 1959.
Kreider, D. L., Kuller, R. G., Ostberg, D. R., and Perkins, F. W., *An Introduction to Linear Analysis*. Addison-Wesley, Reading, Mass., 1966.
Mathews, J. C., and Langenhop, C. E., *Discrete and Continuous Methods in Applied Mathematics*. Wiley, New York, 1966.
Noble, B., *Applications of Undergraduate Mathematics in Engineering*. Macmillan, New York, 1967.
Pease, M. C., *Methods of Matrix Algebra*. Academic Press, New York, 1965.
Pipes, L. A., *Matrix Methods for Engineering*. Prentice-Hall, Englewood Cliffs, New Jersey, 1963.

Directed Graphs

Harary, F., Norman, R. Z., and Cartwright, D., *Structural Models*. Wiley, New York, 1965.
Lorens, C. S., *Flowgraphs*. McGraw-Hill, New York, 1964.
Pullen, K. A., *Topological and Matrix Methods*. Rider, New York, 1962.
Robischaud, L. P. A., Boisvert, M., and Robert, J., *Signal Flow Graphs and Applications*. Prentice-Hall, Englewood Cliffs, New Jersey, 1962.
Seshu, S., and Reed, M. B., *Linear Graphs and Electrical Networks*. Addison-Wesley, Reading, Mass., 1961.
Busacker, R. G., and Saaty, T. L., *Finite Graphs and Networks*. McGraw-Hill, New York, 1965.

APPENDIX Notions of Set Theory

1. TWO PRINCIPLES OF INTUITIVE SET THEORY

For our study of algebra we presuppose a theory of sets called *intuitive set theory*. We use the word "set" as synonymous with "collection" and "class." Although this statement may improve the reader's intuitive understanding of the notion, the term "set" is undefined and takes its meaning from the properties we assume that sets possess.

The properties we assume can be described in terms of the relation of membership. If x is an object and A is a set, then we assume that exactly one of "x is a member of A" and "x is not a member of A" is true. If x is a member of A, we write

$$x \in A$$

and if it is not,

$$x \notin A.$$

The first of our assumptions concerning sets (*the principle of extension*) is this: Sets A and B are *equal*, symbolized

$$A = B,$$

if and only if they have the same members. If A and B are unequal, we write

$$A \neq B.$$

Generally, a proof of the equality of two sets A and B is in two parts: One establishes that if $x \in A$ then $x \in B$ and the other part establishes the converse; for example, if A is the set of all equilateral triangles of the euclidean plane and B is the set of all equiangular triangles of that plane, then, using theorems of elementary geometry, we can prove that $A = B$ in the manner described.

If A and B are sets such that whenever $x \in A$ it follows that $x \in B$, we say that A is *included in* B or A is a *subset of* B and write

$$A \subseteq B.$$

The relation of equality for sets can be formulated in terms of the inclusion relation as $A = B$ if and only if $A \subseteq B$ and $B \subseteq A$. If $A \subseteq B$ but $A \neq B$, we say that A is *properly included* in B or A is a *proper subset* of B and write

$$A \subset B.$$

Suppose that A is a set that consists only of the objects denoted by a, b, and c. We write

$$A = \{a, b, c\};$$

thus A and $\{a, b, c\}$ are names for the same set. If we know names of all elements of a set and these elements are not too numerous, then the above *brace notation* is a convenient way to designate the set.

We turn now to a method for *defining* sets. Our sole offering is based on a method whose roots we discuss in vague terms. Let us agree that a statement about a particular object is a meaningful sentence in which a name of the object occurs and which is either true or false. Some illustrations follow; in cases where there might be ambiguity concerning the object under surveillance its name is underlined:

$\underline{6}$ divides 18 (true)
$\underline{6}$ divides 18 and $\underline{6}$ is a prime (false)
$\underline{4} < 3$ (false)
$\underline{3} < 3$ (false)
the Mediterranean Sea is blue (true)

Now, suppose that in a statement about a particular object we replace each occurrence of its name by a symbol, say x. The result is an instance of what we shall call a *predicate in the symbol* x. The foregoing examples yield, in the manner described, the predicates

x divides 18,
x divides 18 and x is a prime,
$x < 3$,
$x < x$,
x is blue.

The method we now introduce for defining sets is our second assumption concerning them (*the principle of abstraction*). A predicate in x, $P(x)$, defines a set A by the convention that the members of A are precisely those objects a for which $P(a)$ [the statement about a that results on substitution of a for x in $P(x)$] is true. We denote this set by

$$\{x \mid P(x)\}.$$

EXAMPLES

1. The principle of abstraction provides the existence of a set having a specified finite array of objects as its only members; for example,

$$\{x \mid x = 1 \quad \text{or} \quad x = 2 \quad \text{or} \quad x = 3\} = \{1, 2, 3\}.$$

2. $\{x \mid x$ is a positive integer > 1 and x has no positive divisors $\leq \sqrt{x}\}$ is the set of prime numbers.

3. If B is a set and $P(x)$ is a predicate in x, $\{x \mid x \in B$ and $P(x)\}$ is the subset of B consisting of all elements b of B such that $P(b)$ is true. The given denotation of this set is shortened to

$$\{x \in B \mid P(x)\}.$$

4. The unrestricted use of the principle of abstraction can lead to difficulties; for example, if the set R is defined by

$$R = \{x \mid x \text{ is a set and } x \notin x\},$$

it is easily shown that $R \in R$ and $R \notin R$.

5. $\{x \mid x \neq x\}$ has no members. It is called the *empty* or *null* set and denoted by

$$\varnothing.$$

If A is any set, then $\varnothing \subseteq A$ and if $A \neq \varnothing$ (read A is *nonempty*) then $\varnothing \subset A$.

---

2. OPERATIONS FOR SETS

If A and B are sets, the *union* $A \cup B$ of A and B is

$$\{x \mid x \in A \quad \text{or} \quad x \in B\}.$$

Here the word *or* is used in the sense of *and/or*. The intersection $A \cap B$ of A and B is

$$\{x \mid x \in A \quad \text{and} \quad x \in B\}.$$

The *complement of B relative to A*, $A - B$, is

$$\{x \in A \mid x \notin B\}.$$

The operations of union and intersection can be extended to arbitrary collections of sets. Let $\mathscr{C}$ be a collection of sets (that is, a set whose members are sets, $A, B, \ldots$). The *union of* $\mathscr{C}$, symbolized $\cup \mathscr{C}$, is

$$\{x \mid x \in X \quad \text{for some} \quad X \in \mathscr{C}\}.$$

The intersection of $\mathscr{C}$, symbolized $\cap \mathscr{C}$, is

$$\{x \mid x \in X \quad \text{for every} \quad X \in \mathscr{C}\}.$$

The operation of intersection is restricted to nonempty collections $\mathscr{C}$.

As a preliminary for the definition of still another operation for sets, we introduce the *ordered pair* (a, b) of two objects a and b by setting

$$(a, b) = \{\{a\}, \{a, b\}\}.$$

It is left as an exercise to show that $(a, b) = (c, d)$ if and only if $a = c$ and $b = d$. It is this feature of an ordered pair that distinguishes it from a set of two elements. We now define the cartesian product $A \times B$ of sets A and B as

$$\{(a, b) \mid a \in A \quad \text{and} \quad b \in B\}.$$

The *ordered triple* of a, b, and c, symbolized by (a, b, c), is defined as the ordered pair $((a, b), c)$. Assuming that ordered $(n-1)$-tuples have been defined, we take the *ordered n-tuple* of $a_1, \ldots, a_n$, symbolized $(a_1, \ldots, a_n)$, to be $((a_1, \ldots, a_{n-1}), a_n)$. The set of all n-tuples $(a_1, \ldots, a_n)$ whose *coordinates* $a_1, \ldots, a_n$ are members of a set A we designate as A^n.

3. RELATIONS AND FUNCTIONS

If A and B are sets, a *relation from A to B* is a subset of $A \times B$. A relation from A to A we call a *relation in A*. If r is a relation from A to B, we define

$$\text{domain}\ (r) = \{a \in A \mid \text{for some } b \in B, (a, b) \in r\}$$

and

$$\text{range}\ (r) = \{b \in B \mid \text{for some } a \in A, (a, b) \in r\}.$$

Notice that r is a subset of the cartesian product of its domain and its range.

Let f be a relation from X into Y such that

(i) domain $(f) = X$;

(ii) if (x, y) and (x, z) are elements of f, then $y = z$.

We call f a *function from* (or *on*) X *into* Y. The distinguishing feature of a function from X into Y among relations from X into Y is that a function assigns to (or, associates with) *each* $x \in X$ a *unique* element $y \in Y$, namely, that y such that $(x, y) \in f$. Elements of the domain of a function f are called *arguments* of f. If $x \in$ domain (f) and f assigns y to x, we call y the *value* of f at x, or the *image* of x under f, or the element into which f *carries* x. Numerous symbols are used in place of y; some are xf, fx, and $f(x)$. We shall use the first of these in most of our presentation.

The symbolism

$$f: X \to Y$$

is commonly used to convey the information that f is a function on X into Y. If range $(f) = Y$, we say that f is *onto* Y. If f has the property that

$$x_1 \neq x_2 \quad \text{implies that} \quad x_1 f \neq x_2 f,$$

then f is called *one-to-one* (or 1-1). If f is both 1-1 and onto Y, it defines a pairing of the elements of X with those of Y on matching xf in Y with x in X for each x in X. Because of this feature, this function is often called a 1-1 correspondence between X and Y. The 1-1 correspondence between a set X and itself, which carries each x in X into itself, is called the *identity function* on X and symbolized by i_X.

If $f: X \to Y$ is 1-1 and onto, then a 1-1 function, called the *inverse* of f, and symbolized by f^{-1}, on Y onto X is defined by

$$yf^{-1} = x \quad \text{if and only if} \quad xf = y.$$

We continue by defining one further operation for functions. Suppose that $f: X \to Y$ and $g: Y \to Z$. The *product* p of f and g (in this order) is defined by the equation $xp = (xf)g$. It is a function on X into Z and we write

$$p = fg.$$

If $f: X \to Y$ is 1-1 and onto (thus f^{-1} is defined), then $ff^{-1} = i_X$ and $f^{-1}f = i_Y$.

We conclude this section with two further definitions. If X is a set, then a function f on X^2 into X is called a (*binary*) *operation* in X. Thus a (binary) operation in X assigns to each ordered pair of elements of X an element of X. Finally, suppose that g is a function on X into Y and that A is a subset of X. The *restriction of* g *to* A is the function h on A into Y such that $xh = xg$ for each $x \in A$.

4. PARTIALLY ORDERED SETS AND ZORN'S LEMMA

Let A be a set and r be a relation in A. We imitate the notation that is used for such familiar relations as equality and set inclusion by writing "$a\ r\ b$" in place of "$(a, b) \in r$". Suppose now that r has the following properties:

1. For each $a \in A$, $a\ r\ a$.
2. If $a\ r\ b$ and $b\ r\ a$, then $a = b$.
3. If $a\ r\ b$ and $b\ r\ c$, then $a\ r\ c$.

We describe these circumstances by saying that r *partially orders* A or that the pair (A, r) is a *partially ordered set*. It is suggestive to use the symbol $\leq$ for a partial ordering relation. Further, if both $a \leq b$ and $a \neq b$, then we write $a < b$. This is read "a is less than b" or "a precedes b." A partially ordered set $(A, \leq)$ such that at least one of $a \leq b$, $b \leq a$ holds for every pair of elements a, b in A is called a *simply ordered set* or a *chain*.

The third, and final, assumption of intuitive set theory concerns partially ordered sets and is called a *maximal principle*. The formulation that appears below requires three further definitions. Let $(A, \leq)$ be a partially ordered set and C a subset of A such that $(C, \leq)$ is a chain. We call C a *subchain* of A. If C is a subset of A and x is an element of A such that $c \leq x$ for all c in C, then x is called an *upper bound* for C. An element m of A such that for no a in A is $m < a$ is called a *maximal element* of A.

The version of the maximal principle, which we state next, is known as *Zorn's lemma*: Each nonempty, partially ordered set P in which every subchain has an upper bound in P must contain a maximal element.

Index

Index